University Science and Mathematics Education in Transition

Ole Skovsmose • Paola Valero
Ole Ravn Christensen
Editors

University Science and Mathematics Education in Transition

Springer

Editors
Ole Skovsmose
Aalborg University
Aalborg East
Denmark
osk@learning.aau.dk

Paola Valero
Aalborg University
Aalborg East
Denmark
paola@learning.aau.dk

Ole Ravn Christensen
Aalborg University
Aalborg East
Denmark
orc@learning.aau.dk

ISBN: 978-0-387-09828-9 e-ISBN: 978-0-387-09829-6
DOI: 10.1007/978-0-387-09829-6

Library of Congress Control Number: 2008931031

Printed on acid-free paper

springer.com

Preface

Improving the quality of science and mathematics education at universities has been a task to which governments and tertiary education institutions have committed. This was the case in Denmark at the end of the 1990s when the Danish Government, its Ministry of Research, and a network of Universities gathered efforts around the construction and functioning of the Centre for Educational Development in University Science. The centre established collaboration between seven Danish universities around the teaching and learning of science: Aalborg University, Copenhagen University, the Danish University of Education, the Pharmaceutical University, Roskilde University Centre, the Royal Veterinarian and Agricultural University, and the University of Southern Denmark. The centre operated during the period 1998-2001, thanks to the generous funding of 35 millions of Danish Kroner in total.

The Centre for Educational Development in University Science embraced a wide range of educational research and development actives through which the practice of university science education was addressed and improved. Areas such as mathematics, physics and chemistry education were central. The centre ran a Ph.D. programme, which enrolled 12 students who addressed a variety of educational issues in the subject areas of relevance for the centre. The centre also organised a series of conferences and seminars aiming at the professional development of teaching staff in the institutions associated. The centre financed a number of teaching development projects run by university staff in their own institutions and classrooms. Many leading scholars from around the world made important contributions to the work of the centre.

The present book emerged from the wide-ranging network of research and researchers, established through the Centre for Educational Development in University Science. The intention of the book, however, is not to provide any report of the research or developmental activities of the centre, but rather to contribute to the worldwide concern for analysing both challenges and possibilities for university science and mathematics education. Even if the book collects a majority of papers by Danish authors working in Danish contexts, the issues addressed by the different sections and chapters are of a general relevance for tertiary educational environments around the world. Furthermore, the dialogue between the Danish authors and leading international researchers in the field contributes reinforcing the broadness of the book for an international audience, in a changing world were transitions in what is considered to be the core of science and mathematics education in universities are taking place.

We want to thank all the people who have contributed to the completion of this volume. Thanks to the Danish Ministry of Research and to Aalborg University for providing the necessary funding for editing the book. Thanks to Patricia Perry for a careful typographical editing of the manuscript, to Anette Larsen for editorial support, and to Anne Kepple for a language revision of several of the chapters. And thanks to Marie Sheldon and Kristina Wiggings and other members of the staff at Springer for their support and guidance during the edition process.

Finally, we would like to dedicate this collection to the memory of Leone Burton, a remarkable colleague and friend who during very many years supported our work participating in some of the activities of the Centre for Educational Development in University Science, conducting sessions with research students and staff in Denmark, and being a critical partner in our previous work and in an early stage of production of this collection. We are honoured to publish her paper, probably the last printed record of her proliferous and pathbreaking academic carreer.

Aalborg, May 2008

Ole Skovsmose
Paola Valero
Ole Ravn Christensen

Contents

Contributing Authors and Editors

Leone Burton† was Professor of Education (Mathematics and Science) at the University of Birmingham, England . She was a strong advocate for girls/women in mathematics, who wrote widely on gender-based aspects of mathematics education. Burton contributed significantly to developing teaching approaches that promote mathematical thinking. Her book *Mathematicians as Enquirers: Learning about Learning Mathematics* provides a deep insight in to inquiry processes in mathematical research, on the basis of which it is possible to investigate learning processes in a variety of school contexts in great detail. Burton studied the ways in which society and culture have shaped mathematics, examining similarities and differences in mathematical concepts and topics. Leone Burton passed away in 2007.

John A. Bowden is Professor Emeritus at RMIT University and Adjunct Professor at Swinburne University of Technology, in Melbourne, Australia. He has a research and teaching background in chemistry at the University of Melbourne and in educational development at both the University of Melbourne and RMIT University. In 2004 he retired as Professor of Educational Development at RMIT where he had been Director of the Educational Program Improvement Group, Dean of the Faculty of Education and Senior Policy Advisor to the Deputy Vice-Chancellor. The University of Gothenberg awarded him an honorary doctorate in 1998 for his contribution to Swedish university education and Chalmers University, Gothenburg, appointed him William Chalmers Guest Professor, 2001-2003. He has over 150 publications to his name, including *The University of Learning: Beyond Quality and Competence* and *Doing Developmental Phenomenography*. His most recent contributions to research have been to integrate knowledge content, generic attributes and workplace competence into a capabilities-driven curriculum design theory and to establish developmental phenomenography, a modified qualitative research approach.

Ole Ravn Christensen is Associate Professor at the Department of Education, Learning and Philosophy, Aalborg University, Denmark. He obtained a M.Sc. degree in mathematics from Aalborg University (1999), and in 2004 he received a Ph.D. degree from the same institution. Ole Ravn Christensen is Leader of the Board of Studies for Education, Learning and Philosophy. His doctoral dissertation explored the ethical demands faced by university studies in sciences and technology. Through his research he has addressed science and mathematic education, acknowledging the post-modern condition. In particular he has combined these studies with a specific interpretation of Wittgenstein's conception of language and knowledge.

Kathrine Krageskov Eriksen obtained a M.Sc. degree in biochemistry from the University of Copenhagen (1998), and in 2004 received a Ph.D. degree also from the University of Copenhagen. Her doctoral dissertation focused on the reflective dimension university chemistry education. Eriksen has published within the field of molecular biology and on the issues of chemistry, ethics and science education. Her current research interests concern science education, social responsibility, and processes of reflectivity. Eriksen has been involved in the organisation of various groups and conferences promoting the integration of philosophical and ethical considerations in university science curricula. She is leader of the educational programme in medical laboratory technology at University College Sealand, in Denmark.

Niels Grønbæk is Associate Professor at the Institute of Mathematics, University of Copenhagen, Denmark. He obtained his M.Sc. degree in mathematics and physics from the University of Copenhagen (1977), and in 1980 he received a Ph.D. degree in mathematics from the University of California. He has been a visiting scholar at Leeds University, England, Australian National University, Canberra, and University of Newcastle, both Australia. He is member of the steering committee for International Conferences on Banach Algebras and Chairman of Committee on Education of Science Teachers, secondary level, Faculty of Science, University of Copenhagen. Niels Grønbæk has published widely within mathematics.

Tom Børsen is Lecturer at the Center for the Philosophy of Nature and Science Studies, Niels Bohr Institute, University of Copenhagen, Denmark. He obtained his M.Sc. degree in chemistry in 1997 from the Danish University for Pharmaceutical Sciences, and in 2003 received his Ph.D. in University Science Education also from the Danish University for Pharmaceutical Sciences. The observations that 'science and technology should be used to improve the human condition' and that 'this is not always happening today' are basic to Tom Børsen's research and teaching activities. Through a series of publications he is developing the theoretical construct of 'Expert Scientific *Bildung*'. He is a member of the Executive Committee of the International Network of Engineers and Scientists for global responsibility and Treasurer and Board member of The Danish National Pugwash group.

Sandra Harding is a philosopher working at the Graduate School of Education and Information Studies at the University of California, Los Angeles, USA. She is the Author and Editor of many books on issues of epistemology, methodology, and philosophy of science from feminist and postcolonial perspectives. These include: *Is Science Multicultural? Postcolonialisms, Feminisms, and Epistemologies* (1998), *Science and Other Cultures: Issues in the Philosophies of Science and Technology* (2003), *The Feminist Standpoint Theory Reader: Intellectual and Political Controversies* (2004), *Science and Social Inequality: Feminist and Postcolonial Issues* (2006), and *Sciences from Below: Feminism, Postcolonialisms, and Modernities* (2008). She has consulted to several United Nations organizations, including UNESCO, UNIFEM, the Pan American Health Organization, and the United Nations Commission on Science and Technology for Development.

Cathrine Hasse is Associate Professor at The Danish University of Education, University of Aarhus, Denmark. She obtained her M.A. degree in anthropology from the University of Copenhagen (1994), and her Ph.D. in anthropology, also from the University of Copenhagen, in 1999. She also holds a M.A. degree in Communication from The Danish School of Journalism (1984). Cathrine Hasse has been member of evaluation committees at The Danish Evaluation Institute (EVA), and of the boards of The Danish University of Education, and The Association of Gender Research in Denmark. She is a member of the Society for Social Studies of Science, International Society for Cultural Research and Activity Theory, and Psychological Anthropology – American Anthropological Association. She has been visiting research fellow at the University of California, San Diego, at the University of Helsinki, Finland and at La Sapienza University of Rome, Italy.

Sebastian Horst is consultant at the Department of Science Education, University of Copenhagen, Denmark. Since 2003 he has worked with several projects about developing the study programmes and the teaching at the Faculty of Science. He is Author and co-Author of reports investigating results and problems in science education, both at university and primary and secondary level. Since 2006, he has been Editor of the Danish journal for mathematics and science education: *MONA - Matematik- og Naturfagsdidaktik*. Lately he functioned as secretary for a task force preparing a national strategy for science, technology, and health education throughout the entire educational system in Denmark.

Anette Kolmos is Professor at Department for Development and Planning, Aalborg University, Denmark. Her research concentrates on engineering education and problem and project-based learning (PBL). She is holder of the UNESCO Chair in Problem Based Learning in Engineering Education, Aalborg University, Denmark. She has been member of the Advisory Board for the 7th Framework Program, European Commission; Co-Chair for the Danish National Pedagogical Network for Engineering. She is Associate Editor of the *Journal of Engineering Education*, published by the American Society for Engineering Education and Associate Editor of *European Journal of Engineering Education* and Vice-President of the European Society of Engineering Education.

Søren Kruse is Associate Professor in general didactics (Educational Theory and Curriculum) at the Department of Curriculum Research, School of Education, University of Aarhus, Denmark. He is the leader of the Research Program of Teaching and Learning in Higher Education, and head of the Doctoral Program in Didactics. He has written a thesis on the philosophy and didactics of nature experience. He has carried out studies in health and environmental education, teacher education, scholarship of teaching and learning in higher education, and finally conceptual work concerning the sociology of knowledge in curriculum theory. Today his research concentrates on developing a general theory of learning, teaching and curriculum as non-causal relation focusing on empirical studies of teacher education and curriculum in higher education.

Kjeld Bagger Laursen has been a university teacher and researcher of mathematics since the middle 1960s. He has been Head of the Centre for Science Education at the University of Copenhagen, Denmark. He has been Director of Studies in Mathematics, also at the University of Copenhagen. He became Doctor of Science, University of Copenhagen in 1991 and has been an invited lecturer at several international conferences on Banach Algebras and Operator Theory. He was the President of the Danish Mathematical Society 1990–1994.

John Mason is Professor of Mathematics Education at the Open University, England and Senior Research Fellow at Oxford in the Department of Education. Originally a mathematician (Combinatorial Geometry), he was a Co-Founder of the Centre for Mathematics Education at the Open University. He has written hundreds of research papers and a dozen or more books, all aimed at supporting those who find themselves promoting, sustaining, and supporting mathematical thinking in themselves and in others. For more than 45 years, John Mason has been enquiring into what it means to think mathematically, and supporting others in all phases of education who either want to develop their own mathematical thinking, or to support others in doing this. His interests are currently focused on the design of pedagogically effective tasks, and the ways in which the structure of attention of teachers and learners interact to support the appreciation of mathematical reasoning.

Morten Misfeldt is Assistant Professor at Learning Lab Denmark, Danish School of Education, Aarhus University. He obtained his M.Sc. degree in mathematics from the University of Copenhagen, Denmark (2001), and in 2006 he got his Ph.D. from the Danish University of Education. His research concentrates on how new media changes human capacities in relation to thinking and communication, mathematical creativity, and innovative processes in general. He has developed an interest in game technology that supports social learning processes. Currently he is directing the research unit for Science, Technology and Learning. He is member of the program committee of the 4th European Workshop on MathML and Scientific e-Contents.

Kirsten Nielsen is a senior member of the research unit 'Didactics of Mathematics and Science' at the Department of Curriculum Studies, the Danish School of Education, University of Aarhus, Denmark. She graduated as a Geographer but her research and teaching background is mostly in environmental and health education with a great interest in the didactics of fieldwork. Fore some years she has also worked with the Danish program for 'Learning and Teaching in Higher Education' (LTHE program) for new staff members in science faculties.

Wolff-Michael Roth is Lansdowne Professor of Applied Cognitive Science at the University of Victoria, Canada. After teaching high school science, computer science, and mathematics for almost twelve years, he first became Professor of quantitative research methods at Simon Fraser University, Canada before taking his current position in 1997. His research interests concern knowing and learning in science and mathematics from early childhood through school, university, and into the workplace. His interdisciplinary research is published widely, including science and mathematics education, curriculum studies, teaching and teacher education, social studies of science, linguistics, and anthropology. Among his most recent publications are *Being and Becoming in the Classroom* (2002), *Toward an Anthropology of Graphing: Semiotic and Activity Theoretic Perspectives* (2003), *Rethinking Scientific Literacy* (2004, with A.C. Barton), *Talking Science: Language and Learning in Science* (2005), and *Generalizing from Educational Research* (2009, with K. Ercikan).

Camilla Østerberg Rump is associate professor and deputy head of department for teaching at the Department of Science Education, University of Copenhagen. She has a background in engineering and theoretical computer science. She has worked with educational development and research in science and technology education at the tertiary level both at the Technical University of Denmark and the University of Copenhagen. Her current research interests include 1st year physics teaching from the perspective of inclusion/exclusion into the physics community of practice, university teacher training programmes, learning styles seen in a relational perspective, instrumentation and instrumentalisation processes in relation to tertiary science education.

Ole Skovsmose has a special interest in critical mathematics education. He has investigated the notions of mathematics in action, students' foreground, globalisation, ghettoising with particular reference to mathematics education. He is Professor at Department of Education, Learning and Philosophy, Aalborg University, Denmark. He has Authored and co-Authored many books including *Towards a Philosophy of Critical Mathematics Education* (1994); *Educação Matemática Crítica: A Questão da Democracia* (2001); *Dialogue and Learning in Mathematics Education: Intention, Reflection, Critique* (together with Helle Alrø, 2002) and *Travelling Through Education: Uncertainty, Mathematics, Responsibility* (2005). He serves in the Editorial Board of several scientific journals. He has been Co-Director of The Centre for Research of Learning Mathematics, a co-operative project between different universities.

Rie Troelsen is Associate Professor in Teaching and Learning in Higher Education at the Department of Philosophy, Education and the Study of Religions at the University of Southern Denmark. With a M.Sc. degree in chemistry and classical studies and a Ph.D. in chemistry education, she has been doing research and development in various fields, such as a national strategy plan for science education in Denmark, in-service training of university teachers, young people's interest in science and choice of education. Among her present research interests are student attrition in higher education and the professionalization of university teachers. She teaches academic writing and science education to master students, lab work teaching to technical staff and course (re-)design to new faculty members. She is the Editor-in-Chief of the Danish journal of higher education, *Dansk Universitetspædagogisk Tidsskrift,* and serves on Editorial Boards of journals of science education in Denmark and the Nordic countries.

Paola Valero is Associate Professor in Mathematics Education at the Department of Education, Learning and Philosophy, Aalborg University, Denmark. She is leader of the 'Science and Mathematics Education Research Group (SMERG)' and Director of the Doctoral Program 'Technology and Science: Education and Philosophy' at the Faculty of Engineering, Science and Medicine. Her research interests are mathematics, science and engineering education at all levels; in particular innovation and change processes in those fields, curricular development, multiculturalism in science and mathematics education, and science and mathematics teacher education. She is the Editor-in-Chief with Mortne Blomhøj of the journal *Nordic Studies in Mathematics Education*, and serves on the Editorial Board of scientific journals in Colombia, Denmark, Mexico, Spain and South Africa. She is Co-Editor of the book *Researching the Socio-political Dimensions of Mathematics Education: Issues of Power in Theory and Methodology* (2004).

Carl Winsløw's area of research is the didactics of mathematics, and much of his research is drawing on methods from linguistics and semiotics. His background includes studies in French, general linguistics and mathematics. His doctoral thesis and his first 12 papers concerned pure mathematics (von Neumann algebras). From about 1996 he turned to the didactics of mathematics, particularly linguistic and semiotic aspects of mathematics education. He has worked on comparative international studies, studies of ICT-tools in university education, and curriculum theoretical questions. Since 2003, Carl Winsløw holds the first Professorship in didactics at the University of Copenhagen, Denmark, where parts of his work concern development of the teaching at the Faculty of Science. Some of this developmental work is closely linked to his research.

Introduction
The Multi-Layered Transitions of Knowledge Production and University Education in Science and Mathematics

Paola Valero, Ole Ravn Christensen and Ole Skovsmose

More than ever, our time is characterized by rapid changes in the organization and the production of knowledge. This movement is deeply rooted not only in the evolution of the scientific endeavour, but also and especially in the transformation of the political, economic and cultural organization of society. In other words, the production of scientific knowledge is changing both with regards to the internal development of science and technology, and with regards to the function and role science and technology fulfill in society.

Knowledge production has, for some time, stopped being the exclusivity of universities, founded as bastions and guardians of truth and knowledge. This production can now be owned by a variety of institutions and organizations with interests other than the production and maintenance of knowledge for its own sake. This general social context in which universities and knowledge production are placed has been given numerous names: the knowledge society, the informational society, the learning society, the post-industrial society, the risk society, or even the post-modern society (e.g., Castells, 1996, 1997, 1998; Beck, 1992, 1999; Lyotard, 1984).

A common feature of different characterizations of this historic period is the fact that we are living the beginning of its construction. Parts of the World, not only of the First World but also of the Developing World, are involved in the transformations associated with it. There is a movement from former social, political and cultural forms of organization which impacts knowledge production, into new, unknown and uncertain forms. Of course such an observation may be true for any point in time. However, the expansion of information technologies has created global flows of knowledge that implicate experts and non-experts in ways that have not been seen before. The accelerated pace of technological development and innovation opens so many options and possible, unexpected and almost uncontrollable courses of action that change, movements and transformations may go in different directions. Our awareness of the complexity of the changes in our time does not allow us to see a clear end. Somehow it seems that the clear-cut utopias that guided the ideas of development and progress in the past are no longer a strong presence, and therefore the transitions in the knowledge society generate a new uncertain world.

P. Valero, O.R. Christensen, and O. Skovsmose
Aalborg University, Denmark

O. Skovsmose et al. (eds.), *University Science and Mathematics Education in Transition*,

In this context, it is difficult to avoid considering seriously the challenges that such a complex and uncertain social configuration poses to scientific knowledge, to universities and especially to education in the natural sciences and mathematics. It is clear that the transformation of knowledge outside universities has implied a change in the routes that research in the natural sciences and mathematics has taken in the last decades. It is also clear that in different parts of the world these changes have happened at different points in time. While universities in the "New World" (the American Continent, Africa, Asia and Oceania) have accommodated their operation to the challenges of the consolidation of the New World and, thereby, have had a more utilitarian concern, many European universities with a longer existence and tradition have moved more slowly into this time of transformation and have been responding at a slower pace to environmental challenges. The process of tuning universities, together with their forms of knowledge production and their provision of education in science and mathematics, with the demands of the knowledge society has been as complex as the general transformation that society is undergoing. Therefore an understanding of the current transitions in science and mathematics education has to consider different dimensions involved in such a change.

We find ourselves, and our universities, to be in a transitional period. In our choice of the term *transition*, we want to signal the idea that universities and their knowledge production and educational activities are undergoing a particular type of change. We did not want to adopt the term "reform", since we do not want to convey the idea that changes intend clear improvement of the resulting state in relation to the initial state before the implementation of a change. We observe that current changes are happening at a variety of levels, from curricular plans, to administrative levels, and even at international regulations and harmonisations. It is clear that we are leaving "old" practices behind and are entering an era of "new" practices; however, the multiplicity of transformations blur the horizon of when changes will stop and where they will lead to.

Traditionally, educational studies in mathematics and science education have looked at changes in education from within the scientific disciplines and in the closed context of the classroom. Although educational change in the end is implemented in everyday teaching and learning situations, other parallel dimensions influencing these situations cannot be forgotten. The understanding is that the actual potentialities and limitations of educational transformations are highly dependent on the network of educational, cultural, administrative and ideological views and practices that permeate and constitute science and mathematics education in universities today.

1 Multiple Dimensions of Change

This book contributes to understanding some of the multiple aspects and dimensions of the transition of science and mathematics education in the current knowledge society. There is an increasing awareness of the fact that the actual teaching and learning of these subjects are influenced by much more than decisions related to the very same development of the academic disciplines at stake. Social and political changes in recent times have triggered reflections about the role of science in contemporary, post-modern societies. Science and mathematics may not necessarily be conceived as disciplinary ivory towers existing inside the well-protected walls of universities. Such reflections have had consequences for the views underlying the provision of education in these fields.

Universities themselves have been the targets of administrative changes that intend to break the seclusion of knowledge production and put knowledge and education at the service of larger social, political and technological changes. These changes have also impacted the working frame for academic staff in relation to their research activities and, most importantly, in relation to their teaching activity. Science and mathematics education in universities are not immune to neither the administrative nor the cultural changes that may alter the space of teaching and learning.

We will analyze some of the complexities involved in these changes. We have defined four inter-related but still different dimensions of change that we consider necessary to understand the muti-layered transitions of university science and mathematics education.

1.1 Changes in the Processes of Teaching and Learning

First, we consider *changes in the processes of teaching and learning*. University education has symbolized a prototype of the broadcast teaching, where the person "who knows" has to present knowledge to those many students "who do not know". Research in higher education has shown that this is still a dominant form of teaching, but many other ways of organizing educational processes are possible and, in fact, have been established (e.g., Knight, 2002). In opposition to the teacher-researcher as the center of teaching model, new forms of organization of learning advocate having students at the centre with strategies such as, for example, active learning processes around groups of students, or having students involved in the choice of study topics. These new approaches to teaching and learning deal with situations such as learning through communication and cooperation; learning of competencies rather than learning a fixed body of knowledge; the capability to address a broader contextual framework for scientific and technical knowledge as opposed to purely theoretical problems, etc. In Part I, we shall delve deeper into this type of transition in university education and we therefore start this book at ground level where the actual learning and teaching processes are the focal point.

In the late 1960s the students' movement provoked an impact, at least in some universities in Europe. The broadcast teaching model, concentrated on specific content matter issues, was challenged by an anti-authoritarian approach to learning where students claimed the centre of attention and where social problems were the preferred starting point of any learning process. This activist approach to the use and development of knowledge was, in some universities, organized as project work with the problem submerged and the students working in groups with the teacher acting as supervisor or even group member. Problem-based learning was one manifestation of this process and groups of students now produced large reports on real world problems as opposed to the formerly exclusively in depth studies of theoretical constructs. This anti-authoritarian approach also had implications for how the processes of learning and teaching was organized in science and mathematics education. The problem-based approach did not put emphasis on schooling into the traditions of a particular discipline. It was an interdisciplinary approach, which implied that disciplines merged together. Aalborg University, in Denmark, is an example of this development. Kolmos (Chap. 12) presents an overview of the experience of organizing university education in science, technology, engineering and mathematics in this particular institution. Aalborg University was founded on the activist principles mentioned above and based on the problem-based learning approach (PBL). Although the development of these ideas in practice has been very complex – and far from a smooth road – it has opened a space for mixing science and mathematics education with engineering, giving this new university of the 1970s a different profile from the older and often more traditional universities in Denmark.

There have been several phases in the development of the PBL approach. The first can be considered an *exploratory* one, where the original more political and radical view of education opened for new directions in university education. For instance, with respect to mathematics, a guiding idea was to pay particular attention to the applications of mathematics in real-life situations, and to the investigation of the functions of mathematical modelling in different contexts. A second phase was one of *consolidation*. Many routines were established. Examinations took a particular form. New students found both new educational practices that needed to be understood, and it was possible for them to find some traditions they could relate to. Teachers adjusted to play the role of supervisor and found their own way of managing the normally very time consuming supervision processes. Furthermore, the public critique of this "leftist form of university study" faded away. If one could talk about a third phase it could be characterized through the recognition of the *efficiency* of the PBL-approach. Thus broader empirical investigations indicated that candidates that had gone through a PBL-based university education had comparative advantages to their peers who graduated from traditional universities, when observed in complex work situations. A demand for professional scientists, mathematicians and engineers with a PBL-profile developed rapidly. The recognition of the PBL-approach has been further documented through comparative studies showing that education in a topic like mathematics was completed in a more efficient and competent way within a PBL-approach than within a traditional lecture format. The PBL-approach apparently squared nicely with the demands of the labour market in a knowledge society. However, many found that the price of this apparent success was too high, namely the renunciation of the political and critical elements so prominent in the first phase of the PBL-approach. Others, in particular at the early stages of implementation in the 1970s, found that the level of depth in theoretical knowledge dropped. These discussions in favour of and against PBL are all still alive and are part of the make up of institutions where there is an organizational policy in favor of alternative educational forms. In that aspect the changes in teaching and learning practices related to PBL in Denmark do not differ much from other types of initiatives in other countries (Bowden, Chap. 9).

Today science and mathematics education are facing a huge challenge in forming a new idea and vision that can capture the attention of young university students. In some places the influences of the reforms of the 1970s are more vibrant than ever, while in other places they seem to be fading away. Thus the idea of conceiving students as central actors in educational processes in universities is still a very influential perspective. This involves allowing students to take initiative in their learning processes as well as enabling teachers to stimulate such initiative. Mason (Chap. 1) engages this discussion by making explicit the assumptions on which it is possible to envision a teaching and learning situation where people's thinking power can be activated and activated into "productive ways of working".

Another developing theme in teaching and learning processes today is assessment. New and innovative approaches to assessing science and mathematics courses that transcend traditional written or oral exams are studied more than ever (e.g., Poulos and Mahony, 2008). The importance of assessment as a main determinant of learning and teaching processes is illustrated by Grønbæk, Misfeldt and Winsløw (Chap. 4) who analyzed the effect of a teaching intervention, focusing on the impact that the designed assessment had on the evolution of the didactical contract between students and teacher in a third semester mathematics course. The renegotiation of traditional expectations and learning activities – what counts in order to obtain a good grade – is one of the most interesting observations in this study. There is the hope that, as changes in assessment occur in other classrooms, the renegotiation of the traditional university didactical contract will become a noticeable phenomenon beyond the particularity of this classroom.

Troelsen (Chap. 3) and Eriksen (Chap. 2) further exemplify what such engagement could look like. Troelsen explores the meaning that laboratory work in chemistry education could have, theoretically and in practice, if reflections about student competencies are taken into account. In a "reflective society of knowledge" laboratory work should contribute to the development of analytic and cooperative competencies that cannot be secured by traditional, procedural and syllabus-oriented laboratory work. The concept of competencies becomes central in a design of alternative laboratory practices for students. Krageskov Eriksen is concerned with the social role of science in a context of reflective modernity and of a risk society. A key concept for understanding the development needed in science education to match this (changing) social role of science is *reflectivity*. Therefore, Eriksen proposes an operationalization of this concept in a manner that renders it constructive in the tertiary science education context, more particularly, in chemistry education.

Hence, in the first part of this book, different perspectives on the recent and future developments in learning and teaching processes at universities are discussed, hence contributing to the development of new visions for the learning processes in science and mathematics.

1.2 Changes in Academic Cultures

Secondly, we consider *changes in academic cultures* in universities. The notion of academic culture refers to the espoused priorities and discourses which set the scene for educational priorities in relation to the academic communities involved in transformations (Hasse, Chap. 5). Which views of knowledge and science underlie the way academic staff engage in educational processes? What is considered to be relevant to bring into science and mathematics education? What is considered legitimate scholarly criteria for scientific and educational quality? What are considered to be collective principles for teaching and learning? Who defines such priorities? Culture, understood in the broad sense as a set of collective values which are constructed through and which simultaneously emerge from practice, becomes an important category for understanding the transition in science and mathematics education. In university settings, different academic cultures overlap, integrate and also conflict with each other.

According to Snow's *Two Cultures* (published first time in 1959), two cultures of knowledge production seem to drift apart in the Western World. Snow refers, on the one hand, to the culture of the humanities and, on the other hand, to the academic cultures developed around the natural sciences. He describes how this cultural division was acted out even at high tables at the colleges in Cambridge. The topics for the dinner talk were completely different, depending on which of the two cultures was setting the scene. It was not easy to establish any dialogue across this cultural gap. One could assume that this cultural gap was a Cambridge phenomenon, but it appears to be running across all university campuses. Snow sees this cultural gap as a problem of the "entire West" and he believes that "the intellectual life of the whole of Western society is increasingly being split into two polar groups" (Snow, 1969, p.3). As we will see the situation might turn out to be much more complex.

The gap between the way in which the world is conceived from the perspective of the humanities and from the natural sciences establishes a profound condition for today's educational changes, at least in some parts of the world. Hasse (Chap. 5) really goes into the depths of the study of the culture of professional university physics in the Nordic countries and in Southern Europe. She shows how, in Northern Europe – Scandinavia and also the UK – physics is considered masculine and incompatible with human science. The discipline is placed in an untouchable ivory tower of "hard elitist science". This finding resonates very well with Snow's cultural divisions of the corridors in the traditional British universities. The situation, though is quite different the further South and East one goes: "In the Southern part of Europe general cultural conceptualizations of physics seem much less connected to gender. Physics is more integrated with the general cultural history and is not seen as a particularly "hard discipline". Hasse's analysis leads her to wonder whether challenging an internalist view of science and acknowledging its embeddedness in culture could open possibilities for envisioning new more gender-sensitive, responsible ways of doing science. The result of such other forms of science could eventually be breaking the gap of cultures in universities, a point also shared by Harding (Chap. 14).

It is clear that many of the issues related to the cultures of science and mathematics reside in the characteristics of the internal practices of those who do research and teach in these areas: the university staff. An important element of innovations and changes in the way the disciplines reproduce themselves both in research and in the education of the younger generations concerns the education of academic staff. The disciplinary training and expertise of staff in their particular field of research is one of the dimensions at stake here. However, that dimension is not the focus in this book. Rather, we are interested in the qualifications that elate to the staff's competence to lead educational processes for their students, and their ability to engage the breadth and complexity of the teaching situation. This engagement involves discussions on the pedagogical priorities of the academic staff including their conception of good teaching and proper university education. Kruse, Nielsen and Troelsen (Chap. 8) exemplify views among university staff about what constitutes good teaching, and highlight five interconnected points: Mastery of and strong knowledge about the subject matter, mastery of pedagogical skills for communication, management of interpersonal relationships, connection between research and teaching areas, and position of a favorable personality for teaching. An interesting question to pose to this finding is the extent to which these five points reflect the priorities and values of academic cultures, and how these points emerge as a type of shared idea of good teaching among young university staff.

However, there are many more cultural issues to be aware of in relation to university education. One could argue that academic cultures are influenced by external pressure from a business culture that talks the language of "surplus", "productivity" and "effectiveness", while university staff would rather talk in terms of "truth", "insight" and "knowledge". In addition, the culture of globalization has also created a new international cultural mix, which challenges the previously national perspective of universities by the demand for internationally targeted education, and by the need to attract students from all over the world to programmes of study. One could mention the emergence of a new youth culture, which adds to the cultural diversities within universities, when seen from the perspective of the diversity of competencies, interests and expectations that students bring into their studies. It is certainly necessary to talk about different cultures at the university, but hardly about just *two* cultures and *one* gap. It seems to be more appropriate to talk about numerous sub-cultures and several cultural gaps when we engage in university education studies today.

An example of another cultural gap concerns the differences between the science and mathematics learning cultures and the workplace culture that students will later on engage with. Here the central issue might be thought of as differences in general working conditions. The time schedule for studying may be quite incomparable to a work-place schedule. What might be a good solution in the academic culture may be highly problematic in a workplace culture, for instance, with respect to consumption of time, resources used, patents owned etc. Furthermore, the differences in academic and workplace cultures can also be acted out in very specific ways. For instance the ability to use a certain concept or technique may be quite dependent upon the particular context of its use. Some criteria could dominate the academic use while quite different ones may be applied in a workplace context. Roth (Chap. 6) presents an elaborate example of this phenomenon where he addresses the issue of whether it is feasible to suppose that key scientific competences such as reading graphs is a general abstract competence that can successfully be put into operation in any kind of graph-reading situation. His research with scientists in universities and in workplaces shows that graphing competences are highly dependent on the work practices people engage in. Here the term "work" also applies to the work environment of scientists in universities and their practice in a particular field of knowledge. Roth's analysis reveals clearly that scientific work, thinking and competencies are as immersed in cultural practice as any other type of activity and practice one could think of. Roth calls for the need to turn university science education so that it becomes possible for students to develop practice-bounded "knowledgeability".

With respect to a particular scientific discipline it is also possible for cultural gaps to emerge. On the one hand one can consider how researchers within a field of study are formulating new theories and constructing new knowledge. One could talk about "logic of discovery". On the other hand, one could consider how the curriculum is structured within university courses. This structure reveals "logic of representation". These two logics are different and it is often claimed that they have to be so. Thus, it is pointed out that the student must come to master the discipline through "logic of presentation" in order to, later on, be able to master "logic of discovery". This cultural split within a discipline has been pointed out by Burton (Chap. 7) with respect to mathematics. Burton differentiates between the "culture of mathematics" – the aspects of mathematics which are identifiable as discipline-related and that are part of the value set that a participant in that culture needs to be acculturated into – and the "mathematical culture" – the set of socio-political attitudes, values and behaviours that, in situations of communication around mathematics, shape how mathematics practitioners experience mathematics. The very interesting observation emerging from Burton's work is the fact that there seems to be contradiction between what mathematicians express about the culture of mathematics and what can be observed about the mathematical culture. A great challenge for groups of mathematicians in universities is, therefore, to transform their communication practices around mathematics, particularly in the settings of teaching and learning of new apprentices of mathematics, if a more accessible, "humanistic" mathematical culture – to borrow Hasse's term (Hasse, Chap. 5) is to be developed in universities.

1.3 Structural and Administrative Changes

Thirdly, we consider *structural and administrative changes*. Hence, we take a step back and try to understand why and how universities are under structural and administrative transition. Universities have a long history, and, at least previously, they have taken up the position as society's most sublime centre for knowledge production. In the classic university, associated with the Humboldtian tradition, the main aim was to produce knowledge, and to do so independently of particular interests be it religious or economic. The practical aspect of university production was associated with the graduation of candidates, who could, on the one hand, bring knowledge to the world outside the university, and, on the other hand, feed the reproductive system of academia. Today the picture is much more complex, and new demands for the type of knowledge to be developed by universities have come to play a decisive role in the organization of knowledge production and education. Krogh (in press) analyzes the current wave of transitions in European universities, within the context of the general frames for the development of Europe, within the global, informational society. First, Krogh notes that the European Union has given the university a very clear role in relation to economic growth:

> The knowledge economy and society stem from the combination of four interdependent elements: the production of knowledge, mainly through scientific research; its transmission through education and training; its dissemination through the formation and communication technologies; its use in technological innovation. At the same time new configurations of production, transmission and application of knowledge are emerging. [...] Given that they are situated at the crossroads of research, education and innovation, universities in many respects hold the key to the knowledge economy and society (Commission of the European Communities, 2003, p. 4, cited in Krogh, in press)

The Bologna Declaration (European Ministers of Education, 1999) has been preparing the terrain for the realization of the vision expressed in 2003. A reform towards transparency and quality that can secure mobility of students between European universities has been complemented by a series of European initiatives supporting changes in the administration and competitiveness of universities. It is in this scenario that current transitions, at least in Europe, could be understood. One example of these structural changes is the reform of leadership and administration. "Old European" universities have a long tradition for a certain degree of democratic organization. The heads of departments, the Dean and the Rector have normally been elected among colleagues, which establishes a degree of collegial solidarity. However, the trend of new management has established new procedures and accountability systems: the Rector has to be appointed by a board of trustees; the Dean by the Rector; and the heads of departments by the Deans. This top-down structure has serious implications for the organization of both the content and form of the university education, as Horst and Laursen (Chap. 10) illustrate.

This new structure is reflective of a clear alignment with economic and industrial interests, as discussed above. As a consequence, research at universities becomes measured in terms of newly invented scales of productivity. New accountability systems, meant to measure the effectiveness and productivity, are set in operation. These systems identify a way of carrying out input-output analysis. The input can be measured in terms of money allocated to the different institutions, research programmes, research groups, etc. The tricky thing becomes how to measure the output of academic activity. A first manageable but rough measure would be to turn output into numbers of publications written, the number of students going through studies successfully, the amount of Ph.D. degrees conferred on time, the amount of money raised for research purposes from private funds, etc. When one can measure output, then one has established a better grip of managing research and education according to business principles. When research becomes to a larger extent financed by industry and institutions with particular interests, it is not surprising that a management turn is taking place. This turn reveals an intensive interaction between power and knowledge. It is a turn, however, that has not been requested by the research or educational communities.

Science and mathematics education at university level is now operating within a complex set of interests, which really stress what to include in the curriculum. This demand is being set in operation, at least in Denmark, by correlating the funding of the universities with the degree to which the candidates from the universities find jobs. Furthermore, there are huge demands for labelling students and candidates according to their competencies, which should be declared in a way that is transparent for the (international) knowledge market – in the European sphere according to the criteria of the Bologna Declaration.

In his reflections on the structural university reforms that he has witnessed during the last 30 years in Australia and some other English speaking countries, Bowden (Chap. 9) points to the fact that none of these changes have resulted in any substantial improvement of the educational settings that students experience, even though the initial justifications for most of these changes are precisely the improvement of quality of the provision of education. The reason for this might be that many of the external structural demands, which we have described above, obstruct qualitative improvements in university science and mathematics teaching, rather than promote significant betterments.

Laursen and Horst (Chap. 10) present a particular and local example of how a structural reform has been implemented in the Faculty of Sciences at Copenhagen University, and they discuss to what extent its results have been beneficial for creating better frames for student learning. On the other hand, Rump and Winsløw (Chap. 11) argue for the need of educational reforms that have a lasting impact. They consider how tertiary didactics can be beneficially implemented in a Faculty of Science by a combined local and global approach.

1.4 Changes in the Conception of Science

Fourthly, we consider *changes in the conception of science*, in particular concerning the relationship between scientific development and social development in general. From the perspective of Modernity, one could claim that science and mathematics especially stemming from university departments represent genuine knowledge that will emancipate humanity from traditional and religious pre-scientific knowledge and belief systems of the past. This is the narrative told by the Enlightenment

Movement and by social theorists such as Karl Marx and Georg Wilhelm Friedrich Hegel. However, the role of scientific knowledge in society now appears much more complex, and we witness social uncertainties and risks which might be produced by science. We experience how science gives multiple conflicting answers to many questions, and we are constantly faced with scientific inventions that while being effective in one area have less beneficial side effects in others. Many of the chapters in this book address these issues, since they are at the background of the educational, cultural, and structural changes in university science and mathematics education.

Scientific knowledge is in a transitional period from originally being something shared as everybody's property for the common good of humankind to being a protected product, for instance through rules of patenting, in order to be distributed to the market. This transition does not only suggest that university faculties put a price on the knowledge produced. The changes are much more radical as the very organization of university research and education are being transformed. Knowledge that is usable and can perform in practice will win the competitive race for research funding. This means that the emergence of "new sciences" (like nano-science, health-science etc.) finds new prosperous ways of challenging – and sometimes even displacing – the old academic disciplines in university structures and changing them into competitive fields of research – probably engineered by the vision of the EU mentioned above (Commission of the European Communities, 2003). In the early 1970's Jean- François Lyotard was one of the first thinkers to recognize this development. He termed it the "postmodern condition" for knowledge (Lyotard, 1984). Whether or not the change should be termed postmodern, there is no question that we are witnessing a dramatic change in the conception of science. The assumed everlasting knowledge, as developed by the geniuses of the break through of Modern Science like Isaac Newton, Johannes Kepler and Galileo Galilei, is no longer the prototypical form of scientific development. Instead, new products for the market that are made available through technological innovation are what constitute the power of science. The "new sciences" are part of this process and suggest deep conceptual transformations in our understanding of what science and knowledge are (Christensen and Hansen, Chap. 13).

One of the assumptions of Modern Science, which has provided a protecting myth, is that science and mathematics are operating in a neutral domain guided by the criteria of quality of pure research. However, we are in a situation where we must acknowledge the complexities of knowledge fabrication in relation to all kinds of social, political, economic and even cultural interests (Harding, Chap. 14). The classic discourse about the internal qualities of research is challenged by a discourse about research productivity. The big issue at stake is whether scientific cultures within universities are being consciously responsive to the challenges that such change represents to the conceptions that science practitioners themselves hold about their scientific endeavour. One of the points that scientists and researchers on the sociology of science have under-estimated is the consequence of the dominance of Western, androgenic assumptions in science. Gender studies in science – such as Harding (Chap. 14) and Hasse (Chap. 5) – can still interrogate the role of the scientific enterprise in the world today.

Another different but related issue is whether these changes impact educational possibilities and conceptions in mathematics and science studies. Skovsmose (Chap. 15) argues the need for a "critical professionalism" to be part of university studies. Critical professionalism refers to the awareness, on the part of the science practitioner, of the connections between particular fields of knowledge and other fields, of the fact that scientific and mathematical knowledge are bounded to social action, and of the ethical dimensions of producing and applying scientific knowledge. Critical professionalism for Skovsmose is one of the main competences of scientists in a time of uncertainty. This observation resonates with Bowden's observation (Chap. 9) about the need to generate in students a capacity to tackle not what is known and familiar, but what is unknown, uncertain and, probably, unpredictable.

2 About This Volume

As seen above, the current transition in university science and mathematics education is a multi-layered process where changes are happening simultaneously across several dimensions and in several spaces. Covering such a vast landscape exhaustively would be almost impossible. Therefore, when organizing this volume we did not envision an organization that could represent a logical flow of argumentation through the different chapters. We collected a series of papers that based on particular experiences in particular contexts, open critical points of debate about the multiple predicaments that university science and mathematics education face at the moment.

The fact that most of the authors belong to Danish universities does not bind the discussions raised to the context of Denmark. We have invited authors to transcend national boundaries and to link their experiences to more general trends that are common to many other countries in the world, and to many other university environments. The conversation with experienced researchers in science and mathematics education in Australia, Canada, the United Kingdom and the United States broadens the perspectives presented in the collection.

Finally, we have spoken generally about science and mathematics education. Nevertheless in this book we address more directly three classic domains of university education, namely, mathematics, physics and chemistry. We let "science" function as a general label, being well aware that one could claim the importance of making a distinction between, on the one hand, mathematics as being a formal science, and, on the other hand, physics and chemistry as belonging to the natural sciences. The considerations and reflections in this book might in many cases be more general, and therefore be pertinent to other domains such as biology, computer science, statistics, bio-chemistry, etc. We are well aware that a claim of generality can always be contested by other observations. But we take that fact as a compliment if, reading this book generates further reflections about the transitions in other fields.

References

Beck, U. (1992). *Risk Society: Towards a New Modernity*. London: Sage.

Beck, U. (1999). *World Risk Society*. Cambridge: Polity.

Castells, M. (1996). *The Information Age: Economy, Society and Culture. Volume I: The Rise of the Network Society*. Oxford: Blackwell.

Castells, M. (1997). *The Information Age: Economy, Society and Culture. Volume II, The Power of Identity*. Oxford: Blackwell.

Castells, M. (1998). *The Information Age: Economy, Society and Culture. Volume III, End of Millennium*. Oxford: Blackwell.

Commission of the European Communities (2003). *Communication from the Commission: The role of the Universities in the Europe of Knowledge*. Brussels: Commission of the European Communities. Retrieved May 2008.ftp://ftp.cordis.europa.eu/pub/documents_r5/natdir0000038/s_1940005_20030317_144432_GERC031941en.pdf.

European Ministers of Education (1999). *The Bologna Declaration on the European space for higher education*. Brussels: European Ministers of Education. Retrieved May 2008. http://ec.europa.eu/education/policies/educ/bologna/bologna.pdf

Knight, P.T. (2002). *Being a Teacher in Higher Education*. Buckingham: SHRE – Open University Press.

Krogh, L. (in press). Nye udfordringer i undervisningen ved universiteterne - en professionel strategi for undervisningen, set i lyset af aktuelle samfundsmæssige krav [New challenges for Teaching at Universities – A Professional Strategy for Teaching in the Light od Current Social Demands]. In P. Rasmussen & A.A. Jensen (Eds.), *Læring og forandring*. Aalborg: Aalborg University Press.

Lyotard, J.-F. (1984). *The Postmodern Condition: A Report on Knowledge*. Manchester: Manchester University Press.

Poulos, A. & Mahony, J. (2008). Effectiveness of Feedback: The Students' Perspective. *Assessment & Evaluation in Higher Education*, 33(2), 143–154.

Snow, C.P. (1969). *The Two Cultures: And a Second Look*. Cambridge: Cambridge University Press.

Part I
Changes in Teaching and Learning

Chapter 1
From Assenting to Asserting

Promoting Active Learning

John Mason

The issue addressed in this chapter is age-old: How can learners be stimulated to move from *assenting* (passively and silently accepting what they are told, doing what they are shown how to do) to *asserting* (actively taking initiative, by making, testing and modifying conjectures, and by taking responsibility for making subject pertinent choices). How can learners be provoked into actively working on and making sense of the ideas and techniques that they encounter, and how can this cultural ethos be fostered and sustained?

I use the term *asserting* because of the assonance with *assenting*, but also because it signals that the learner is taking initiative and making significant choices. It is not intended to indicate that learners become either arrogant or garrulous. Much of the most desirable *assertive* behaviour is internal, and need not have visibly overt external behaviour. It involves taking initiative, taking control, making choices, and becoming independent.

One of the reasons often given for assenting or passive stances taken by students is the fact that they are immersed in a culture of testing, so that their focus is on being told what they have to do to pass the next test. In other words, it is to be expected that in a culture of testing you get test-oriented (assenting, passive) behaviour. However, in this chapter the claim is made that it is perfectly possible to develop a culture of seeking to understand, and more, a culture of enquiry, within any testing regime. The way to do this is to evoke, support and develop learners' use of their own natural powers of sense making.

The chapter describes some of the most fundamental powers which all learners possess and have already used before coming to class, but which may have been put to one side due to previous experiences of being taught which failed to make use of these powers. Identification of these powers comes from reflecting on personal experience informed by the seminal insights of Pólya (1962) and Gattegno (1970) and developed over many years since Mason et al. (1982). The chapter also suggests tactics which have been used by teachers to get learners to make use of those powers, in contrast to teachers unwittingly doing the real work for learners, whether in plenary sessions, in tutorials or in informal learner discussions through an unconscious

---

J. Mason
Open University, United Kingdom

O. Skovsmose et al. (eds.), *University Science and Mathematics Education in Transition*,

decision to "push things along a bit", "get to the end of the lesson on time", or "helping learners to understand". The tactics are equally applicable to lectures, tutorials, problem-based sessions and the use of Information Communication Technology (ICT). In each case there are forces acting to increase passivity that need to be countered through suitable choice of teaching tactics and the development of a culture that overtly values learners taking initiative.

Promoting active rather than passive learning can be expressed in terms of the development of personal agency and identity. The chapter draws upon studies of similarities and differences between Confucian and Western views of education to try to locate a common core of personal agency, which straddles the two traditions.

1.1 Assumptions

Before describing powers and strategies, it may be helpful to articulate some salient assumptions on which the chapter is based.

1.1.1 Assumptions About Human Beings

Human beings are seen here as naturally active construers of the sense impressions which they experience. They possess not only fundamental powers for dealing with these impressions, but also a fundamental curiosity or drive to want to make sense of them. When learners are encountered who seem to want to sit passively and learn by rote, then it is because they have been trained or acculturated to act this way. Nevertheless it is still possible to engage such learners, to provoke them into activity, by refusing to amplify their tendency to want things packaged and painlessly injected. Instead of trying to "do the learning for them", teaching can be exciting and stimulating by taking as its core purpose the provocation of learners to activate, become aware of, and develop their natural powers through encounters with pervasive mathematical and scientific themes.

This chapter takes an overtly constructivist stance, combining aspects of both psychological and social constructivism. Learners are seen as active agents (though the level of activity may be below desired thresholds!) whose psyche consist of the interweaving of behaviour (enactive), emotional (affective) and intellectual (cognitive) strands through the exercise of will, and who are embedded in a social milieu in which colleagues and teachers display practices which are adopted and adapted by learners.

1.1.2 Assumptions About Cultural Heritage

Covert activity, in which the learner is assembling and internalizing what is being offered so as to engage deeply with it subsequently is all too easily confused with actual passivity, in the sense of "waiting to be told what to do and explicitly how to do it", and "taking as little initiative as possible". For example, in the Confucian tradition, "human self-perfection [is] pursued as the highest purpose of life through personal commitment to learning" (Lin, 2004, pp. 129–130). Whereas in the West, rote memorization is seen as an end in itself, an attempt to minimize effort and simply regurgitate what has been "taught", the Confucian heritage sees commitment to memory as simply the first of a sequence of four phases. Memorized passages are interrogated for intention, style and meaning, and then applied to other situations in order to test it out in experience. This moves into critique and modification so that what was "memorized" is internalized and integrated into the learner's functioning (Lin, 2004, p. 131).

At the heart of both Western and Confucian heritage approaches to teaching and learning lies initiation into productive "ways of working" on science and mathematics, whether in the classroom or in the world outside. The tactics proposed in this chapter serve to activate and intensify activity corresponding to the later phases of the Confucian approach, but right from the beginning without depending on an initial phase of memorization.

1.1.3 Assumptions About Mathematics and Science and Their Didactics

Mathematics and science, as bodies of knowledge occupy what Popper (1972) called the "third world" of accounts lodged in libraries. However, these accounts are at best the expression of how someone else's attention was altered in its structure. They indicate subject specific sensitivities to notice and dispositions to act, and characteristic ways of acting. What the authors attended to, and how they attended, are evidenced only through what they were then able to express as a result, through the manifestation of their behaviour, itself usually refined and distilled. To make use of these accounts it is necessary for readers to experience transformations in the structure of their own attention, even if only vicariously and in their imagination.

Mathematics and science are both seen here as activities, as ways of thinking and acting in the world. Facts and figures are accumulated and rehearsed while engaging in pertinent activity. While learners sit passively, accepting and assenting to what they are told or shown, taking notes to "learn" later, they are not actually doing mathematics or doing science. Certainly they are not developing their powers of thinking mathematically and scientifically, to make mathematical and scientific sense of the world.

How then can learners be provoked into taking initiative, into making use of and developing their natural powers, into anticipating what is coming rather than rushing to catch up with what is being written on a screen in front of them by an expert? How can they be acculturated into developing both disposition for, and familiarity with productive ways of working on scientific and mathematical phenomena?

1.2 Phenomena

An incident or situation can only become an instance of a phenomenon when some boundary is drawn around it, when some aspect is discerned and fore-grounded. This stressing and consequent ignoring is the basis for abstraction, as Gattegno (1987) pointed out. It is one of the basic powers which human beings possess, but it can be all too easily subverted and parked if not used effectively. Mathematics, physics and chemistry all arise from becoming aware of characteristic situations as a phenomenon, of discerning details and relationships which comprise that phenomenon, and seeking underlying mechanisms to explain or account for the apparent relative invariance which constitutes that phenomenon.

1.2.1 Phenomena Encountered in Teaching

The following generalized incidents are intended to be generic examples of experiences in teaching, which, if they speak to experience, constitute phenomena encountered in teaching.

> Some learners attend a laboratory and follow the instructions as a recipe, finally writing up what they measured or observed. They can only assent to what they are told to do, and try to do it as faithfully as possible.

- For example, I discovered as a first year undergraduate that my father had 25 years earlier attended the same physics laboratory sessions and written up the identical experiments. Indeed neither the descriptions nor even the equipment had changed! Consequently I was able to find out in advance what each experiment was supposed to show and what the results were supposed to look like. I learned nothing about doing or constructing experiments.

In some cases, however, learners are required to design (at least parts of) a laboratory experiment for themselves in order to resolve some stated question. They are not permitted into the lab until the description is deemed acceptable from a safety as well as a scientific standpoint. This is becoming standard practice in science, and supports a move to asserting rather than mere assenting.

> Some learners sit in rows waiting to be told what mathematical or scientific procedures and facts they need to know. They naturally want to minimise their investment of energy, particularly when the 'subject' is peripheral or seen as at best a necessary tool rather than as a central element of their discipline. They are in a transition phase waiting to acquire a 'license to practise' in their discipline in the world beyond education. They assent to what they are given and asked to do, but have no thought to take initiative or to do more than requested.
>
> In some cases, however, learners are actively engaged in trying to make sense of phenomena. They see their courses as a means to achieve this, and they apply themselves actively to everything they encounter. They assert themselves in and through their engagement. Nevertheless, sometimes the going gets tough, the ideas complex, and the techniques unclear or complicated and multi-phased. Learners revert to minimising effort, looking for 'what they need to do' to pass the course. They assent to the practices of the system.

Underlying these phenomena is a pervasive theory of learning: if learners complete most of the tasks they are set by the teacher, to a reasonable standard, then the requisite learning will have taken place. Educationalists have railed against this attitude for as long as teaching has been written about. For example:

- Plato (Laws VII 819b-c, in Hamilton and Cairns, 1961) complained about the teaching of arithmetic, and many of his dialogues are explorations into what learning is and how it can be promoted.
- Spencer (1878, p. 28) complained about generalizations such as definitions and rules preceding exposure to cases, and that telling or showing does not teach how to observe.
- Whitehead (1932, p. 7) coined the expression *inert knowledge* to describe the central problem of education, which he saw to be the problem of keeping knowledge alive (assertive).
- Brousseau (1984, 1997) elaborated the notion of an implicit *didactic contract* which produces an inescapable tension: the more precisely the teacher indicates the specific behaviour being sought, the easier it is for learners to display that behaviour without actually generating it from understanding.
- Harré and van Langenholve (1999) used *positioning theory* to articulate the ways in which people are both positioned and position themselves into taking different roles at different times, through sometimes delicate social interaction.

The task of teaching is to create conditions in which the doing of set tasks will lead to activity that promotes active construal rather than simply "getting answers". Inner pondering and solitary work, discussion occasions this construal with colleagues, and interaction with a teacher (relative expert). The teacher's task is to call upon and make use of learners' natural powers of sense making in any or all of these milieux. The next section elaborates some of those powers and proposes ways of constructing tasks, which can lead to fruitful and pedagogically effective activity and interaction between learners and between learners and teacher. The suggestions sketched briefly here are elaborated in Mason (2002a).

1.2.2 Learning Phenomena

One of the observable outcomes of effective learning, and hence of effective teaching, is an enhanced and enriched disposition to notice phenomena in the material world to which mathematical or scientific thinking can be applied. For example:

- People often beat egg whites to create a foam, but what happens when you beat them, and why does it sometimes collapse and sometimes not?
- People often take showers where there is a soft curtain, and as the water is turned on, the shower curtain moves inwards towards the water flow and the person's legs, but why not ask yourself why?
- People are aware that the moon is sometimes visible and sometimes not, but when can you see a vertical half-moon (the terminator is parallel to your body as you stand and look) and can you ever see a horizontal half-moon?

- People often go through double doors, but why is it that if you open both doors at the same time for someone, the gap takes longer to become wide enough for them to get through than if you only open one door?
- People expect businesses to have pricing policies, but themselves are only interested in the price they are offered: the entrepreneur uses algebra, the customer uses arithmetic.

Incidents like these only become phenomena when someone becomes aware of a (potential) repetition and then isolates those features that are common to various incidents. Becoming sensitized to notice such opportunities, even if they are not taken up, is to become more mathematical or scientifically assertive, in the sense of engaged with and contributing to the community. Some forms of teaching are more effective than others at promoting such a disposition (Boaler, 1997). Lin (2004, pp. 134–135) points to differences between Western and Confucian heritage learners in this respect. The former emphasize activity as the origins of learning, through problem solving, and depend on suitable and compatible teaching resources. The latter value and emphasize "heart and mind wanting to learn", "seeking knowledge" and opting for both breadth and depth. Both are aimed at active agentive conditions rather than passivity, but they are expressed differently.

1.3 Powers

The idea of people possessing powers has ancient roots in Indian psychology. In the west, Spencer (1878), Montessori (1912), Gattegno (1970) and Krutetskii (1976) among others all referred to them in the development of their specific approaches to teaching and learning. My observations have been inspired and informed by Pólya (1962), and confirmed and developed by my own observations and reflections over a period of nearly 40 years. The powers elaborated here have been chosen because they are, in my experience, the core powers that are used both to make sense of mathematics and science and to make mathematical and scientific sense of the world. My aim is not to be comprehensive, because such a list would then become unwieldy. Rather my aim is to draw attention to key powers, which, though on the one hand perfectly obvious to any mathematician and probably to any scientist, are on the other hand, often overlooked and not made use of fully in pedagogic and didactic settings.

- Imagining and Expressing
- Specializing and Generalizing
- Conjecturing and Convincing
- Focusing and De-Focusing Attention
- Organizing and Classifying

Gattegno (1988) proposed the notion that, in a manner of speaking, by responding to sense impressions the soma constructs its brain, or as others might say, brain structure co-emerges as sense impressions and soma co-operate. Connections are made as the soma attends, and attention becomes possible due to brain connections forming. Mason (2007) elaborates the claim that these powers are demonstrated by newborns, and form the basis for effective participation in and action upon the material world. The aim here is to elaborate how these powers are the key to getting learners to engage with mathematics and science.

Sense-making powers are activated when learners experience some sort of disturbance: expectations are revealed when something happens which produces surprise or when some task presents a challenge because it cannot be accomplished through the routine use of habits. This notion of disturbance as the impetus to learning also has a long history. Piaget (1971) used a biological metaphor (assimilation, accommodation and equilibrium seeking), Wertheimer (1945) recognized disturbance explicitly as the source for activity leading to learning, and Festinger (1957) made disturbance the core of his notion of *cognitive dissonance*.

1.3.1 Imagining and Expressing

The power to imagine and to express what is imagined through movement, pictures, words and symbols is fundamental to human functioning. This power permits us to imagine what is not present even when not being physically possible. It also permits the use of sophisticated language. Mental imagery is also the means we use to harness the energies produced by our emotions, by mentally imagining ourselves in future situations and carrying out actions as we intend. Indeed Norretranders (1998) concludes that consciousness lags behind and interprets behaviour rather than directing it, reinforcing ancient psychological insights that mental imagery as a planning device acts as the reins to direct emotions. It is what we use in order to plan and prepare for the future. It is how we tell stories and how we communicate with others. "Imagining" is taken here to encompass virtual sensations of any mode or combination of modes, including sight, sound, taste and smell, touch, and a more generalized "sense-of" which is not directly attributable to any particular mode.

Expressing what we imagine produces communication of many different forms, including both science and art. Education is constantly confounded by trying to identify what is imagined (understood, appreciated) with what is expressed, for example in the act of assessing learners. Just because someone does something in one situation (say on a test), it does not follow that they either can or will think to do it, much less do it, in another context. Hence the notion of situated cognition (Lave and Wenger, 1991), and situated abstraction (Noss and Hoyles, 1996), in which emphasis is placed on the important role of the situation or situation-type in which encounters and hence learning take place. Conversely, just because someone does not do something in a situation, it does not follow that they could not, or that they will not in some other situation. As archaeologists are fond of saying, "absence of evidence is not evidence of absence".

Language contributes to imagination (you can strengthen some images by repeating inner speech as reinforcement of images) but imagination also provides the source of what is expressed in language. Becoming articulate in speech and writing, whether in primary or in tertiary level, is as much about connecting language with inner experience and thought as it is about learning the relevant and appropriate format. Language also strengthens the power to imagine, because when one person expresses what they are imagining they can stimulate imagination in others. For example diagrams are static images, which provide a stable background on which to superimpose variation and further construction. A diagram or photograph of a cell, a picture of a moment during an experiment, or a geometrical figure can all serve in this way. The learner can then imagine further details, other aspects, and what happens before and after. Similarly dynamic images, such as a video or animation, provide what Salomon (1979) called *supplanted images:* images that can be used as eidetic components in the construction of further mental images.

However, images, like text, can be interpreted in many ways. There are likely to be relationships which are particular to the specific diagram and not intended to be included as part of the generality being illustrated. Fischbein (1987, 1993) coined the expression *figural concepts* to describe the concepts constructed by learners from attending to unintended features. The issue of how people learn general concepts from particular examples has exercised psychologists for some time (Bruner et al., 1956; Rissland, 1991; Renkl et al. 1998) and is of abiding interest in computer science and A.I. (Winston, 1975) as well as in education (Sweller and Cooper, 1985).

Experience with dynamic geometry software has also highlighted the fact that in a diagram or picture there are local relationships, which are not directly observable (Laborde, 1995). In geometry it is vital to become aware of these relationships, which are often what the diagram is really about. Something similar applies to any picture or diagram that is supposed to illustrate some relationship, property, some process, principle or technique. Furthermore, relationships may be detected among discerned elements in a particular case without being perceived as properties, which can apply in many different situations, or in the case of figural concepts, properties may be inappropriately or unintentionally assumed to be essential. This is what gives rise to undesirable situatedness. Thus some subtle shifts in the structure of attention may be required, from discerning relevant items to recognizing relationships, to perceiving these as properties, before learners can be expected to engage in reasoning on the basis of those properties, and so displaying understanding (Mason 2003, 2004).

In science as in mathematics, being invited to imagine (guided by the teacher) the steps in carrying out a task or in thinking about a phenomenon, which the topic can help explain, can be a powerful introduction to a lesson or suite of lessons. For example:

- Imagine a beaker, half filled with pure water. Put it in a beaker of ice. Insert a thermometer and watch the temperature…
- Imagine the graph of a function. Imagine a point not on the curve. Imagine a line passing through that point and slowly rotating. Sometimes it intersects the curve, sometimes it does not, and if these are both possible, somewhere in between it touches the curve at two coincident points (it is tangent there) as well as, perhaps crossing the curve elsewhere…

A learner who is trying to follow what they are being told or shown is at a disadvantage, because they are always one step behind, reacting to what has happened rather than participating in real time. This is often evident when learners who are copying notes from a board or screen fall behind the production of those notes. They cannot therefore be concentrating fully on what is being said, so much of it goes over their heads, literally and figuratively. By contrast, learners who are able to anticipate what is about to happen are actively engaged and in a much better position to experience disturbance or at least contrast and surprise when their expectation is contradicted. This is the moment when a crucial decision can be made, for if the exposition flows ever onward, the disturbance may only be suppressed or parked. The learner who falls behind is in a much less favourable position to anticipate, and is driven into a position of severely restricted and situated assertion.

1.3.1.1 Making Use of Imagining and Expressing

The power to imagine can be strengthened and developed, through use. But that use has to be active rather than passive. One of the dangers of the sea of images imposed on our senses by television and earphones is the pollution and over stimulation of our senses. Bombarding people with sense impressions in a competitive free-for-all is unlikely to result in the development of learners' powers. Rather, learners can be called upon to anticipate, to imagine what something will look like or what will happen next, before being shown.

Say What You See

Learners can be called upon to *Say What You See* to others in order to discover when listening to others that there are other details to discern, other relationships to recognize, other properties to perceive than the ones that may come to attention immediately. A variant is to be asked *what is the same and what is different?* Or *what is changing and what is invariant?*

- For example, observing a fountain, it can be seen simply as "fountain". However, if someone endeavours to say what they see, trying not to make unwarranted assumptions, they will find their attention directed to what is changing (the apparent path of the water droplets, the movement on the surface of the water) and what is invariant (the basin etc.). In the case of the second fountain shown, the spout is itself rotating. Nevertheless, even though individual drops follow their own unique path, there is the appearance of a path being formed by the water (a relative invariance or an invariant relationship). *Say What You See*, informed by seeking what is (relatively) invariant and what is changing is likely to invoke curiosity: what is the actual path? Are the apparent paths actually paths, and are they also parabolae?

Walker (1975) provides descriptions of numerous everyday phenomena, which can be explained by relatively elementary physics. If learners are not acculturated into seeing the material world through the eyes of a curious physicist, then they are not learning to be physicists, and arguably, not even learning physics. They are merely being trained in the use of formulae. The same applies to chemistry. Why is it difficult to open a freezer door immediately after closing it (assuming it is functioning!)? How does soap work? What happens when you heat up ice and it turns into water or heat up water and it turns into steam? Curiously, phase change can be a helpful metaphor for teaching, in which effort sometimes goes in to explaining and correcting with no visible effect, and then suddenly the learner acts as if they have actually learned something.

Say What You See is an effective pedagogic device for provoking learners to probe beneath the surface of situations, to detect an underlying phenomenon, and to locate details that might be used to explain or account for what is observed. *Say What You See* can also be used with abstract entities such as appear in mathematics. For example, *Say What You See* can be used with either of the following objects or both together for comparisons, in order to draw attention to different component elements and what they signify, and to indicate which features are irrelevant and which are essential to some problem or task type.

$$\int_{\frac{1-\sqrt{5}}{2}}^{\frac{1+\sqrt{5}}{2}} (x(x^2-1)-x)dx$$

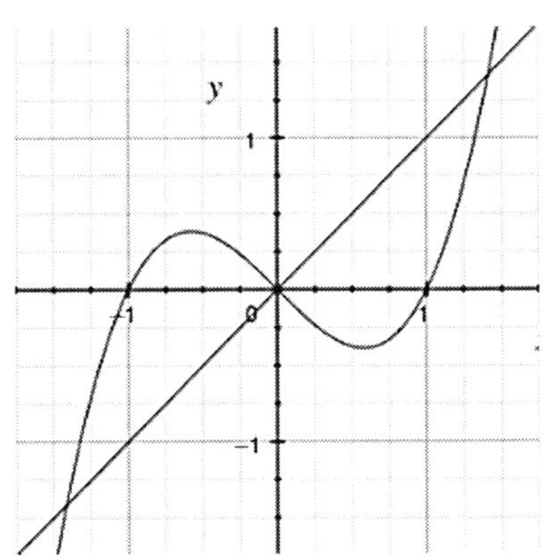

Through being exposed to how others attend, learners can discover that they too can choose to direct their attention, to perceive differently, and to make different sense.

Indicate When You Can See …

Learners can also have their attention directed explicitly so that they discern relevant details (*look at this; hear that*) using verbal and physical pointing. More subtly, they can be invited to recognize significant relationships through being invited to indicate when they can detect a particular relationship (Thompson, 2002). For example, after gazing at this diagram,

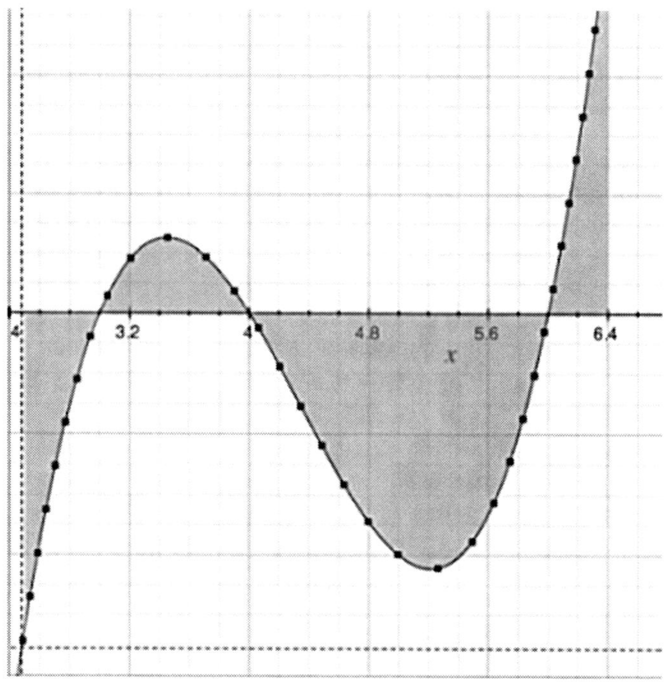

can you see:

area under a curve?	four areas?	an area-so-far function?
an integral to calculate a mathematical area?	four integrals to calculate a total area?	An instance of the fundamental theorem?

Indicate When You Can See … is useful for prompting learners to become aware that it is both necessary and useful to see one object from multiple perspectives. Of course the question *"Can You See"* on its own has only a "yes/no" answer, so it is much more effective to ask learners to indicate when they can see. The time delay then may suggest whether more time could usefully be spent on the particular issue. The question is also useful simply for directing attention to relevant details before drawing attention to relationships that are instances of general properties.

It is vital for a teacher to be able to direct learner attention appropriately, and this requires becoming aware that learner attention may not spontaneously be directed as the teacher expects. Not only can attention be focused on different features, causing break down in communication, but, even where there is a common focus, learners and teacher may actually attend differently. Some may be occupied by discerning details, others may be seeking relationships between particular discerned components, yet others may be perceiving relationships as instances of more general properties, and some may be prepared to reason about the phenomenon only on the basis of explicit and agreed properties, and not just on anything and everything they think they know.

1.3.2 Specializing and Generalizing

Pólya (1962) used the term *specializing* to refer to all aspects of trying out a particular case of something in order to "see" what is going on, to experience the underlying structure, and hence to re-generalize for oneself. Specializing includes ordinary familiar examples as well as extreme or "special" cases. People do this entirely naturally, as when they ask for an example, or resort to using diagrams or physical objects as a model of a more complex situation. The interesting thing from a teacher's point of view is that when learners get stuck on a problem or with a concept, they often do not think of using their natural power to specialize for themselves (Mason et al., 1982). Consequently it is important pedagogically to display and enact specializing and generalizing, to call upon learners to use those powers themselves, and to draw learner attention to the fact that they have used those powers. These are the three aspects that contribute to noticing opportunities to act in the future and so to inform practice (Mason, 2002b): acting, reflecting, imagining.

1.3.2.1 Making Use of Specializing and Generalizing

Halmos (1994, p. 852) stresses concentrating "attention on the definite, the concrete, the specific". As Whitehead (1932, p. 4) put it, "To see what is general in what is particular and what is permanent in what is transitory is the aim of scientific thought". I prefer a rephrasing as "to see the general through the particular and the particular in the general" and "to be aware of what is invariant in the midst of change" is how human beings cope with the sense- impressions that form their experience.

Scaffolding and Fading

Wood et al. (1976) introduced the metaphor of *scaffolding* as a description of the way in which a teacher can act as "consciousness for two", augmenting the learners' awareness and reminding them of things that are not coming to mind because their attention is fully taken up with matters which will ultimately become integrated and subordinated into their functioning (Gattegno, 1970). For example, when a learner is stuck on a problem, asking them to specialize (give me an example ...) can be seen as an act of scaffolding. However, teacher intervention can create dependency as well as independence. It all depends on *fading*, that is, reducing the support being offered (Brown et al., 1989; Love and Mason, 1992). A useful way to do this is to move from direct prompts (have you got an example?) to increasingly more indirect prompts (What did you do last time? or What question do you think I am going to ask you?) until learners have taken up those prompts and internalized them.

Learner Constructed Examples

Given the problems which arise when learners try to use examples to work out what features are intended to be exemplary and what features are in fact specific or special, a useful additional strategy turns out to be to get learners to construct their own examples (Mason and Watson, 2001, 2005). Not only does this reveal some of the richness or poverty of what learners have access to, but as an action it actually serves to extend and enrich the space of examples to which learners have access in future.

For example,

Can you construct a cubic whose inflection tangent does not cross the curve?

Sketch a cubic graph;
and another;
and another

Construct a cubic;
Construct a cubic which has only one real root;
Construct a cubic which has only one real root and which has a local minimum;
Construct a cubic, which has only one real root, which has a local minimum and whose inflection slope is positive.

Can you find a situation involving a gas in which temperature changes but neither pressure nor volume change?

Can you find a situation in which change of temperature produces a change of pressure?
And another?
And another?

Find a situation in which change of pressure produces a change of temperature;
A change of volume produces a change of pressure;
A change of volume produces a change of temperature

The structure of the second task is intended to illustrate how being asked for another and then another (not all at once, but in sequence) directs most people to begin to explore the full range of generality from which to make a choice, rather than sticking with the first idea that comes to mind. In this way their personal example space becomes enriched (Watson and Mason, 2005).

The third task is intended to illustrate how treating constraints one at a time, and paying attention to the full generality at each stage can be more instructive than trying to meet all constraints at once. The constraints are chosen to try to force people to reconsider their current example. At each stage choices have to be made, but sticking with a generality is more fruitful than trying to modify a single example.

Example Spaces

The central idea being explored herehas been articulated by Ference Marton and colleagues (Johansson et al., 1985; Marton and Booth, 1997; Marton and Trigwell, 2000) as *dimensions of variation*. To understand or appreciate a concept means to be aware of what aspects (dimensions) can vary (change) in an example and still it remains an example. A mathematical version of this is the theme of invariance in the midst of change, which lies at the heart of mathematics and science (Whitehead, op. cit.) and perhaps of all learning. Marton goes further, proposing that learners become aware of a dimension of possible variation through exposure to variation in sufficiently quick succession to be attended to *as* variation. Watson and Mason (2005) go further in drawing attention to the importance of becoming aware of the permissible range of change in which the invariance is preserved. Very often learners are aware of a subset of the dimensions of possible variation, and even when they are aware of a particular feature that can be varied, their sense of the range of permissible variation is often attenuated. For example, they may think only of integers instead of real numbers. Consequently when the teacher uses a concept, the learners' sense of what is being said may be so greatly restricted as to nullify its significance.

A classic example in mathematics is the function $x \rightarrow |x|$ which is the standard example of a continuous function differentiable everywhere except at one point. It is well known that many students simply monster-bar this example and continue to think of it as merely an aberration (MacHale, 1980). Until learners think of exploring possibilities such as functions which are non-differentiable at 2, 3, ... points, and even the set of points at which functions could be non-differentiable, they have not appreciated what the counterexample is telling them. By becoming familiar and confident with example tinkering, they enrich both their appreciation of a concept and their accessible example space.

In science, a diagram of a cell, no matter how beautiful the graphics, suffers from the possibility that learners will attend differently to the ways intended. They

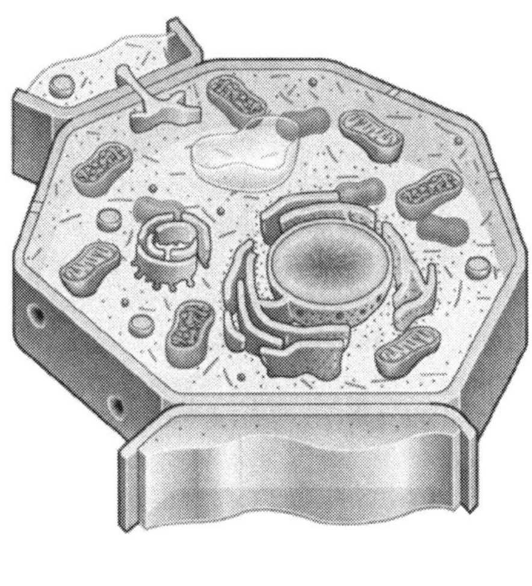

may take some specific details as generic when in fact they are arbitrary, and some details as arbitrary when in fact they are specific.

The nucleus need not be in the position shown, nor even near the centre; other components displayed three times appear hundreds or thousands of times, and so on. A learners' sense of what can vary and what is invariant nor necessary can be highly attenuated and yet they can answer many questions correctly and appear to understand the topic.

Watson and Mason (2005) have found that there is a tendency for learners of mathematics to have a restricted range of examples to use as touchstone and on which to experiment. Their example spaces may be much more restricted and impoverished than teachers realize. Asking learners to construct one example, then another, and then yet another often has the effect of expanding learners' awareness of the range of possibilities from which a choice can be made when looking for a suitable example. It is this sense of choice, of potential (if not actual) infinity that enriches and educates their awareness and provides access to an expanded example space in the future. The confidence gained by having a sense of control over the potential complexity, which comes from being able to construct your own variations, also enriches the possibilities when a learner wants to specialize in order to make sense of a generality.

1.3.3 Conjecturing and Convincing

Mathematics and science proceed and develop through conjectures, proofs, and refutations (Lakatos, 1976). Convincing themselves, friends, and finally, other sceptics is what mathematicians and scientists do. But so does everyone, although in less formal and explicit ways. Every action in and on the world is a conjecture; when the response is unexpected, that disturbance generates energy, which can be used to try to return to equilibrium, to make sense, to accommodate or to find awareness and sensitivity enriched.

Human beings are also natural story-tellers, or as Bruner (1996) said, *narrative animals*, for we try to account for our experience, weaving it into a story which justifies our actions and explains why things happen the way they do. diSessa (1987) uses the label *phenomenological primitives* for core explanations that people use to account for phenomena, to justify their conjectures about why things happen. The term *primitive* is used to indicate that the explanation itself is not probed any further. For example, when asked what happens to the sound of a vacuum cleaner motor when the tube is blocked, "working harder" is often used to account for claims both that the pitch rises and that it falls.

Scientists and mathematicians have developed formal practices in order to try to convince each other, although these practices do change and develop over time as both questions and reasoning become more sophisticated or political. Learning to adopt these practices, to take up the relevant discourse, is one of the challenges for learners in any topic.

1.3.3.1 Making Use of Conjecturing and Convincing

If learners are to become mathematicians and scientists, they too need to engage in overt conjecturing and convincing. They need the time, opportunity and impetus to try to articulate their reasoning to others in order to convince themselves. They also need to learn to be sceptical about someone else's reasoning, developing the disposition to look for counter-examples, to test out special cases in an effort to find flaws. In this way they develop their power to convince not only colleagues, but also a more sceptical "other" or "enemy" (Mason et al., 1982). In science, many learners arrive with assumptions and partial stories to account for observations about the material world that need to be challenged in order that more appropriate scientific perspectives are adopted. For example, as diSessa (1987) showed, even very intelligent students can adopt a split view that Newtonian mechanics is what you use in class while retaining an Aristotelian view of forces. He asked undergraduates to predict the path of a particle emerging from a circular track at speed, and found many considered the circularity to continue after release. It is vital that learners become aware of assumptions they are making, convert these into conjectures, and begin to construct (even if only mentally) experiments to test out those conjectures.

Because of the strong effect of years of schooling, conjecturing does not just "happen" in classrooms. Establishing a conjecturing atmosphere or ethos takes time. Attention needs to be drawn to the status of assertions and assumptions as conjectures needing confirmation and justification. Those who are uncertain need to be encouraged to try to articulate their conjectures and be overtly respected for modifying their conjecture as a result; those who are confident or certain need to be encouraged to ask helpful questions of others, to try to find counter-examples to other people's conjectures, and to probe themselves and others for justifications and reasoning. Most importantly, conjecturing is most effective as a collaborative rather than a competitive interaction with others. Everybody learns when someone modifies a conjecture or proposes a counter-example.

Overtly calling upon learners to make a conjecture before they test it on an example or against a phenomenon puts them in a more active stance. When a conjecture is refuted, learners have more personal commitment to sorting it out, modifying, augmenting or even radically changing their conjecture. For example:

> For what conjectures could the function $f(x) = \begin{cases} x^2 \text{ when } x \geq 0 \\ -x^2 \text{ when } x < 0 \end{cases}$ be used as a counterexample?

> For what conjectures could the fact that the ratio of the masses of a substance before and after burning is independent of the sample size be used as a counterexample?

It is so easy to sit back passively and wait for the "correct" answer to emerge, when in fact "learning" is the result of struggling to articulate ideas and reasons rather than actually getting the particular answer. Most lectures have a few people attending who fool themselves into thinking that "I could have got that, done that, thought of that" when they don't make an act of commitment through formulating a conjecture. Pólya (1962) recommended making conjectures and then disbelieving them. As long as ideas and possibilities toss around in your mind like clothes in a tumble drier, they are hard to pin down and likely to dissipate. By recording a conjecture and then immediately looking for counter-examples the learner engages in mathematical and scientific thinking through taking an active stance. There is of course a close connection between anticipation and conjecturing, setting up expectations which might then be confounded by examples or by the phenomenon, but which can then lead to scientific advance.

Legrand (1990, 1998/2000) has developed *scientific debate* as a format or structure within which learners are encouraged to conjecture and justify, challenge and try to convince each other, in order to develop their understanding of a concept or theorem, as well as to develop their powers to engage in mathematical-scientific practices (Warfield, 2005). Mathematical thinking in its full sense can really only thrive in a conjecturing atmosphere, where everything said is taken as a conjecture, and everyone is encouraged to test it and to offer potential modifications where necessary.

To establish an appropriate atmosphere and collection of practices the teacher needs to be able to hold back and listen (Davis, 1996, p. xxiv). Disturbance felt by learners is not often assuaged simply by assertions from teachers, especially when the teacher jumps in at the first opportunity to correct what a learner has offered. It does not matter that wrong conjectures are made. What does matter is the establishment of a conjecturing atmosphere, an ethos of enquiry; what does matter is that learners encounter counter-examples, and that they modify their conjectures until they are either put to one side for further exploration some other time, or justified using mathematical or scientific reasoning. Effective teaching has taken place when learners are moved (motivated) to go away and re-construct for themselves mathe matical objects as examples of concepts, applications of theorems, and illustrations of techniques.

1.3.4 Focusing and De-Focusing Attention

Attention is partly under conscious control, and partly not. People can have their attention attracted by sudden movement or sound, by the appeal of something with particular meaning for them, by having attention attracted or directed by a respected "other" such as a teacher, or by direct application of their own will-power. Careful observation of your own attention soon reveals that noticing begins with a sudden rush of concentrated attention, a sharpening of awareness, which then gradually dissipates until it is either refreshed and renewed or overlaid with some other noticing.

Attention can be highly focused on some detail, but it can also be very diffuse as when gazing at a diagram or into space waiting for inspiration. Attention can be uni-focal but also multi-focal as when you think about preparing supper while listening to a colleague and looking at something out the window.

Teaching can only be effective when learners and teacher are attending to the same thing and in similar ways. There are subtle but important shifts not only in what it is worth attending to, but how, which learners need to experience repeatedly as they begin to perceive and think in ways characteristic of the discipline and the particular topic. Sometimes it is useful to *hold wholes*, that is to gaze at the whole of a phenomenon, situation, object or experience. Sometimes it is important to work at *discerning details*, which includes locating boundaries that distinguish and define sub-wholes. Sometimes it is relevant to *recognise relationships* between particular details, either in the same whole or between wholes. Often it is valuable to *perceive properties* as generalizations of specific relationships, which might hold amongst other objects or other details elsewhere in the current situation or in the future. This is part of the move to generalization and abstraction. Relationships may arise as particular instances of perceiving a property to be relevant, and sometimes perceiving properties arises as an abstraction from relationships between particular details. This is another way of thinking about "seeing the general through the particular, and the particular in the general". Justification and explanation is usually based on articulating *reasoning* (based on intuition arising from gazing), which is only possible when it proceeds *on the basis of properties* that are accepted, and acknowledged, and agreed. These five subtly different structures of attention lie at the heart of what learners are doing when being taught. Although there is considerable overlap and consonance with the van Hiele levels articulated by van Hiele (1986) there is a significant difference when applied to attention itself. Rather than forming levels through which a learner might be expected to move, self-observation reveals that attention often shifts very quickly between structures, back and forth, round and round, while at other times one or other structure becomes quite stable.

One of the values of sensitizing yourself to these subtle shifts in structure is that they can be used to account for learner difficulties. For example, during exposition, learners gazing at a whole may not even hear distinctions being drawn by a teacher or be aware of relationships being identified. If they are occupied with discerning details they may not be able to make sense of descriptions of relationships amongst those details. If they are seeking out relationships in the situation before them they may not be aware of those relationships as instances of more general properties, or even the potential for those relationships to be generalized and perceived as properties. If they are coming to grips with properties as general qualities which may explain, capture, or indicate specific relationships, they may not be in a position to benefit from the teacher's use of certain properties as the basis of reasoning.

One of the reasons that scientific and mathematical reasoning might be so difficult to teach is that learners are rarely in an appropriate state with their attention encompassing the perception of properties, which can be used as the basis of reasoning. There is a subtle difference between using language (which is inherently general) to talk about specific relationships recognized in a situation, and reasoning about that and similar situations on the basis of properties. The same language is often used in both cases, but learner experience is quite different. This would account for the otherwise mystifying experience of learners who seem to be reasoning quite generally but who suddenly resort to particulars or fail to use that reasoning in a new situation. For example, Rowland (2001) reports the case of students claiming to follow and appreciate a number theory proof carried out in a particular case, but unwilling to accept the implied generality of the proof.

1.3.4.1 Making Use of Attention

Becoming aware of the structure of one's own attention, of not only what is being attended to but in what manner, improves sensitivity to learner experience. Awareness of one's own attention makes it more likely to notice mismatches with that of learners, and then to be motivated to construct tasks which provide the time necessary for learners to engage with different structures of attention, and which prompt learners to shift from one structure to another. Awareness of one's own attention also makes it possible to refer explicitly to the fact that it is possible and desirable to attend to certain things in certain ways. A number of frameworks have been developed which help as reminders when planning a session or when in the midst of interacting with learners (Floyd et al., 1981; Mason and Johnston-Wilder, 2004, 2004/2006).

See, Experience, Master

It is rare for learners to make a great deal of a concept or technique the first time they encounter it. With continuing exposure they develop experience, and as the practices and ways of thinking and perceiving become integrated into their functioning, they move towards mastery. The label *See–Experience–Master* serves as a warning, a trigger not to expect too much, and to construct tasks that provide requisite time and experience.

Manipulating – Getting-A-Sense-Of – Articulating

As many educators have suggested since Plato, having something tangible to manipulate can often assist in recognizing significant relationships and so making sense of some generality. Specializing or particularizing is another way to refer to the entirely natural act of seeking out a familiar and confidence-inspiring example or object through which to try to see the general. Getting learners to construct their own examples is one way to support learners in developing a rich repertoire of objects to which they can refer when they want to test out a conjecture or try to get a sense of some underlying structure or phenomenon. Attempting to articulate that "sense" can assist with the sense making, especially when it takes place within a supportive conjecturing atmosphere.

1.3.5 Organizing and Classifying

There is a sense of satisfaction and energy release when some sort of order or orderliness is brought to a collection of disparate and scattered items. The whole effect of language is to provide labels for collections of objects according to shared properties to support organizing and characterizing. Thus language has the effect of organizing experiences. To recognize suddenly that two or more apparently different situations are actually examples of one phenomenon makes future sense making much easier on the one hand, while on the other hand it sets up the possibility of failing to attend sufficiently to differences. Thus every act of organization has the potential to inform but also to mislead.

Much of mathematics and of science can be seen as a process of classifying phenomena, characterizing the conditions under which the phenomena come about or can be recognized. Theorems asserting that such and such a condition is equivalent to some other condition, or that all objects of one sort also have some additional property all serve to organize example spaces and characterize objects in those spaces.

1.3.5.1 Making Use of Organizing and Classifying

What Is the Same, What Is Different About?

Offering learners two or more objects, situations or phenomena and asking them what is the same and what is different about them directs their attention from specific details of individual cases to relationships which are common to several instances and helps create in the learner a sense of phenomenon (an identifiable repeatable experience) and hence generality. It also adds to their enculturation into becoming aware of what features are worth attending to and are of significance in the particular discipline, for each discipline consists of a collection of characteristic awarenesses and sensitivities to notice. Learners are most likely to become attuned to these when their attention is suitably directed and when they are in the presence of a relative expert whose attention is similarly directed.

The notion of invariance lies at the heart of elementary science. If you burn a quantity of substance and compare the masses before and after burning, there will be an invariant ratio. This is due to the fact that it is the nature of the substance, not its size, which matters in terms of physical and chemical properties.

Sorting and Ordering Tasks

Offering learners collections of objects (descriptions of specific experiments, states or phases in a single experiment, a collection of mathematical exercises or other objects) and asking them to sort them in some way that seems sensible, or in some way which illustrates a collection of concepts, or to order them in some suitable fashion invites learners to stand back from doing calculations and to think about structure, relationships and properties. Once they start thinking in terms of "types" they are organizing and classifying, as well as preparing themselves to be tested on an examination by recognizing the type of a question.

1.4 Concluding Comments

Teaching consists of acts which take place *in* time, but learning takes place *over* time, usually during sleep when the impressions of the day are selected, rejected, and organized. Where learners have been actively engaged, taking initiative and asserting rather than merely assenting, where they have engaged all five of the aspects of their psyche: enaction, affect, cognition, attention and will, they are more likely to have something salient to process during sleep, to link to other experiences, and so, to learn.

To promote an active rather than a passive stance, to get learners asserting themselves and becoming more intentionally agentive rather than merely assenting requires the teacher to establish an ethos, to set up conditions (the tasks set), and to interact with learners in ways which evoke and promote the use of learners' own powers. It requires developing confidence that learners will indeed make use of their powers if given the opportunity. It requires not so much "getting the sage off the stage" (King, 1993) as sensitizing yourself to notice opportunities for learners to take responsibility, to make subject-significant choices, to become active learners rather than mere receivers.

The powers suggested here play a central role in any mathematical or scientific activity. They can be used to inform the planning and conduct of workshops, lectures and tutorials, the design of course curricula, and the writing of texts. Adopting an approach to pedagogy and didactics which aims to evoke and develop learners' use of their own powers frees the teacher to work on the direction of learner attention rather than on rehearsing well-known content in front of assenting learners. It opens the way to having fruitful interactions with learners about the core aspect of mathematical and scientific thinking.

Constructs and frameworks such as those suggested here are only of value if they are integrated with personal experience. Integration is achieved through looking for examples in your own experience which highlight or illuminate the general, and mentally imagining yourself in a future teaching situation with a label coming to mind which triggers you to act in a fresh rather than habitual manner. It all depends on noticing an opportunity in the moment, whether when planning or when teaching, and the possibility of noticing freshly is enhanced by preparing for the future, based on reflecting on the past (Mason, 2002b).

References

Boaler, J. (1997). *Experiencing School Mathematics: Teaching Styles, Sex and Setting*. Buckingham: Open University Press.

Brousseau, G. (1984). The Crucial Role of the Didactical Contract in the Analysis and Construction of Situations in Teaching and Learning Mathematics. In H. Steiner (Ed.), *Theory of Mathematics Education*, Paper 54 (pp. 110–119). Institut fur Didaktik der Mathematik der Universitat Bielefeld.

Brousseau, G. (1997). *Theory of Didactical Situations in Mathematics: Didactiques des mathématiques, 1970–1990*, N. Balacheff, M. Cooper, R. Sutherland & V. Warfield (Trans. & Eds.). Dordrecht: Kluwer.

Brown, S., Collins, A. & Duguid, P. (1989). Situated Cognition and the Culture of Learning. *Educational Researcher*, 18(1), 32–41.

Bruner, J. (1996). *The Culture of Education*. Cambridge: Harvard University Press.

Bruner, J., Goodnow, J. & Austin, A. (1956). *A Study of Thinking*. New York: Wiley.

Davis, B. (1996). *Teaching Mathematics: Toward a Sound Alternative*. London: Garland.

diSessa, A. (1987). Phenomenology and the Evolution of Intuition. In C. Janvier (Ed.), *Problems of Representation in the Teaching and Learning of Mathematics* (pp. 83–96). Mahwah: Erlbaum.

Festinger, L. (1957). *A Theory of Cognitive Dissonance*. Stanford: Stanford University Press.

Fischbein, E. (1987). *Intuition in Science and Mathematics: An Educational Approach*. Dordecht: Reidel.

Fischbein, E. (1993). The Theory of Figural Concepts. *Educational Studies in Mathematics*, 24(2), 139–162.

Floyd, A., Burton, L., James, N. & Mason, J. (1981). *EM235: Developing Mathematical Thinking*. Milton Keynes: Open University.

Gattegno, C. (1970). *What We Owe Children: The Subordination of Teaching to Learning*. London: Routledge & Kegan Paul.

Gattegno, C. (1987). *The Science of Education Part I: Theoretical Considerations*. New York: Educational Solutions.

Gattegno, C. (1988). *The Mind Teaches the Brain*. New York: Educational Solutions.

Halmos, P. (1994). What is Teaching? *American Mathematical Monthly*, 101(9), 848–854.

Hamilton, E. & Cairns, H. (Eds.) (1961). *Plato: The Collected Dialogues Including the Letters. Bollingen Series LXXI*. Princeton: Princeton University Press.

Harré, R. & van Langenhove, L. (1999). Introducing Positioning Theory. In R. Harré & L. van Langenhove (Eds.), *Positioning Theory* (pp. 14–31). Oxford: Blackwell.

Johansson, B., Marton, F. & Svensson, L. (1985). An Approach to Describing Learning as Change Between Qualitatively Different Conceptions. In L. West & A. Pines (Eds.), *Cognitive Structure and Conceptual Change* (pp. 233–257). New York: Academic.

King, A. (1993). From Sage on the Stage to Guide on the Side. *College Teaching*, 41(1), 30–35.

Krutetskii, V.A. (1976). J. Teller, Trans.) In Kilpatrick, J. & Wirszup, I. (Eds.), *The Psychology of Mathematical Abilities in School Children*. Chicago: University of Chicago Press.

Laborde, C. (1995). Designing Tasks for Learning Geometry in a Computer Environment. In L. Burton & B. Jaworski (Eds.), *Technology in Mathematics Teaching: A Bridge Between Teaching and Learning* (pp. 35–67). London: Chartwell-Bratt.

Lakatos, I. (1976). *Proofs and Refutations: The Logic of Mathematical Discovery*. Cambridge: Cambridge University Press.

Lave, J. & Wenger, E. (1991). *Situated Learning: Legitimate Peripheral Participation*. Cambridge: Cambridge University Press.

Legrand, M. (1990). Débat scientifique en cours de mathématiques. *Repères-IREM*, 10, 123–158.

Legrand, M. (1998/2000). Scientific Debate in Mathematics Course. *La Lettre de la Preuve Novembre/Décembre 2000* on-line text. Retrieved May 2008 Http://europa.eu.int/comm/research/conferences/2004/univ/pdf/univ_sciencediss_mathematics_290704_en.pdf.

Lin, J. (2004). A Chinese Cultural Model of Learning. In F. Lianghuo, N.-Y. Wong, J. Cai & S. Li. (Eds.), *How Chinese Learn Mathematics: Perspectives from Insiders* (pp. 124–156). Singapore: World Scientific.

Love, E. & Mason, J. (1992). *Teaching Mathematics: Action and Awareness*. Milton Keynes: Open University.

MacHale, D. (1980). The Predictability of Counterexamples. *American Mathematical Monthly*, 87, 752.

Marton, F. & Booth, S. (1997). *Learning and Awareness*. Hillsdale, USA: Lawrence Erlbaum.

Marton, F. & Trigwell, K. (2000). 'Variatio est Mater Studiorum'. *Higher Education Research and Development*, 19(3), 381–395.

Mason, J. (2002a). *Mathematics Teaching Practice: A Guide for University and College Lecturers*. Chichester: Horwood Publishing.

Mason, J. (2002b). *Researching Your Own Practice: The Discipline of Noticing*. London: RoutledgeFalmer.

Mason, J. (2003). On the Structure of Attention in the Learning of Mathematics. *AAMT*, 59(4), 17–25.

Mason, J. (2004). Doing ≠ Construing and Doing + Discussing ≠ Learning: The importance of the structure of attention. *Regular Lecture at the 10th International Congress on Mathematics Education*. Copenhagen, July 2004.

Mason, J. (2007). Making Use of Children's Powers to Produce Algebraic Thinking. In D. Carraher, J. Kaput & M. Blanton (Eds.), *Algebra in the early grades*. Hillsdale: Erlbaum.

Mason, J., Burton, L. & Stacey, K. (1982). *Thinking Mathematically*. London: Addison Wesley.

Mason, J. & Johnston-Wilder, S. (2004). *Fundamental Constructs in Mathematics education*. London: RoutledgeFalmer.

Mason, J. & Johnston-Wilder, S. (2004/2006). *Designing and Using Mathematical tasks*. Milton Keynes: Open University, republished (2006). St. Albans: Tarquin.

Mason, J. & Watson, A. (2001). Stimulating Students to Construct Boundary Examples, *Quaestiones Mathematicae*, 24 (Suppl 1), 123–132.

Montessori, M. (1912 reprinted 1964). (George, A. Trans.) *The Montessori Method*. New York: Schocken Books.

Norretranders, T. (1998). (J. Sydenham, Trans.). *The User Illusion: Cutting Consciousness Down to Size*. London: Allen Lane.

Noss, R. & Hoyles, C. (1996). *Windows on Mathematical Meanings: Learning Cultures and Computers*. Dordrecht: Kluwer.

Piaget, J. (1971). *Biology and Knowledge*. Chicago: University of Chicago Press.

Pólya, G. (1962). *Mathematical Discovery: On Understanding, Learning, and Teaching Problem Solving*. New York: Wiley.

Popper, K. (1972). *Objective Knowledge: An Evolutionary Approach*. Oxford: Oxford University Press.

Renkl, A., Stark, R., Gruber, H. & Mandl, H. (1998). Learning from Worked-out Examples: The Effects of Example Variability and Elicited Self-explanations. *Contemporary Educational Psychology*, 23, 90–108.

Rissland, E.L. (1991). Example-based Reasoning. In J.F. Voss, D.N. Parkins & J.W. Segal (Eds.), *Informal Reasoning in Education* (pp. 187–208). Hillsdale, NJ: Lawrence Erlbaum.

Rowland, T. (2001). Generic Proofs in Number Theory. In S. Campbell & R. Zazkis (Eds.), *Learning and Teaching Number Theory: Research in Cognition and Instruction* (pp. 157–184). Westport, CT: Ablex.

Salomon, G. (1979). *Interaction of Media, Cognition and Learning*. London: Jossey-Bass.

Spencer, H. (1878). *Education: Intellectual, Moral, and Physical*. London: Williams and Norgate.

Sweller, J. & Cooper, G.A. (1985). The Use of Worked Examples as a Substitute for Problem Solving in Learning Algebra. *Cognition and Instruction*, 2, 59–85.

Thompson, P. (2002). Didactic Objects and Didactic Models in Radical Constructivism. In K. Gravemeijer, R. Lehrer, B. van Oers & L. Verschaffel (Eds.), *Symbolizing, Modeling and Tool Use in Mathematics Education* (pp. 191–212). Dordrecht: Kluwer.

van Hiele, P. (1986). *Structure and Insight: A Theory of Mathematics Education*. London: Academic Press.

Walker, J. (1975). *The Flying Circus of Physics with Answers*. New York: Wiley.

Warfield, V. (2005). (accessed June 05). Calculus by Scientific Debate as an Application of *didactique*. On line document. Retrieved May 2008. Http://www.math.washington.edu/warfield/articles/Calc&Didactique.html

Watson, A. & Mason, J. (2005). *Mathematics as a Constructive Activity: The Role of Learner-Generated Examples*. Mahwah: Erlbaum.

Wertheimer, M. (1945). *Productive Thinking*. New York: Harper.

Whitehead, A. (1932). *The Aims of Education and Other Essays*. London: Williams and Norgate.

Winston, P. (1975). Learning Structural Descriptions from Examples. In P. Winston (Ed.), *The Psychology of Computer Vision* (pp. 157–210). New York: McGraw-Hill.

Wood, D., Bruner, J. & Ross, G. (1976). The Role of Tutoring in Problem Solving. *Journal of Child Psychology*, 17, 89–100.

Chapter 2
A Reflective Science Education Practice

Why, What, and How?

Kathrine Krageskov Eriksen

2.1 Reflectivity: A Key Concept

When the Danish University Law (Videnskabsministeriet, 2003) was reformed, a component was added stating that Danish universities should not only do research and offer education that meet the highest international standards, but that they should, moreover, convey to society research and educational results for the further prosperity of the latter. In the legal text it is further explained that universities shall be obliged to actively exchange knowledge and competence with society, and that this includes an obligation to engage in public debate about important social issues. Apparently, something has changed – a change that seems to include a closer entanglement of academic science and society. In the context of science education at least two questions are thus raised: How is the social role of science changing? And what (if any) implication does this bear for science education?

The argument made here is that a change is indeed taking place; from traditional (academic) science which celebrates disinterestedness and unblemished quest for objective knowledge to contemporary science infused with socio-cultural and political-capitalist interests and influencing societal development profoundly. A change, which makes the scientific community a much-needed participant in social reflectivity. However, studies encountered below suggest that contemporarily the scientific community tends to cling to the old celebrated virtues and it only reluctantly enters its new role as participant in societal development and debate. The problem is not only that science apparently "chooses" not to participate in the social reflective processes, but more significantly that this choice is grounded in the traditional academic ethos according to which scientists of today have been trained. Science education, the entry point of the scientists of tomorrow, hence becomes extremely important as a place for inducing changes to the institution of science and the scientific self-perception.

I suggest that as a response to this, tertiary science education must focus on the development of students' scientific reflectivity, and that this concept includes the

---

K. K. Eriksen
University College Sealand, Denmark

O. Skovsmose et al. (eds.), *University Science and Mathematics Education in Transition*,

development of contextualized awareness and professional humility, as well as gap sensitive interactive skills. Further, I launch and develop a teaching method, anchored ethical dialogue, through which one may stimulate the development of all three notions. Hence, anchored ethical dialogues could become fix points for the development of scientific reflectivity and thus, ultimately, for raising the level of social reflectivity. I have attempted to depict this in Fig. 2.1.

However, before considering the various concepts and notions included in Fig. 2.1 in more detail, I will take a closer look at the development postulated above.

2.1.1 Science and Risk Society

Modernity can be seen as an era in which modern scientific thinking and technological development created a profound belief in progress – a belief that was inevitably linked to a trust in the ability of science and technology to further free and enlighten humankind. However, since these heydays of the modern breakthrough the role of science within society has changed.

> In developed civilisation, which had set out to remove ascriptions, to evolve privacy, and to free people from the constraints of nature and tradition, there is thus emerging a new global ascription of risks, against which individual decisions hardly exist for the simple reason that the toxins and pollutants are interwoven with the natural basis and the elementary life processes of the industrial world.
>
> (Beck, 1992, p. 41)

As the quote illustrates, to Ulrich Beck the central concept for describing this change is *risk*. It is the production of risks and the attention subsequently ascribed to these risks by society that constitutes the driving force in societal transformation.

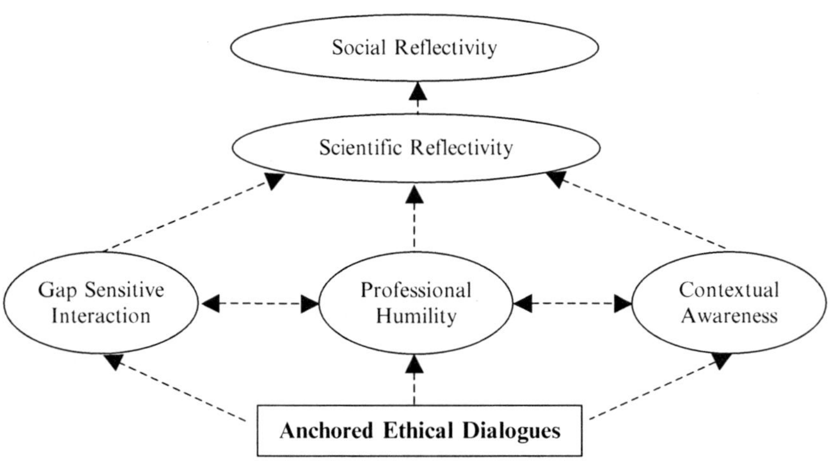

Fig. 2.1 A reflective educational practice

Beck understands risks as something potentially concrete. Risks are the potential results of the activities of humankind. The main difference between the risks Beck describes and the risks of preceding societal orders is thus their relation to modernity. Whereas the risks of previous ages were stemming from nature (or from the scarcity of resources), the risks marking present society are somehow related to the modernization processes that had the liberation of humankind from these restraints of nature as their goal. More precisely, they are the results of the very same modernization processes: The notion "risk society" refers to "a stage of radicalized modernity" (Beck, 1997, p. 20), where it is the success of modernity, the liberation of humankind from poverty, hunger and so forth that produces a new social order.[1] The unintended consequences or "hazardous side effects" (Beck, 1992, p. 20) of modernity, that is of the scientific and technological development constituting its hallmark, the production of risks, become central to the development and the focus of our attention. Thus according to Beck what we are experiencing is a break inside modernity where society removes itself from traditional industrial society and assumes a new constitution, but where the break is a result of the processes of the very same modernity: "Modernization has consumed and lost its other" (Beck, 1992, p. 10), that is nature and tradition, and now turns toward itself – hence the designation reflexive modernity.

It is exactly this attention given to the observed and potential effects of our own endeavours that characterizes risk society and becomes a new social driving force. The concept of risk does not (only) refer to damage or destruction of nature in itself, it refers to potential destruction, the potential time bomb against our health and reproduction that may and may not have been created: "The concept of risk thus characterizes a peculiar, intermediate state between security and destruction, where the *perception* of threatening risks determines thought and action" (Beck, 1999, p. 135).

Beck's analysis thus points to important differences between the emerging and the preceding social orders in relation to the perceived risks/hazards. One of the most important points is the involvement of science and it is exactly Beck's pinpointing of the intimate relationship between science, technology and social development which makes the risk society perspective a useful one to apply in the present context. Thus, the central trait of risk society, the characteristic differentiating it from pre-risk societies, Beck stresses, is not the occupation with risk, nor is it the increased extent of these risks. The crucial feature is the scientific constitution and the society-changing scope of the risks (e.g., Beck, 1992, pp. 153ff).

However, the role of science in risk society is not merely that of an enfant terrible, producing all the risks we are encountering. The situation is far more complex. It is not the failure of science but its success that constitutes the basis for the up-growth of risk society. Scientific and technological development is not simply the background for the very improved conditions for humankind of today, for we are also dependent on science for detecting and dealing with the risks of this risk

[1] Again, we are talking about the Western world.

society. Thus, science takes at least three social roles, that is the role of risk creator, risk detector, and risk analyser (Beck, 1992, Chap. 7).

The role of science as risk creator of course refers to the crucial involvement of science and scientific thinking in all aspects of modern society and specifically in the creation of solutions to the problems of humankind. Solutions, the unintended consequences of which, we are now facing in risk society. This, the societal transforming role of science has become visible. Further, we are absolutely dependent on science for the process of detecting and analysing risks. The risks of risk society are intangible; the human sensory apparatus often cannot directly experience them. Moreover, there is an enormous gap in time and space between impact and the manifestation of a risk. Risks instead come to life only in the form of scientized probabilities, which speak in "the language of chemical formulas, biological contexts and medical, diagnostic concepts" (Beck, 1992, p. 52). The extent of endangerment of the individual cannot be estimated by the individual him – or herself. Essentially, we all have to rely on external, scientific knowledge and thus we are becoming "*incompetent* in matters of [our] own affliction" (Beck, 1992, p. 53). This development leaves science in a vital social position and it exposes exactly how crucial it is for society that the scientific sphere is aware of its own position, abilities, and limitations.

2.2 Scientization

This development has of course affected science and the view on the scientized approach to life in the modernity. Beck describes this development by employing the notion of *reflexive scientization*. Initially man's relations to the surrounding, given world was scientized (*simple scientization*). However, as scientization has gradually changed this world, science is being confronted with itself, that is its own products and defects, and as a result our relation to science becomes scientized (reflexive scientization) (Beck, 1992, p. 155).

Beck goes on to say that reflexive scientization also results in the dogmatization of science being abolished, in that science has turned towards itself and scientized the scrutiny of its own foundations, products, and effects. It may be argued, however, that Beck's use of the term reflexive scientization in the latter sense becomes somewhat confusing. Who exactly is reflecting on whose foundations, products, and effects? If the answer, as indicated by Beck (1992, p. 156), is that the study (sociology, philosophy etc.) of science is scrutinizing the foundations of science scientifically, then the problem is that the concept of science is employed too broadly. This is not science scrutinizing its own foundations. Rather, it is one sphere of science scrutinizing the foundations of another sphere of science, for example natural science. So, seen in this light it is indeed *reflexive* scientization in that scientization is being reflected scientifically – but more so in the sense of a mirror being held up by others than in the sense of self-reflection. And crucially, it does not mean that (natural) science cares to look into that mirror: it may carry on its endeavours unaffected! So, at least in this interpretation, I see some problems in Beck's line of argument.

Instead, I suggest the introduction of an additional notion that is *reflective scientization*. I suggest replacing the use of reflexive scientization in the meaning outlined above (in the mirror sense), and instead use it to designate scientific self-reflection, that is the (natural) scientific sphere reflecting on its own foundations, limitations and so forth and working actively to change these (and in this process the insight gained from for instance science studies is of course vital). I believe this separation of notions (and of scientific spheres) makes the analysis from Beck more clear and fruitful to employ in the present context. And actually, according to Beck, it is exactly such a reflective science that is needed:

> To be sure, risk cannot be banned from modern life, but what we can and indeed should achieve is the development of new institutional arrangements that can better cope with the risks we are presently facing; not with the idea in mind that we might be able to regain full control, but much more with the idea in mind that we have to find ways to deal democratically with the ambivalences of modern life and decide democratically which risks we want to take.
>
> (Beck, 1999, p. 108)

The point is not merely that science should raise its voice in the political debate about the application of its results. Rather, Beck implies a changed approach to doing science, in which science is capable of self-criticism concerning all aspects of its own endeavours, including "what it considers noteworthy or not, how it asks questions and casts the "nets" of its causal hypotheses, how it decides on the validity of its conjectures [...]" (Beck, 1992, p. 180). Only in this way can we try to discover and thus avoid the unintended consequences, which are adding to the risks of risk society. In short we can say: To deal with reflexive modernisation (in the shape of a risk society), we need reflectivity within the institutional setting of science!

This realization provides us with a link to the overall topic of the present study, science education. Development of science education could provide one way of promoting reflectivity. To add to our understanding of the development needed, I will briefly turn to a number of other scholars and their interpretation of the relationship between science and society in the emerging social order.

2.2.1 Post-Academic and Mode II Science

Traditionally, science, or maybe rather the scientific archetype or "the stereotype of science in its purest form" (Ziman, 2000, p. 28), has been academic science, that is the social institution of science as one has been able to find it within academic institutions for the last century or so. It is John Ziman's claim that although the term academic science by no means covers all scientific work going on[2]; in many respects

[2] Science within the setting of industry accounts for up to 90% of all research taking place (Ziman, 2000, p. 16).

it does, however, epitomize the traditional scientific self-understanding. Academic science of course includes many disciplines with separate sets of rules and epistemic values. However, "the sub-tribes of academia span a common culture" (Ziman, 2000, p. 31) and to describe this culture in terms of its ideal values Ziman refers to the set of norms defined by the sociologist Robert Merton (1968). The norms include communalism, universality, disinterestedness, and organized scepticism and together they describe an ideal scientific endeavour that shares knowledge publicly, is culturally unbiased and free of material interests, and in which critical examination of published findings through peer-review provides the framework for systematic testing of claims and thus the basis for the reliability of scientific findings. Put together, the initial letters of the Mertonian norms spell out the acronym CUDOS meaning acclaim or prestige, and ideally it is by performing and contributing to science according to these norms, the scientific ethos, that academic scientists earn recognition among their colleagues in the form of citations, titles, and employment (Ziman, 2000, Chap. 3). The extent to which science has, and possibly can, actually be carried out in compliance with this ethos has been much debated, nevertheless the ethos influences scientists' behaviour and general attitude. For example, the norm of disinterestedness has been supportive of the view that the sphere of science can been seen as separated from "real life" (Ziman, 2000, pp. 53ff).

A crucial point made by Ziman is, however, that academic science is changing and giving way to a new *post-academic* science governed by a different ethos and occupying a different social role. The main characteristic of post-academic science is that it is turned outwards to a degree unknown to academic science. This is not only with reference to a growing openness between various disciplines inside science. The focus of scientific attention is becoming increasingly influenced by outside forces. Post-academic science is to a large degree oriented towards complex conglomerates of problems that require multi-disciplinary approaches and thus force scientists to work together across disciplinary boundaries. Financial limits to growth of the scientific sector paired with an increasing instrumental sophistication and costliness in many scientific disciplines also force scientists to work together. Moreover, the limited financial sources have precipitated a hitherto unknown competition for research funding. This development in turn influences scientists' choice of research objects, since the funding bodies give away money according to priorities influenced by current political and economic interests. Consequently, the scientific focus shifts from the general quest of knowledge towards potential applications and exploitation of the knowledge generated according to current, and therefore necessarily limited, vision (Ziman, 2000, pp. 66ff).

However, the traditional academic scientific ethos is not compatible with this development and the changed requirements. Thus, the emergence of post-academic science signals the beginning of a novel scientific culture. In the new culture, for instance, the commitment of making public the knowledge acquired is likely to often conflict with commercial interests. The focus on specific problem solving may link research and technological development even more closely than is presently the case. In evaluating research results peer review will thus be supplied or replaced by quality control of people, projects, and performance. Consequently, post-academic science will be infused with social values and interests. This may have consequences for the reliability of scientific knowledge – negative as feared by Ziman (1996) or positive as postulated by Nowotny et al. (2001).

Gibbons et al. (1994), and Nowotny et al. (2001) provide related accounts of contemporary changes to science. They propose that the "closer interaction of science and society signals the emergence of a new kind of science: contextualized, or context-sensitive, science" (Nowotny et al., 2001, p. vii) in which the value of science becomes more connected to its concrete use and the social impact of this use. The critique of scientific results is hence no longer primarily an internal scientific matter; various social actors with various social interests now join in. The authors suggest that a broad discussion of novel scientific development will test more aspects of this development than did traditional scientific discussions and therefore in a sense make it more reliable or *socially robust* (Nowotny et al. 2001, Chap. 11). Ziman, on the other hand, points out that this means that society loses science as *the* impartial expert.

In all circumstances it means that future demands put on scientists will be revolutionarily different to the demands put on scientists in the traditional academic setting. Scientists must be aware of aspects of their research that are very different from the aspects relevant in the academic scientific culture. Instead of praising detachment from outside interests and the remoteness to practical utility of research results, scientists must become capable of anticipating potential applications and consequences, whether environmental, social, ethical and so forth, of their work. And they must become capable of discussing potential future scenarios with actors from outside the scientific sphere. Or, using the terminology developed above, they must become able to act reflectively.

2.3 Reflectivity as an Educational Concept

A next question arises: May education actually influence individual scientists and, crucially, science as an institution to act reflectively? In exploring an answer it is difficult to avoid the concept of *Bildung* from the German educational tradition.

2.3.1 Science Education in a Bildung Perspective

Bildung has classically been interpreted as education or maybe, rather, formation of the individual to become able to reflect upon and act to change the common conditions. I will not go into a further presentation of the Bildung idea here, but focus specifically on the distinction adaptation/Bildung launched by the Norwegian philosopher Jon Hellesnes (1976). As already noted, in the classical Bildung tradition Bildung is formation to become able to better the common conditions, not merely to become able to enter a community. Nevertheless, the former necessitates the latter! Thus, the formation processes undertaken by the individual, and in the present context this equals education, will always be a form of socialization. As Hellesnes accentuates, the crucial issue is what kind of socialization is taking place. Hellesnes sees Bildung as the antithesis to adaptation and hence as a *reflective* socialization in which "the rules of the game" are uncovered and critically evaluated – and talking general Bildung this game is society at large.

However, if we transfer this interpretation of Bildung to the tertiary science education context, some fundamental questions arise: Which game and rules are we talking about? On the face of it, the answer is of course straightforward. We are talking about science education, thus the game is surely science and the rules are obviously scientific ways of dealing with reality. However, what we learned from the analysis of the relation between science and society is that the answer is *not* straightforward! Science can no longer (if ever) be seen as an isolated entity. Rather, science is to a hitherto unknown degree interwoven in the societal development and in pending social problems, both so in the form of risk creator, risk detector, and risk analyzer and in relation to the structures determining the focus of scientific research. Moreover, it was suggested that science and the conventional scientific self-perception is inadequate for dealing with this changed role of science. It is against this background that the answer is clearly not merely science. The answer is to be found in the connection of science as a knowledge generator and as a social actor. The game new scientists are to become socialized into is thus this complex social institution of science, and the rules they are to learn concern scientific research as such but also the relationship between science and society. Further, if this socialization is to amount to Bildung and not merely adaptation, this complex setting has to be somehow unveiled to the students *and* the students must become capable of critically reflecting upon it. The next question is of course, how can this be done? If education is socialization and what is desired is reflective socialization, then how is this acquired?

2.3.2 Science Education as Socialization

One proposed link between education, the socialization/Bildung of coming scientists, and in turn societal development may be extracted from a Danish study of higher education by Bo Jacobsen (1981). Jacobsen hypothesizes a relationship between educational structure and personality types. Employing Jacobsen's study, it may be proposed that socialization processes in higher education depend on at least two structural levels: An upper level of organization and general content of the education in question and an intermediate level of the concrete pedagogical interaction taking place at the educational institution. Further, that it is the "tightness" or "looseness" of these different structural levels of the educational system that influence the socialization of students, so that tight systems tend to produce more tight-minded individuals than loosely structured ones do.

However, to claim that these structural levels determine the socialization of students is a simplification. First, students are not empty jars upon their arrival; they have already been socialized through their previous experiences, educational and others. Second, throughout their higher education students will receive inputs from sources other than the educational system per se; inputs that will also influence their socialization. And third, students are not simple mechanisms that can be expected to react in a certain way to a given stimulus.

Thus, I believe that Jacobensen's approach, which may be read as being solely structuralistic, needs modification. In doing this, I will lean on Anthony Giddens' theory of structuration.[3] The theory of structuration is Giddens' contribution to resolving the agent-structure dualism. He views the relationship between agent and structure as a coherent relation or duality where structures are seen as both the means for and the results of the actions of individual agents (Fig. 2.2). In Giddens' optics, structures do not exist per se; rather they are constantly created and re-created through our actions. Since we base our actions on structures, we continuously reproduce them. The link between agent and structure is thus the social practice (Kaspersen, 1996, pp. 398–400).

Employing Giddens' theory of structuration, socialization can be comprehended not merely as the product of structures but also as the basis for the reproduction of structures and thus ultimately for the structures themselves. Therefore, the adoption of this perspective can help us see beyond the existing structures and highlight the potential for change. Giddens gives part of "the responsibility" back to the individual agents since structures in his point of view cannot be comprehended as something that *is* but only as something that *becomes* through reproduction (by agents). Thus, in contrast to a one-sided structure perspective Giddens' theory provides a link between changes to the socialization of agents and changes to overlying structures, in turn society – or in the present context between science education and the role of science in society. In other words, in Giddens' perspective working with the lower structural levels (as we in educational work do most of the time) or trying to "do something" as an individual agent, does posses potential for change.

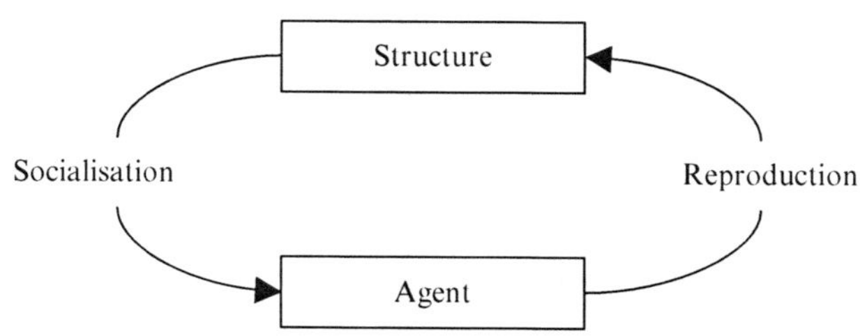

Fig. 2.2 Giddens's structuration theory – the social practice[4]

[3] Giddens' structuration theory can be found in an elaborated version in Giddens (1984). However, I base my description on Kaspersen (1996).

[4] The figure is based on the description of Giddens's theory of structuration from Kaspersen (1996).

If we apply Hellesnes' distinction between two forms of socialization, Bildung and adaptation, to Giddens' theory of structuration then, depending on the socialization type, the actions of the agents, that is the reproduction of structures, becomes not merely reproduction but either a "blind" reproduction (simply following the game and the rules) or a reflective reproduction (questioning the game and the rules and if necessary working actively to change them). To illustrate this point I have elaborated Fig. 2.2 into Fig. 2.3.[5]

At this point, I will again consider the concept of reflectivity and use the term to designate both a reflection (cognitive action) and the action for change of the practices (active transformation). Against the background of the preceding analysis of education as socialization, a higher science education that matches the social challenges identified earlier may thus be conceptualized as a socialization process allowing for reflectivity.

Thus, this analysis suggests that science education could be one place to initiate changes to the scientific self-perception. And further, that an educational practice which acknowledges the social need for a scientific community prepared to confront the problems and risks connected to scientific and technological development (and is thus developed as a response to the social challenges analyzed above) may be characterized as an educational setting in which the socialization of students is considered explicitly and measures are taken to ensure that the socialization processes are open for reflectivity.

But what does this imply in more concrete educational terms? On a very general level it implies that scientists need to develop additional abilities, sensitivities or competences. In addition to traditional academic scientific virtues (e.g., possess a broad professional base; skilled in laboratory techniques; the ability to find necessary information, and the ability to make new discoveries), students must develop

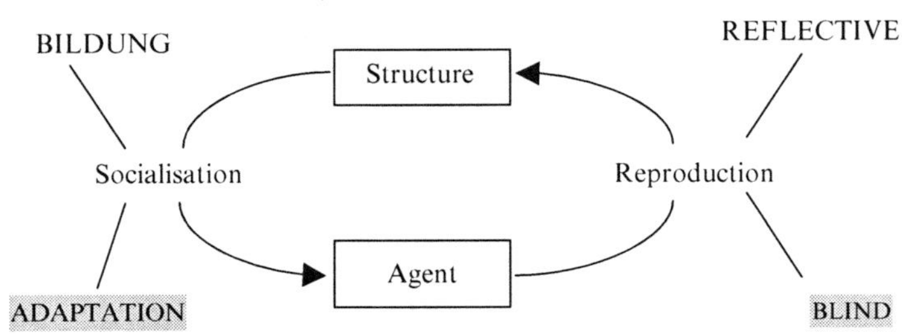

Fig. 2.3 Giddens' structuration theory in a Bildung perspective

[5]Giddens works with a related distinction when differentiating between practical and discursive consciousnesses. Our practical consciousness guides our actions – provides the background for "playing the game" – but most of the time we do not make explicit our knowledge. In contrast, our discursive consciousness enables us to present explanations for our actions and thus it also provides the basis for changing these actions (Kaspersen, 1996, pp. 400–404).

a reflective approach to the scientific endeavour. The final questions that this study aspires to answer hence become the following: What does development of reflectivity imply? And how may we change higher science education as to put more focus on development thereof?

2.4 Reflectivity and Pedagogical Interaction[6]

In Bo Jacobsen's study (1981) two structural levels influencing students' socialization are identified: The upper level of organization of the studies and the level of pedagogical interaction. Concerning the upper level, one may tentatively say that organizational structures that endorse an integrative approach and the development of own structures must be promoted if the socialization process of students is to be open to reflectivity. I will, however, not engage in further analysis of this structural level – but focus instead on the second level of pedagogical interaction, which is of particular relevance in the present context.

2.4.1 Reflectivity and Content

In relation to the level of pedagogical interaction at least the content, the format, and the social relations of teaching and exam situations need to be considered. Concerning content let me start by introducing an example: teaching university chemistry students about halogenated organic compounds usually involves an introduction to the spatial structure of the compounds, ways of synthesizing the compounds and concrete examples of compounds from the group, DDT being one. This type of knowledge relates to chemistry as a product and it includes knowledge about chemical compounds, concepts, and laws. Borrowing from Leif Östman (1999), for the present purpose I will refer to this type of knowledge as *ontological* chemical knowledge. However, further aspects of chemical knowledge exist. First, in the teaching of halogenated compounds the historical background to the synthesis of these compounds or perhaps a discussion of the synthesis and testing procedures linked to the development of new chemical compounds could be included. All these aspects are linked to another sphere of the subject of chemistry, which I will refer to as the *epistemological* sphere (Östman, 1999) or the understanding of chemistry as an activity and as a scientific community. Second, the subject of chemistry also consists of a third sphere. This could be referred to as the social or ethical sphere (Östman, 1999) and it contains knowledge of chemistry in a social context,

[6] This interpretation of scientific reflectivity is partially based on experience from an experimental ethics teaching sequence and subsequent interviews with chemistry students at the University of Copenhagen (see Eriksen, 2003).

including questions of how chemistry is part of society and which considerations (ethical) could be made in this regard. In the case of halogenated hydrocarbons a discussion of the use of DDT as an insecticide and the consequences now being linked to this use could be a way to include this third sphere of the subject of chemistry in the actual teaching.

The explicit incorporation of all three spheres of chemical knowledge into tertiary chemical education could help ensure reflectivity at the subject content level – the constant reflection on this content knowledge: What is chemical knowledge? How is it produced? Is it true? How is it used? What are its limitations? What are the benefits and dangers connected to this use? Do we as chemists have a responsibility for this use? Traditionally, much chemistry teaching at university level has primarily been linked to the ontological knowledge sphere of chemistry, carrying with it a tendency to treat the subject of chemistry as a collection of factual information that should be learned as well as possible (Eriksen, 2003). But as discussed above this is not an adequate approach; all three spheres of chemical knowledge must be included in the teaching in order to explicate and open the rules of the chemistry game for reflection and debate. Paraphrasing Kant we can say: reflectivity without content is empty, content without reflectivity is blind!

2.4.2 Reflectivity and Format

In the interpretation launched here, reflectivity is connected to both the understanding of rules and of the development of abilities to reflect upon and actively change these rules. This accentuates the role of the actual teaching, and it seems obvious that forms of teaching where the teacher presents accepted knowledge to the students who are then supposed to internalize this knowledge is not compatible with this aim. Wolfgang Klafki (2001, pp. 162–184) as the teaching basis for his *categorial Bildung* suggests *exemplary learning*. Exemplary learning, basically, employs the idea that working with the particular case as starting point, students can gradually develop more or less generally applicable insights. By means of these general insights students should become able to comprehend structurally related phenomena; that is to develop categorial insight, and thus ultimately to critically evaluate and act upon future categorically related challenges. Employing an exemplary learning approach, the teaching process is therefore not seen as the presentation of pre-determined knowledge and skills for the students to acquire, rather it is focused on pedagogical assistance of the students' own active learning. Central formats for exemplary learning are hence experimental work and project work in which the acquisition of basic, categorical insights and abilities are prioritized. However, students cannot acquire all the knowledge they need through time-consuming exemplary learning. It must be supplemented with orientation-like teaching. In Klafki's point of view exemplary learning provides precisely the necessary ballast for extracting meaningful messages from a more informative (fact-presenting) teaching. Thus, exemplary absorption and the acquisition of broad orientation-like knowledge are not merely supplementary; they actually provide the necessary basis for each other to occur.

Accepting Klafki's analysis of exemplary learning we can now undertake a more nuanced discussion of teaching formats and socialization of students. If the aim is opening for reflectivity, the teaching can of course not be carried out solely in transfer-oriented settings. However, from Klafki's analysis it becomes clear that the crucial factor is not that all teaching activities are based on project work and room for students' own discovery and categorization of reality. Rather, the idea must be to secure anchorage points in the subject matter. At crucial "points of impact" in the content of a given course students must be given the opportunity to work independently and in depth and thus to develop their own categories or structures for comprehending this particular exemplary piece of knowledge.

2.4.3 Reflectivity and Social Relations

According to Jacobsen's (1981) subdivision of the intermediate pedagogical structural level, a third dimension has to be taken into account when the influence of this level on the students' socialization is discussed: the social relations. And it seems obvious that adaptive socialization and socialization as Bildung are correlated with very different forms of social interaction between both teachers and students and in-between students. If students are to develop the ability to criticize the rules of the game they cannot solely find themselves in situations where someone else has the authority to determine right from wrong. Instead, at least some of the time they must become engaged in situations where their own interpretations and opinions are recognized and valued. In other words, the teaching situations must sometimes be based on dialogue – a dialogue in which students and teachers engage on (to the extent possible) equal terms.

2.4.4 Reflectivity and a Note on Exams

My reflections on the structural level of pedagogical interaction have primarily been concerned with the teaching setting. However, as Jacobsen also stresses in his study, the settings of teaching and exams act in concert. Parallel to the above argued effects of the explicit and implicit messages sent to students through the teaching, the organization of exams can be said to profoundly influence the students' enculturation into accepted attitudes and priorities and thus ultimately their socialization. To exemplify, the teaching may very well be based on project work and aimed at the development of insight into the underlying principles. If exams test and reward only the acquisition or rote learning of facts and figures students will of course adjust to this and focus on these aspects in their studies, despite the intentions behind the teaching.

I will not, however, more explicitly consider the various exam formats. The reflections on exam content, format and social relations do not differ significantly from the reflections on the teaching setting already encountered and I will leave the transfer of arguments to the reader.

2.5 Reflectivity Operationalized

As already revealed the present study suggests that education into a reflective scientific practice may be operationalized through the concepts of *contextual awareness, gap sensitive interaction,* and *professional humility*. These concepts grew from experimental teaching of ethics at the University of Copenhagen and subsequent evaluation in collaboration with students and teachers (Eriksen, 2003).

2.5.1 Contextual Awareness

As discussed above the concept of reflectivity entails an aspect of *contextual awareness*. If scientists are to contribute to the social reflectivity they must be(come) aware of the contextualized nature of science. That is, they must develop an understanding of the cultural and philosophical context in which science is embedded. I suggest that the development of this contextual awareness requires that the curriculum as a whole reflects the contextualized nature of science. Otherwise the discrepancy between for example one course in philosophy of science and the "hidden curriculum" behind the composition of the rest of the courses may lead (some) students to, at best, view philosophical reflections on science merely as a spice on the "science dish" or, at worst, as a waste of time. Exemplary case based projects could act as possible anchorage points for unfolding of the contextualized nature of science.

Taking chemistry education as an example, this idea of contextual awareness and a case based development thereof can be illustrated. Developing contextual awareness in relation to organic chemistry means that students must develop an appreciation not only of different types of organic compounds, typical reactions, and synthetic pathways but also of the contextual embedding of organic chemistry. In other words the need to develop knowledge from all of the three spheres presented above, that is the ontological, the epistemological, and the ethical spheres. The field of organic chemistry is embedded in a historical, disciplinary and societal context. Developing an awareness of the two former contexts means that students must gain insight into the historical development of the chemical subfield of organic chemistry: what is the basis for the emergence of such a field? Why is it called organic chemistry? How has it developed? What are its aims? What are its constraints? How is it connected to other chemical disciplines? And developing an awareness of the latter context means that students must learn about the applications of organic chemistry; about the panoply of societal contexts dependent on organic chemistry (food industry, medicine, agriculture to mention a few) and about the enormous impact which the explosive development of our ability to synthesize new organic compounds has had on society.

Exemplary learning based on a case centred on one organic compound like DDT may easily exemplify many of these aspects (Simon, 1999). Working with a DDT case along different lines could for example illustrate the emergence of an organic chemical subfield (based on the growing understanding of the structure of organic compounds, chemists in the last part of the nineteenth century became able to analyze compounds and to suggest methods for their synthesis and DDT was first synthesized in this exciting time). Furthermore, the DDT story illustrates a growing application oriented and commercial interest in organic synthesis (the insecticidal properties of DDT were discovered during a large screening project for potential insecticides); the widespread use of organic compounds in society (DDT has been used as an insecticide in private homes, the military, and in agriculture); environmental effects (the story of DDT in nature illustrates food chain accumulation, degradation, or lack thereof, spreading from the point of impact, effects on living organisms etc.); as well as social effects (the development of a large industry-like agriculture in the West has been dependent on fertilisers and insecticides and it has transformed many aspects of society, environmental issues influence a wide range of societal actions from political decision making to grocery shopping). Thus, by working with a case like the DDT story students' contextual awareness of organic chemistry could be developed. The idea is not to turn science students into historians or sociologists but to develop their awareness of the contextual embedding of their own discipline in order to develop their ability to co-create social reflectivity in their role as scientists.

2.5.2 Gap Sensitive Interaction

As outlined above much more than good rhetorical skill is an issue when developing an interactive competence suited to partaking in social reflectivity. Students must become able to interact in what we might refer to as "a gap sensitive way." That is, students must become able not only to present research results in an easily comprehensible way, but also to engage in discussions of the implications, at several levels, of these research results, and, in doing this, being sensitive to several potential gaps. Previously in this study, we have identified at least one gap, which can now be considered, that is the gap between contemporary contextualized science and the conventional, academic scientific self-understanding. Furthermore, awareness of the gaps between science and ethics; between different ways of perceiving the world, nature, and man's right to "fiddle" therewith; and between a scientific and a "lay" interpretation of central concepts such as risk are central for gap sensitive interaction.

Again referring to the DDT story, we may exemplify some of these gaps: To decide whether the use of DDT should be prohibited when currently many developing countries have no realistic alternative in their fight against malaria is a question that spans the gap between science and ethics. We need science to estimate the risks connected to the use of DDT as a preventive measure in the fight against malaria and we need science in order to develop safer alternatives. However, deciding whether to continue the use of DDT in this relation is not a scientific question and science cannot provide an answer, only necessary background information. It is possibly less obvious that the issue of risk estimation spans the same gap. As discussed by Beck, the use of science to analyze risks and the use of scientific language when risks are described can make us blind to their real nature. So when discussing the risks connected to the use of DDT and thus parameters such as "acceptable values" and "average exposure," we do not always realize that we are really allowing "a permanent ration of collective standardized poisoning" (Beck, 1992, p. 65). When employed in a gap insensitive way, the use of science and of scientific language can be a way to obscure that risk estimation is also a matter of ethics, not merely chemistry – we should not poison each other (completely...).

2.5.3 Professional Humility

Closely connected to interacting in a gap sensitive way is the development of what we may label professional humility. In this concept I include the idea that in order to communicate genuinely with critics of their research scientists must develop a humble attitude towards their own professional knowledge. I do not mean to indicate that scientists should abandon their understanding of science as an outstanding knowledge generator, but merely that they must develop the awareness that nothing is certain, that science engineered to solve one problem may generate new and worse problems, and that science can not answer all types of questions; some for instance are questions of ethics. If scientists are to become productive partakers in social reflectivity, science students can not leave university with the attitude that "ordinary people" are simply too stupid or frightened to understand anything or that science "knows" all the answers. Instead, they must appreciate the fact that everything is more or less uncertain and that attention must also be given to these uncertainties and to potential "blind spots" in our knowledge. A crucial aspect of scientific reflectivity is the realization of the extremely difficult position in which science and society are caught; while we do not know everything, we have to make decisions. And development of professional humility may be one way of attempting to minimize the risks we inflict on ourselves.

For example, in the case of DDT the discourse in the community debating and influencing the use of DDT and other pesticides remained dominated for a long time by a focus on the actual, immediately recognizable damage; this was despite the fact that in debates on pesticide use an emerging focus on potential damages as opposed to immediate ones date back to the late 1920s (Böschen, 2002). A more pronounced professional humility, that is more explicit awareness of the limitations of our knowledge and of the blindness that goes hand in hand with any focused attention, might have given the potential damages connected to the widespread use of DDT more weight at an earlier stage in the discussion and decision-making process.

2.6 Anchored Ethical Dialogues

Having now, through the concepts of contextual awareness, gap sensitive interaction, and professional humility interpreted and developed aspects of scientific reflectivity, let us turn instead to the process: how exactly are we supposed to develop students' contextual awareness and so forth? I have suggested that the answer may be case based exemplary work and ethical dialogue. Below I will take a closer look of a teaching and learning "method" that aspires to combine the two things.

2.6.1 A Note on Anchorage

I suggest the introduction of "anchorage points" on a regular basis (e.g., twice or thrice a year) into the planned study programmes. That is, opportunities for the students to reflectively relate to the bigger picture. Further, I suggest that these anchorage points may take the form of *anchored ethical dialogues*. By this notion I mean ethical dialogues anchored to a concrete case (with a basis in students' course work) that students work with for some time. It has been suggested here that engaging in dialogue and being challenged by different viewpoints on the situation in question is crucial for a reflective educational practice. And here the anchored ethical dialogues enter the stage. The students' in-depth work with a concrete case provides an excellent opportunity for discussing ethical questions connected to the case. In this way the ethical dialogue becomes anchored to concrete circumstances and to the other aspects of the case on which the students have worked extensively.

2.6.2 A Note on Ethics

In a science teaching context ethics is often interpreted as good scientific conduct. It is my claim that this interpretation of ethics in the university chemistry teaching setting is inadequate. The difference between my idea of the ethical sphere introduced previously and good scientific conduct boils down to the following "learning how to do science the right way" versus a critical approach "what is the right thing to do and why?" Contained in this latter is a broadening of the students' world perspective and the ability to see the relatedness of various spheres, whereas a limited, internal perception of ethics leaves out reflections on science as a social actor. At its extreme, the limited, internal perception of ethics teaching could convey to the students the idea that when everything is being performed according to the internal ethical guidelines, it constitutes "good science," the responsibility of the scientist is fulfilled. This leaves out the social dimension and as Ziman (2002, p. 43) points out:

> [T]he scientist who takes a job doing research on Napalm on the grounds that it is 'good chemistry' is almost as much a pervert as the medical researcher who experiments on patients without their informed consent. Doing 'good science' is not synonymous with being a good person.

So, we can add that this view on science is definitely not the answer to the call for increased reflectivity. In continuation of this, I suggest perceiving ethics in relation to chemical education as *contextualized ethics* and to include under this heading reflections via for example historical, philosophical, and sociological analysis on the chemical enterprise and the values governing this endeavour, including the discussion of the adequacy of these values. Ultimately, the ethical sphere thus refers to the question of how we want to live and it encompasses reflections on scientific development and its consequences for humankind.

In order to make this idea of ethics operative in teaching, some additional reflections are needed. The kind of ethics we are talking about is obviously a sort of practical or applied ethics (Thomassen, 1997, p. 38). Bent Flyvbjerg's (1993, pp. 20–22) interpretation of a phonetic idea of an applied ethics focuses, for the choice of a line of action, on studies and analysis of the concrete case within its context, not on the application of a general ethical theory on specific cases. This understanding of applied ethics does not exclude the use of ethical theory in decision-making, it merely emphasizes that the basis is always the specific case, not the application of a specific ethical theory. Thereby, the phronetic idea of an applied ethics is in accordance with the methodological idea of *contextualism* as introduced by Earl Winkler (1993, p. 344):

> [Contextualism] is the idea, roughly, that moral problems must be resolved within concrete circumstances, in all their interpretive complexity, by appeal to relevant historical and cultural traditions, with reference to critical institutional and professional norms and virtues, and by utilizing the primary method of comparative case analysis.

2.6.3 A Note on Dialogue

As the notion "anchored ethical dialogue" indicates an important aspect of the teaching method I advocate here is dialogue. It was argued above that engaging in dialogue and being challenged by different viewpoints on the situation in question is crucial for a reflective educational practice. As discussed, it is important that this dialogue is anchored in a concrete case. Further, it is crucial that a genuine dialogue is taking place, and that the case work thus represents an opportunity for breaking the sender-recipient relationship between teachers and students. One may easily imagine that students who "come back" from the mental journey into an exemplary case, present their results, and in class discuss with the other groups ethical questions related to the case, will experience a very different meeting with their teacher.

Suddenly students will have "been to places" that the teacher has not necessarily "seen." I believe such experiences could be important for students' ability to develop an own attitude to the scientific game they are becoming partakers in.

Other, more practice tied, considerations must be made: These include "dialogue catalysts" and "dialogue shapers." In order to create a catchy dialogue in a group of students, a large number of who are likely to share professional viewpoints, the discussion material or the discussion format must somehow act as a catalyst. And in order to create a constructive dialogue that will actually enrich and refine students' viewpoints and dialogical approaches, the students must be introduced to various tools for shaping the dialogue. Relating to the catalysts, several suggestions may be proposed: the case that provides the basis for the ethical dialogue could include materials that provoke firmly cemented attitudes. Or, as part of the project students may be asked to produce arguments for opposing viewpoints (not necessarily their own ones). More interesting – when also considering the development of professional humility and gap sensitive interaction – is the idea that the anchored ethical dialogues should include people from "outside," for example students from other faculties than that of science. For example, biology, medicine and psychology students discussing the brain and the experience of self, or biochemistry and theology students discussing the concept of life would, given the proper guidance, probably develop students' comprehension of both their own discipline and of differences and similarities between this and other disciplines, their basic assumptions, and approaches to the material world. In this way students could become sensitized to other ways of comprehending the world.

About the discussion shapers, we may propose that students need to be introduced formally to ways of building an argument; to an overview of different ethical theories; and to epistemological reflections on science if the ethical dialogues are to become sufficiently concrete to the students.

To put it briefly, we may say that in anchored ethical dialogues one may at the same time stimulate several aspects connected to the development of students' reflectivity. Thus, the combination of casework and anchored ethical dialogues seem to tie together the development of contextualized awareness, professional humility, as well as gap sensitive interactive skills (see Fig. 2.1).

2.7 Conclusion

In conclusion, the present study suggests that one challenge science education is currently facing is to prepare students for the changed societal role of science, and that this implies that future scientists must be prepared to see science as an integral part of societal development and for partaking in social reflectivity. Further, that this necessitates a reflective educational practice which aspires to develop (at least) students' *contextual awareness, professional humility*, and their ability to *interact in a gap sensitive way*. Finally, the idea has been developed that one road towards this goal could be to include in the curriculum, project work and *anchored ethical dialogues* that "force" students to contemplate the contextualized nature of science, to engage in discussions of ethical questions, and to become challenged by different viewpoints on the case, science, and the world in general.

While the introduced notion of social reflectivity relates to general socio-political considerations, the presented operationalization of reflectivity, and the development of the idea of anchored ethical dialogue, transfers the discussion to the micro-level and thus focus directly on changes to the processes of teaching and learning.

A further refinement of our understanding of development of students' scientific reflectivity and more concrete suggestions for teaching "approaches" await more empirically based data. Currently Danish higher educational institutions are working to implement mandatory philosophy of science courses into their study programmes. Gathering experiences from the science departments could be an interesting place to start.

References

Beck, U. (1992). Risk Society – Towards a New Modernity. London: Sage.
Beck, U. (1997). The Politics of Risk Society. In J. Franklin (Ed.), The Politics of Risk Society (pp. 9–22). Cambridge: Polity.
Beck, U. (1999). World Risk Society. Cambridge: Polity.
Böschen, S. (2002). DDT and the Dynamics of Risk Knowledge Production. HYLE – International Journal for the Philosophy of Chemistry, 8(2), 79–102.
Eriksen, K. (2003). Reflectivity and Tertiary Chemistry Oriented Education, Doctoral Thesis. Copenhagen: University of Copenhagen.
Flyvbjerg, B. (1993). Aristotle, Foucault and Progressive Phronesis: Outline for an Applied Ethics for Sustainable Development. In E.R. Winkler and J.R. Coombs (Eds.), Applied Ethics: A Reader (pp. 11–27). Oxford: Blackwell.
Gibbons, M., Limoges, C., Nowotny, H., Schwartzman, S., Scott, P., and Trow, M. (1994). The New Production of Knowledge. London: Sage.
Giddens, A. (1984). The Constitution of Society – Outline of the Theory of Structuration. Cambridge: Polity.
Hellesnes, J. (1976). Socialisering og teknokrati. Copenhagen: Gyldendals pædagogiske bibliotek.
Jacobsen, B. (1981). De højere uddannelser mellem teknologi og humanisme. Copenhagen: Rhodos.
Kaspersen, L.B. (1996). Anthony Giddens. In H. Andersen and L.B. Kaspersen (Eds.), Klassisk og moderne samfundsteori (pp. 397–412). Copenhagen: Hans Reitzels forlag.
Klafki, W. (2001). Dannelsesteori og didaktik – nye studier. Århus: Forlaget Klim.
Merton, R. (1968). Science and Democratic Social Structure. In R. Merton (Ed.), Social Theory and Social Structure (pp. 604–616). New York: The Free Press/Macmillan.
Nowotny, H., Scott, P., and Gibbons, M. (2001). Re-Thinking Science. Cambridge: Polity.
Östman, L. (1999). Science Education Versus Science – Conflicts Within the Academic Setting. Presentation at the Second International Conference of the European Science Education Research Association. Kiel, Germany, September.
Simon, C. (1999). DDT – Kulturgeschichte einer chemischen Verbindung. Basel: Christoph Merian Verlag.
Thomassen, N. (1997). Etik – En introduktion. Copenhagen: Gyldendal.

Videnskabsministeriet. (2003). The new Danish University Law (law no. 403 May 28, 2003). Retrieved 14 October 2003. http://www.videnskabsministeriet.dk.

Winkler, E.R. (1993). From Kantianism to Contextualism: The Rise and Fall of the Paradigm Theory in Bioethics. In E.R. Winkler and J.R. Coombs (Eds.), Applied Ethics: A Reader (pp. 343–365). Oxford: Blackwell.

Ziman, J. (1996). Is Science Losing its Objectivity? Nature, 382, 751–754.

Ziman, J. (2000). Real Science – What It Is, and What It Means. Cambridge: Cambridge University Press.

Ziman, J. (2002). Getting Scientists to Think About What They are Doing. In T.B. Hansen (Ed.), The Role of Philosophy of Science and Ethics in University Science Education (pp. 23–44). Göteborg: NSU Press.

Chapter 3
The Development of a New Laboratory Course in Chemistry

Rie Troelsen

> It is especially important that we develop new perspectives for laboratory instruction in science. It is a widely accepted credo that experiences in the laboratory are essential to effective science programs. Occasionally this tenet is questioned, however, and, although laboratory experiences survive, the answers offered to critics are not overwhelmingly impressive. Certainly laboratory work is not as efficient and effective a mode of instruction as it might be, and it is high time that we take a hard look, with a fresh perspective, at learning in the laboratory.
>
> (Anderson, 1976, p. vi)

This opening quotation is taken from a book entitled *The Experience of Science: A New Perspective for Laboratory Teaching* in which it is advocated that given the massive expansion in the amount of scientific knowledge, it is crucial, with regard to lab work, to give a great deal of consideration to what kind of information is presented to the students and what approach is used. The book and the recommendations are more than 30 years old but there is still a need to develop new perspectives for laboratory instruction in science.

In this chapter I will try to respond to this need by describing the development of a new laboratory course in introductory chemistry at university level. Society has changed dramatically over the last 40 years and with that,students, educational goals and the amount of professional knowledge have also changed. These are all changes that require a different approach to teaching and learning at universities. The key concept in this different approach is teaching with a competence focus rather than a syllabus focus. But how are the visions of teaching with a competence-based focus implemented in a "real" university teaching sequence? This is the question I try to answer in this chapter by describing the developmental process involved in transforming a traditional lab course, which concentrates on the learning of factual knowledge and craftsmanship for its own sake, into a course which aims at competence-based learning of factual and procedural knowledge.

As the first step in the development of a new course, however, I will briefly outline some reflections on the nature of contemporary society and the essential

---

R. Troelsen
University of Southern Denmark, Denmark

O. Skovsmose et al. (eds.), *University Science and Mathematics Education in Transition*,

abilities this society demands, which form the basis of the course. Previous research and developmental projects focusing on competence-based teaching in a lab work setting is another kind of basis of the course. Hence, an overview of former research in the area follows, serving as a catalogue of inspiration for the development of the first version and trial run of the laboratory course. This is described and analyzed in the second part of the chapter in terms of the relationship between the teacher's intended objectives and the students' perceived learning outcome. A range of problems and questions arise from this analysis, and with these problems in mind – as well as findings from former research on teachers' views on laboratory teaching – I will move on to the third step in the development of a competence-based course: a proposal for an improved version of the same lab course.

3.1 The Basis for the Course

In order to determine what types of knowledge and skills students need to develop during the course of their university education, the nature of the society in which graduates are going to work and live must be considered. This must be kept in mind when developing a new course at tertiary level (and all other educational levels).

3.1.1 The "Reflective Society of Knowledge" and the Essential Abilities Needed in This Society

Based various sociologists' and philosophers' (Bauman, 1992; Beck, 1992; Giddens, 1990; Qvortrup, 2000; Simonsen and Ulriksen, 1998 and Ziehe, 1989) thoughts and descriptions of contemporary society's features, I suggest that contemporary society should be regarded as a "reflective society of knowledge". This designation implies that knowledge and lifelong learning are essential driving forces in future societal development. Knowledge is here not to be understood in a narrow sense as purely intellectual and academic knowledge but also as encompassing practical and social knowledge. Furthermore, knowledge of knowledge, that is, acknowledgment and verbalization of personal knowledge and ways of obtaining more knowledge, also becomes – or may already be – a crucial factor in future society and possibly the most important individual qualification for ensuring the competence of future citizens.

Today, all this knowledge and learning is permeated by a reflectivity that is far deeper and broader than ever before. All knowledge is subject to reflection, and today, being able to reflect, put into perspective and choose are necessary and important skills in a society that no longer provides secure points of reference.

For these reasons I find the concept of a "reflective society of knowledge" useful in describing the challenges and pitfalls that characterize contemporary society. These characteristics are manifested in the way hierarchies are crumbling while networking as a concept is emerging as the new form of cooperation, in the way that moral and ethical values are experiencing a renaissance in which citizens seek new, fixed standpoints, and in the speed at which development is increasing and knowledge is being outdated. At the same time, and as a consequence of this, new class barriers are arising between those who know and are able to acquire new knowledge and those who cannot.

In a reflective society of knowledge in which the all-important factors increasingly are knowledge, the acquisition of knowledge, and the application of knowledge, educational institutions, including universities, play a central role. A greater amount of learning will undoubtedly be moved into organizations and companies but universities have a unique opportunity as well as a societal obligation to take the shift in societal structure seriously and equip the ever-increasing number of students with the abilities necessary to succeed as citizens.

Precisely what abilities are necessary for the present and future citizen? What abilities and skills are required to succeed in a reflective society of knowledge? In the field of education – as in many others fields – there is a growing recognition that society's need for citizens who are ready to change, are in possession of analytic and cooperative abilities and are capable of engaging in lifelong learning processes fail to be satisfied by the previous focus on syllabus mastering as the dominating goal of education. Hence, in the field of education the concept of competence is seen as a means to help the system on its way to an increased focus on how the learner is learning a given subject rather than how the teacher is teaching the subject.

One answer to the above questions seems to be that the concept of competence has to be implemented in tertiary education in order to educate future academics and citizens. Their implementation implies designing teaching that insists on developing competencies. This can be formulated as a range of principles for all teaching situations and for education in general:

- a social dimension
- a value orientation in terms of both the choice of content and the communication of this content
- participant responsibility for learning as well as for teaching
- a new professional teacher role based on pedagogy and subject-related didactics[1].

These principles apply to every teaching situation and to every form of science education at universities, but in this chapter I will focus mainly on how to apply them in a chemistry laboratory-teaching context.

[1] 'Didactics' is here and throughout the chapter used as the very broad sense of theories of teaching and learning in all circumstances and in all forms. I adhere to the wider European use of the term and not the English use of it.

To sum up, the development of society increasingly forces the educational world to regard the development of competencies rather than the development of factual knowledge as the goal of education. This is, of course, also the case in the area of chemistry laboratory teaching. Content knowledge is not disclaimed but the approach to teaching and learning of the content must be renewed.

3.1.2 An Overview of Previous Research and Developmental Projects Focusing on Competence-Based Teaching in a Laboratory Setting

This renewed approach to teaching and learning chemistry content implies, of course, renewed approaches to designing actual teaching situations. In an actual teaching situation it is not sufficient to simply change some isolated exercises into more open-ended exercises while maintaining the traditional form of evaluation, the teacher's role, and the expectations of student engagement. The idea behind a course, its aims and objectives, and its form and evaluation must be changed if everyone involved in the teaching situation, including students, teachers, and laboratory technicians, is to contribute to carrying out an optimal course, which effectively combines aims with outcome.

For these reasons, special attention is now drawn to previous literature that describes alternative forms of laboratory teaching aimed at promoting the development of competencies relevant in present and future society. The aim of describing the following alternative forms of lab teaching is to shed light on possible ways of designing lab courses where students work with scientific problems in order to strengthen their problem solving skills and to develop different social, personal and scientific competencies. Furthermore, the selection of literature builds on the criteria that the alternative teaching should be, as a rule, tertiary, within chemistry or chemistry related subjects and based on goals of teaching students something more than craftsmanship skills and training in following instructions.

Arce and Betancourt (1997) describe one piece of research that meets these criteria. The researchers follow a first-year university chemistry laboratory course. As a background for this research, the writers state that they want to get first-year students to engage in scientific laboratory investigations. Hence, they have developed simple projects that present the students with a problem to solve without the help of detailed information on experimental procedures, but with the guidance of a range of questions aimed at pointing the students in the right direction. The projects are developed on the basis of six general elements: the learning cycle, the challenge level, factual knowledge and guiding questions, time to think and to repeat, a safe environment, and evaluation with emphasis on thinking. *The learning cycle* consists of three phases: exploration, interpretation and application, and the projects guide the students through a complete cycle. Focus on *the challenge level* implies matching the student's skill level and the level of difficulty of the experiment to avoid frustration as well as boredom. In order to help the students stay focused, every project begins with a review of the necessary *factual knowledge* needed to work with the problem and also provides *guiding questions*. It is important to devote enough *time to think and to repeat* in the projects to promote reflection and scientific thinking. *A safe learning environment* is necessary if the students are to develop thinking skills rather than focusing on reaching a correct answer. Last, but not least, it is crucial that the *evaluation emphasizes thinking* in order to direct the students' attention to what is important.

In the same way, an introductory analytical chemistry course described by Wright aims at allowing the students to actively experience the feeling of being practicing professionals (Wright, 1996). The course is based on an active learning perspective and is designed to involve a cooperative learning style, spreadsheet programmes and open-ended laboratory projects. The open-ended lab projects have well-defined goals, but students are free to use their creativity in reaching these goals. Many of the projects are based on traditional lab experiments and are conceptually simple but nevertheless rich in detail and many unanticipated problems may arise during the process. The students must cooperate in groups of four to solve every problem in the project. This means that one student concentrates on measuring methods, one on concepts and theories, one on statistics and error analysis, and one on group management. Each member of the group meets with a teaching assistant and other students with the same tasks from other groups to discuss, for instance, measuring methods, and is then responsible for passing that knowledge on to the rest of the group. The project finishes with a group report including descriptions of the experimental design, the results and a discussion. The students in the group meet with the teacher and teaching assistants to discuss the report on a relatively high scientific level.

Experiments with laboratory teaching have also been made in a first-year university chemistry course in Holland (Biessels, 2000). The course is called *From Monoester to Poly-ester* and its main goals are to introduce students to a multidisciplinary problem approach involving the obtaining of information through literature searches, the construction of an approach to solving chemical problems through discussions and cooperation with colleagues, the production of written and oral reports, and the mastering of the basic skills necessary for working in a chemistry lab. The course begins with an introduction to the rest of the course and concludes with student reports and evaluation sessions. The in-between part of the course is organized in five phases. In *Phase 1* the students are lectured on safety, ester syntheses in theory, library searches, and how to cooperate. The students are lectured on measuring methods in *Phase 2*, and are required to synthesize, purify and analyze 6 esters. *Phase 3* consists of physicochemical experiments and the oral presentation by students in their subgroups of six of their results from Phases 2 and 3. During this phase the students have to prepare several "products" for the teaching assistant, including the oral presentations and the evaluation of these presentations, protocols for the determination of the physicochemical parameters, and a list of aims and a time schedule for this phase. In *Phase 4* the students are lectured on polyesters and their syntheses, and synthesize and characterize a polyester. Here too, protocols for the synthesis and characterization of the ester, and time schedules are to be handed in to the assistant. The last phase, *Phase 5*, includes an oral presentation of all experiments to a subgroup of colleagues and is followed by handing in a list of aims and a time schedule for this phase to the assistant as well as the oral presentation and the evaluation of this presentation. This form of organization of a lab course depends heavily on well-educated teaching assistants in terms of both experimental overview and pedagogical knowledge.

Another example of teaching in a chemistry lab, which does not prompt the students to "follow a recipe", also starts with the students solving a problem through lab teaching. In this case the problem is characterized by having a closer connection to "the real world" than the teaching projects described above. Ram (1999) describes a second-year university analytical chemistry lab course in which problem-based learning (PBL) has been successfully implemented. There are five well-defined stages in the PBL process: introduction, inquiry, self-directed study, revisiting the hypothesis, and self-evaluation. Students concentrate on solving one problem during a semester. In this case, the problem is to identify, understand and run analyses on water samples collected in a local polluted river. The students are organized in groups of two or three that are responsible for one or two water quality tests. The groups' duties include collecting equipment for their specific test(s), running the standards, helping their fellow students conduct the test, and finally, collecting everyone's data and performing the data analysis. In this way all students perform nine water quality tests. Ram concludes that PBL is an effective way of motivating students by allowing students to experience authentic problems and carry out self-directed studies. The course has, however, been very time-consuming, a fact that needs to be taken into consideration by students and educational planners.

The previous four lab courses have to varying degrees tried to involve "the real world" by transferring problems from the surrounding environment to the laboratory for closer analysis. This last example of an alternative teaching sequence goes one step further into the real world by making collaboration between students and local or regional companies on the premises of the company mandatory, albeit adjusted to the students' educational level (Henderson and Buising, 2001). The overall aim of this lab course is to be research based, student centred and cross-disciplinary by attempting to mimic real-life experiences in research laboratories. The research company, which is a cornerstone of the course, provides relevant experiments, mentors, materials and apparatus. In return, the scientists at the company receive useful data from the students, gain teaching experience, and establish relationships with the community and future scientists. The course consists of the following components: *A technique symposia*, where the students present techniques to each other, *preparation for research*, which involves students meeting company staff, *the research problem,* which the students define in groups, *grant writing*, which requires students to describe the problem and experimental plans, to prepare a literature review, and to draw up a budget, *research,* involving the conducting of the actual research, *peer reviews*, through which students evaluate each other's participation, *research reports*, in which students present their work in poster and journal format, and finally, *peer evaluation of reports*, whereby the students evaluate each other's written and oral performance.

As mentioned before, my aim in describing the research projects above was to shed light on possible ways of designing lab courses where students work with scientific problems in order to strengthen their problem solving skills and to develop different social, personal and scientific competencies instead of working with scientific problems for the sake of the problems themselves. There are many different ways to do this but some general guidelines may be suggested:

- Start out with traditional teaching to prepare for project work
- Ownership is important, so let students pick their own problem
- Let students work in groups
- Let students present their results and processes to each other
- Create a form of evaluation that focuses mainly on students' work processes and to a lesser extent on experimental results.

There are two main problems associated with these courses. First, they are extremely time-consuming for students. The students are expected to work far more outside the scheduled classes than they are in traditional courses. Second, the teacher role changes. In addition to being a good researcher and scientist, the teacher must act as a facilitator of the students' learning processes, and be able to identify and analyze group dynamics.

3.2 Description and Analysis of the First Version of the Course

In the following I will describe and analyze the first version of a chemistry lab course that is developed with the guidelines above as a starting point. The term "first version" relates to the developmental process from visions of competence-based teaching to implementation in an actual lab course. This first version of the course should be regarded as an attempt, by means of a "real" university teaching sequence, to be specific about the ideas of a kind of teaching that meets the challenges of contemporary society and students and focuses on the development of competencies in a sound professional context.

The first version of the course is based on the above guidelines suggested from reviewing the literature. By reviewing the literature one can learn from a range of experiences and presumably avoid many practical and communicative pitfalls in developing one's "own" lab course. Nonetheless, it is necessary to gain personal experience because every lab course is embodied in a specific context featuring both practical, staff related, subject related, institutional and student related elements.

In the following I will describe a lab course in general chemistry at the University of Southern Denmark, aimed at gaining that experience. In addition to the communication of chemical knowledge, the aim of the course is to help students attain more than an ordinary mastery of the syllabus. This is achieved in part by restructuring the course into a more project-oriented form.

In order to learn, in a further developmental process, the course is also analyzed in relation to the intended goals and the outcome of the course. This is done by regarding to the three different levels of the teaching situation: what the teacher intends, what the teacher presents through his teaching, and what the learner learns. An identification of possible matches or mismatches between these three levels leads in a constructive way to the development of further and more effective versions of the lab course.

When the following analysis was conducted, the course was held for the first time and still regarded as an experiment.

3.2.1 Brief Description

The lab course is part of the overall course in general chemistry, which also includes lectures and tutorials. The course takes place at the end of the first semester and is mandatory to all students studying science at the university. This means that students following this course are on different levels in terms of both their interest in chemistry and their knowledge of chemistry.[2] The lab course consists of five sessions, each of four hours' duration, during which the students work together in groups of two or three. In the first session, all students conduct a titration experiment in which the decalcification agent ÷*kalk* ("minus lime") is titrated with sodium hydroxide. The aim of this exercise is to introduce the students to the laboratory, its equipment and practices. During the following four sessions the students work with one of four chemicals known from everyday life.

- *Nonoxal* (an ordinary household chemical used e.g., to bond oxalic acid in stewed rhubarb). In this exercise the concentration of calcium ions, chloride ions and lactic acid must be determined through freezing point reduction and titration
- *Easysept* (a disinfectant used for disinfecting and storage of soft eye lenses). The aim of this assignment is to determine the concentration of hydrogen peroxide and chloride ions by means of titration and to determine the order of reaction for the decomposition of hydrogen peroxide
- *CMA-plus* (a salt used for de-icing the Great Belt bridge in Denmark during winter). In this exercise the concentration of magnesium, calcium and acetate ions must be determined through titration and the freezing point for CMA-plus
- *Mineral water* (both Ramlösa and Evian are used). The amount of carbonic acid and the concentration of chloride ions in Ramlösa are determined. For comparison the concentrations of hydrogen carbonate, calcium and magnesium ions in Evian mineral water are determined.

[2] Students studying biology, for instance, are not required to have the same level of chemistry knowledge from high school as students studying chemistry, but at the University of Southern Denmark students studying any science subject begin their studies with a one-year basic science course.

Hence, all of these four eligible exercises have a consumer product as their object of inquiry. The exercises also have in common the fact that the concentration of the chemical's component must be determined, mainly through titration experiments. All methods for measuring different ion concentrations are described in the lab manual.

This course in general chemistry is, as noted above, divided into three parts: lectures, tutorials and laboratory teaching sequences. As with other first-year courses in which about 150 students are taught over a semester, ... the teacher responsible for the course lectures all students and teaches a few tutorials and laboratory classes. The remaining laboratory classes are taught by other faculty members or teaching assistants.

Another practical aspect of conducting the lab course that is worth mentioning is the fact that the teacher in charge of the course neither wrote the new laboratory manual nor was responsible for the final testing of the experiments involved in the course. A teaching assistant carried out this work.

3.2.2 Analysis with Regard to the Relation Between Intentions and Outcome

As mentioned above, I analyzed the lab course in terms of the teacher's intended goals, the students' expectations, the actual conducting of the course, and the students' actual learning experiences. This kind of analysis is based on the assumption that every teaching situation can be described at three different levels (Bauersfeld, 1979):

- the matter meant (what the teacher intends)
- the matter taught (what the teacher through her actions presents and represents)
- the matter learned (what is learned by the learner).

The content of and the relations between these levels are indicative of the consistency of every teaching process. If a closer analysis of the relations between the three levels reveals an inconsistency, the identification of these mismatches can be regarded as a constructive way of developing a more effective teaching process (Argyris, 1992).

As part of the relational analysis I conducted a series of interviews with students before the course started and again after having attended the lab course (group interviews). I also interviewed the teacher in charge (in an individual interview) and analyzed the lab manual.

3.2.2.1 The Matter Meant

The first level considered is the level of the matter meant. The teacher in charge was interviewed about the intended goals of the course. In the faculty's official course catalogue the goals of the course are described as "to impart knowledge of and practical experience in conducting simple, quantitative analyses to students". The teacher elaborated on this goal by noting that the course mainly aimed at imparting an independent way of thinking as well as an independent way of working to the students. The teacher repeatedly pointed out that the most important learning outcome was the independency:

> I believe that this course should try to teach them (the students) to think independently, to plan things independently, to give them some working methods.

However, he had little or no confidence in the students learning new chemical concepts through lab teaching – instead, they would become acquainted with the scientific way of working. It was also important to the teacher that the students were given personal space and leeway in terms of time to make mistakes. The teacher stressed this message to teaching assistants, stating the students should become "motivated to make some decisions themselves", and thereafter experience the consequences of these decisions. According to the teacher the students would also gain other skills:

> [To] be able to gather relevant data and to write a decent report, to be able to work together in pairs. Some of these abilities cannot be directly measured, but hopefully, they [the students] will be able to use them in other situations.

The teacher chose consumer products rather than more traditional chemical solutions on which to conduct the chemical analyses. This was to create an interesting and motivating frame for chemical analysis – a frame that would contribute to the engagement of the students due to the fact that they knew the products beforehand:

> I think that some of the students like to work with things they know. They can see that the products they buy at the grocery store or wherever they buy them are just ordinary chemical solutions, which actually are analyzable.

3.2.2.2 The Matter Taught

The second level considered is the level of the matter taught. As one way of gaining insight into how the course is actually conducted I have chosen to analyze the lab manual. The lab manual is analyzed with special focus on the function of verbs and their contribution to the general impression of the manual gained by students (Parkinson and Adendorff, 1997). The lab manual consists of three parts. In the first part, where techniques and apparatus are introduced, verbs indicating action are, of course, overwhelming. The second part, which gives an overall description of the five exercises, starts with an explanation of the chemical's function in everyday life. The explanation consists of verbs that partly describe physical phenomena (the effect of the chemical in everyday life) and partly clarify scientific facts (what the chemical consists of). The tasks of the exercise are then listed using the verb "determine", which indicates both action and thinking. The third part, which is the most exhaustive part, is characterized by concise descriptions of measuring methods.

As an example, I have analyzed the verbs' function in one of these descriptions of measuring methods, the potentiometric titration of chloride ions, in more detail. The potentiometric description consists of a theoretical part (the chemical principles in the titration method), a precondition part (calculations that are preconditions for further investigations), an implementation part (how to conduct the experiment), and finally, an equipment part (glass equipment, apparatus and chemicals used in the exercise).

Table 3.1 clearly shows that almost one third (12 out of 38) of the verbs in this description have the function of instructing the student to act, while half (19 out of 38) establish universal facts or describe a physical reality. Only on one occasion does the description require students to think about and explain more than what concerns the procedure of the titration, namely when students are asked: "What would the titration curve look like if Nonoxal [the consumer product] contained both chloride and bromide ions?"

From this linguistic analysis it can be concluded that the lab manual is primarily aimed at teaching the student well-known techniques based on well-known facts. Only to a far lesser extent does the manual signal to the students that they must think or plan themselves, be critical of well-known facts, and pose questions.

Of course, this analytical method is to a certain extent normative, given that the judgment of the verbs' meaning in relation to specific categories is subjective. Nevertheless, the overall impression of a lab manual that instructs the student to act without posing any critical questions is to a certain extent validated by the traditional structure of the manual, which consists of precise and strictly organized descriptions of various measuring techniques, although an attempt is made to signal a more project orientated approach in the second part. Furthermore, the students' interview statements about their experiences with the course validate this picture.

Table 3.1 The verbs' function in the potentiometric description

		Theoretical part	Pre-condition part	Implementation part	Total
Instruct the student to:	Act	4		8	12
	Analyse via plots and calculations			1	1
	Think and explain			2	2
Thinking functions	Establish universal facts, describe physical properties	18		1	19
	Reason	1	3		4
Organizes the exercise[a]	Predict the actions of the student				
	Comment on the process				
	Comment on the expected results				
Total number enumerated		23	3	12	38

[a] This category is included despite the lack of this kind of verbs in the manual. The reason for the inclusion of the category is to show the wide span of the linguistic analysis method

3.2.2.3 The Matter Learned

The third and last level considered is the level of the matter learned. A way to measure what students have learned is to interview them about their experiences with the process of a teaching sequence and their views on its learning outcome. Regarding this particular course the students initially expected that the exercises would be more project-oriented, and that they would find out themselves what solution should be used to titrate the consumer product. In this aspect the course did not live up to their expectations. A student told me about his high school experiences, where the students planned the experimental set-ups themselves in a physics course. According to his experience, you learn more or gain a better understanding of a topic when you yourself have been involved in the planning stage:

> "It is exciting that way[...] It seems more scientific or research-related when you try to figure out some things for yourself instead of just doing what you are told. Actually, I think you learn more from it ".

On the other hand, another student argued that she did not have sufficient theoretical knowledge to conduct project-oriented exercises where the students were supposed to figure out what solution to use in the titration:

> "If you don't have any idea of what to titrate with, then I don't think that I would be able to proceed at all".

To sum up, the students belong to two groups: one group has self-confidence and experience in chemistry and hence likes the challenge of an independent planning phase in the exercises, while the second group feels insecure in the field of chemistry and therefore likes a very specific lab manual. The self-confident group has by virtue of their starting level not learned anything new regarding craftsmanship and general laboratory practice but has received practical training, whereas the insecure group felt they had learned a lot in both respects.

The students found it exciting to work with consumer products and to be given the opportunity to work in depth with the product. Likewise, they were glad to work in groups as they felt secure and less burdened with the responsibility of knowing everything beforehand.

3.2.2.4 The Relational Analysis

The relational analysis takes up the relations between "the matter meant", "the matter taught" and "the matter learned" and it indicates the existence of three mismatches: the scientific working method, independence, and accuracy.

Mismatch one regards the scientific working method. For the teacher the most important outcome is the students' development of a scientific way of thinking or ways to work scientifically in the laboratory. What is understood by the term "working scientifically in a laboratory"? If it involves what might be termed *inquiry tactics* – "toolkits" of strategies and approaches that can be considered in planning an investigation – it could be argued that the students learned or became more familiar with them during the lab course. By repeating a measurement until the results do not differ from one another by more than 1%, the students may learn which situations require accuracy. Moreover, when writing the final report they may learn to select relevant data and present it in a graph or a table to show trends and abnormalities. On the other hand, if the term "working scientifically in a laboratory" is regarded as equivalent to the working method of scientists, the students do not learn anything about it. The students are not forced in any way to think about relevant or proper circumstances and design in connection with the particular experiment or to consider other alternatives to doing an experiment. The distinction can be seen as the difference between *learning to practice science* and *practicing science* (Millar, 1991). Teachers must keep this distinction in mind when structuring a lab course, but it is also important that the students are aware of the difference. Furthermore, the students are evaluated on the basis of a written report on their scientific results and not on the process they have been through or how they have worked in the laboratory. This brings into focus the gap between the teacher's goals with the course and the chosen form and content of evaluation of the learning outcome.

The second mismatch regards the notion of *independence* as it is also very important for the teacher that the students learn to work and think independently. It matters to the teacher whether the students are able to plan, work and think for themselves by the end of the course. However, neither the lab manual nor the structure of the course gives the students a fair chance to develop *independence*, at least not with regard to working methods. Because all working methods are thoroughly described in the lab manual the students are only allowed to be independent about their attitude to work; they decide for themselves in what order they want to conduct the measurements. The students do not mention anything about independence or scientific thinking when asked about what they think they have learned from the course. They feel that they have become more familiar with the laboratory settings and remember some chemical concepts better than before.

The third mismatch deals with the need for *accuracy* in the course. Accuracy is *not* important, the teacher claims. In spite of this, in several places the lab manual demands an accuracy of 1% for the results of two measurements made by the student:

> The experiment is done one more time, in order to carry out a double determination. These two determinations must not differ more than 1% from each other. If the difference is larger, the experiment must be done again.

From comments like these, the student gets the impression that precision and accuracy are important factors in titration, as of course they are. However, because the majority of practical exercises in the course focus mainly on titration, the student can easily get the impression that precision and accuracy are important factors in *chemistry*. In fact, almost 70% (67 students out of 97) of all students taking the lab course agreed or strongly agreed with the statement: "A good chemist is one who works precisely and without errors in the laboratory". The students in the interviews also talk about good laboratory practice as being precise and accurate. Some of them even equate being accurate with being scientific. This is only partly true. A good chemist is *not just* one who works precisely and without errors in the laboratory. The ability to work creatively with complex and unfamiliar problems, the ability to communicate effectively in both speaking and writing, the ability to work in teams and the ability to assess and reflect on one's knowledge are also important qualities for the future chemists to possess (Tolman and Parshall, 1999; Troelsen, 2001). The students are not directly introduced to this broad range of characteristics of a good chemist through this lab course.

In conclusion, the students do not obtain the intended skills and competencies from the course. The reasons for the mismatches between "the matter meant", "the matter taught" and "the matter learned" are many and various, and are related to different practical, organizational and human circumstances. These barriers to the development of a competence-based lab course in which intended goals are implemented and acknowledged by teachers, students and organizations are described further in the following section.

3.2.3 Barriers to Development

The analysis of the new lab course at the University of Southern Denmark has pointed out some barriers to development. In addition, findings from research on teachers' views on laboratory teaching and research on laboratory teaching in general indicate barriers. In the following I will point to three barriers to development partly learned from the present analysis of the first version of a chemistry lab course and partly learned from former research in the field. These two sources to the localization of hindrances to an ideal lab course are, so to speak, validated by each other.

The first barrier to consider is that changing a course demands changing all parts. It is not enough for the teacher in charge of a course to believe in a new teaching strategy if the traditional learning and teaching style of the course as a whole is to be changed. Many different people must have the courage and the ability to carry out the changes. For instance, the teacher's intention with the course is that the students learn to think, plan and work independently in a laboratory. These intentions are to a certain extent counteracted by the design of the manual, which signals that none of the proposed methods can be discussed and urges the students to "follow a recipe" without reflecting on isolated steps, for example, by mostly using verbs that establish universal facts or describe a physical reality. Another barrier is the lack of change in evaluation form. The actual evaluation form still sets the stage for the final report to be an accurate reproduction of the manual's specification of methods and theory without any reflection or critical assessment. This kind of evaluation does not help students develop skills in cooperation and processing information.

Secondly, another barrier to development is the lack of a didactical contract. The concept of didactical contract is to be understood as a common agreement between all people involved in the teaching process on the learning and teaching goals of the course, the outcome for the students, how this outcome should be achieved, the kinds of knowledge and learning addressed in the course, and so on. A didactical contract is not an actual signed, binding contract, but more like a mutual understanding of the aims and content of a course, which all parties are aware of and obliged to act in accordance with. The students in this research project are not completely aware of the aspects of the course that are in focus after the reorganization. The lab manual does not help them, and the teaching assistants do not draw their attention to the learning goals of independence and a scientific working method, probably because they are themselves unaware of them. Lewis (1999) has carried out case study research on the implementation of mini-projects in a second-year university course. The students found the mini-projects demoralizing and frustrating because they were not aware that the process in conducting the mini-projects was more important than the product. In order to implement such projects successfully, Lewis points out that the aims and objectives must be accessible and explicit. Second, all involved partners in the course must make clear (or have made clear to them) what the mini-projects demand from students and to what extent these requirements are fulfilled. These two aspects are part of an acknowledgment of the didactical contract as an important tool in insuring that everybody in the teaching situation is working towards the same goals. Lewis also points out that a third aspect must be considered: sufficient time, help and support to ensure that students have the possibility of developing the intended skills and understandings.

Last, a third barrier to the development of a new laboratory course is the lack of education of the teaching personnel. In courses aimed at developing students' independence in working experimentally, it is essential that the teacher plays the role of a catalyst for the students' own learning processes. The teacher should not give students the solution to a problem, but rather give advice and suggestions for solutions. The teaching assistants (and in some cases, lab technicians) must also be involved in a didactical contract and acknowledge their new role as teachers. Previous research on teachers' views on the conditions and possibilities of laboratory teaching in tertiary education (Troelsen, 2003) indicates that many teachers regard the employment conditions of teachers, and especially the merit rating of employees at universities, as a potential barrier to the development of new courses. The employees' research skills and results are more highly valued than their teaching skills and results, even though in principle these two parameters should be valued equally. If a university employee is both to keep up with the latest research (in order to write the research articles necessary to keep her job) and to develop new teaching forms and methods, she needs considerable time, energy and resources. Experience from research conducted by van Keulen (1995) also concludes that full implementation of new ideas in a teaching setting cannot be achieved by simply handing out guidelines to teaching assistants and students. Instead, both groups must be introduced to the course and its intentions, and be offered support and help to fulfil these intentions throughout the course. Buning and Thijs (2001) also emphasize, in their research on different teacher roles, that a well-educated teaching staff is absolutely necessary in order to successfully implement new teaching ideas in the laboratory.

3.3 Suggestion for an Improved Version of a Lab Course

Having been convinced of the necessity of teaching with a competence focus rather than a syllabus focus in order to meet the demands of the reflective knowledge society, and having been inspired by reviews on former research and developmental work regarding competence-based lab courses, I will now suggest a realistic alternative to the form of lab teaching generally pursued at Danish universities today. I use the term *realistic* to indicate that the suggested outlines are based on actual experiences from the first version of the new lab course and previous research on barriers to implementing new teaching strategies in lab settings. The need for realism also connects with the discussion of the gap between educational research and educational practice. It is important that research in alternative, innovative, and relevant teaching reaches the teachers who are to implement all the new and groundbreaking research results.

The lab course consists of a range of sessions throughout a semester and the students are grouped in twos or threes. The first session is an introduction to the subject area in question and the techniques, methods and apparatus involved in the sense that the student is introduced to the physical, technical and theoretical "tools" which are necessary for the rest of the course. After the first session the groups choose one of five possible exercises, which they then work on during the remaining sessions. The five project-based exercises are all relatively open-ended exercises with fixed goals that can be reached in many different ways.

In order to develop different competencies it must of course be made clear to the students what kind of expectations there are regarding the acquisition of knowledge, subject related and social reflection, communication of personal views and interpretations, and respect of other's views. Likewise, it is necessary to make clear to the students that cooperation between groups and exchange of experience and knowledge are allowed and in some cases even necessary. Finally, it is important that the students are presented to the different elements of the total and final assessment at the start of the course so that they can form a picture of the learning goals characterizing the course in question.

The assessment of the course consists not only of the traditional written report but also of a range of partial assessments that are incorporated in the final assessment with different weightings. The final report must as usual be of good written quality. However, the report should also demonstrate the student's ability to put things into perspective, for example, by discussing why the consumer product is put together exactly that way or whether the declaration of the product is in agreement with the result of the experiment. The group hands in the report to the teaching assistant, who, after having read and evaluated it, discusses the report and its content with the students. After that, the group is given the task of presenting the most important features of their project on a poster. On the poster the students must, in a well-organized and informative way, account for the goals, experimental design, methods, and results and suggest possible changes to the project design, keeping a potentially more precise and elaborate fulfilment of the project's goals in mind. All posters are presented in a joint session for the entire class and five groups are picked out to present their posters orally to the rest of the class. The selection is made in such a way that each of the five exercises is presented. The oral presentations are judged partly by the teaching assistant, who considers the visual and oral presentation skills of each of the group members, and partly by the other groups working with the same exercise. To ensure that the rest of the groups are impartial and constructive in their criticism, the teaching assistant also takes their criticism into account as part of her assessment of them. In this way, the final assessment of the student consists of an evaluation of the written report, of her ability to express herself orally and visually, and of her ability to give and receive constructive criticism from peers.

Courses containing such elements point to an increased focus on and development of some of the competencies described earlier. The open-endedness prompts the student to practice making her own choice, arguing for this choice and finally, if the consequences of the first choice dictate it, choosing again. The independent working form sets the stage for an on-going reflection on one's own learning and dealing with knowledge while at the same time facilitating the student's ability to think and act creatively and trust her own judgment. The group work and evaluation form promote cooperation competence both internally in the group and between groups, which involves the acknowledgment of the importance of being able to cooperate while being true to one's own views and values.

It is, of course, vital for the successful implementation of this kind of course that lab teaching is not given low priority for economic or staff-related reasons. Lab teaching demands large staff resources compared to, for instance, lectures. Hence, a great deal of the teaching staff is normally instructors (mainly students studying for their Master's degree) who are relatively inexperienced when it comes to teaching. These instructors keep abreast of the pedagogical and didactical intentions of the course by attending weekly meetings between all instructors, more experienced teaching assistants, and the teacher in charge. In such forums questions can be raised about pedagogical consequences, and guidelines for the evaluation of reports, posters, and criticism can be discussed along with subject related didactical questions. The instructors and the teaching assistants as well the regular teachers should not feel left behind or caught between the actual teaching situation and the overall didactical intentions of the teacher in charge.

3.4 Concluding Remarks on Lab Work in Particular and Science Education in General

Now that my proposal has been put forward for a lab course focusing on imparting chemical knowledge through competence-based teaching rather than for the sake of this knowledge itself, one obvious objection arises. As the proposal does not claim to describe the ideal lab course but rather represents a step toward a better lab course, the question is whether this developmental process is hindered by the rather detailed description of a particular lab course given above. And yes, there is a possibility that the level of description and the focus on general chemistry and first-year university students can block future development of other science courses in other subject areas with other students. It is hence very important to underline the key features of an alternative lab course that may have a direct transfer value to other teaching situations:

- The course is structured with a theoretical and technical introduction, followed by the students working on their own projects in groups of two or three
- The students and the involved teachers negotiate a "didactical contract", which points out preferred teaching forms and the important learning goals of the course
- Peer reviews and student talks are added to the traditional assessment forms
- The pedagogy of teaching in the lab is given high priority. All members of the teaching staff must at least be aware of the elements in the didactical contract and try to teach accordingly.

Nevertheless, it is important to stress that this proposal for further development of lab teaching and its possible innovative potential are not to be regarded in isolation from the *subject-related* context. Actually, subject-related issues are imbedded in the concept of competence – you cannot just be competent, but must achieve competence in *something*. Therefore, transforming a laboratory course in chemistry into a competence-based course is not synonymous with being indifferent to what content the students are presented to in their learning processes. It is still necessary for the teacher to give some serious thought to the content by means of which the relevant competencies are to be developed.

In other words, competence is a hollow shell without *content*. In a similar manner, considerations of competence-based teaching cannot exist without pedagogical reflections on which *teaching form* is relevant for developing the actual competence. In fact, if the concept of competence is to be taken seriously it is not sufficient to transform a lab work course from concentrating on the learning of factual knowledge and craftsmanship for its own sake to aiming at a competence-based learning of factual and procedural knowledge. This is a dangerous statement to make, especially after having campaigned for exactly this aim in all of the above. However, in referring to this insufficiency, it is not my aim to disprove or claim that any of the changes that have been made so far or will be made in the future in the chemistry lab course described above are unimportant. My aim in these concluding remarks is to draw attention to the fact that macroscopic changes in society ought to lead not only to microscopic changes in a small lab course but to macroscopic changes in science education as a whole. If the road to better science education that trains students to become adequate citizens goes through competence-based teaching, then the interplay between competence and teaching form must be viewed from a reverse angle: considerations on which competencies to develop must have precedence over the choice of teaching form.

In many cases in current teaching, syllabus and teaching forms predict the teaching objectives. This ought to be reversed: first, as a teacher (or as a teaching and learning institution) one must reflect on which competence a specific teaching sequence should aim at developing. Then follow considerations on the content by means of which this competence can be obtained and which teaching form is appropriate. This inversion demands to a very high extent the belief in and commitment to the concept of competence-based teaching not only on the part of teachers in higher education but also on the part of the educational leadership at universities. It demands a total reorganization of university teaching where the development of competence comes first, and teaching forms and syllabus have to fit in.

Another change in university teaching as we know it, which is due to a commitment to competence-based teaching, is the stipulation of putting the learner in focus. This means that teaching to a higher extent should be based on the student as active and responsible, which can be put into practice by more project-oriented teaching but also, and more generally, in an organization of teaching which hands over more responsibility to the students. Such a change in organization may mean taking into account the ratio between teaching and learning. It is seldom the case that the ratio between confrontation time and the time used by students for acquiring knowledge is other than 1:1 – sometimes the ratio is even bigger in favour of the teaching part! A new organization of teaching that hands over more responsibility to students would imply that the amount of learning exceeded the amount of teaching in such a way that the ratio between teaching and learning became, e.g., 1:4. Changing the ratio entails thorough reflection on the teaching sequence and what the students are really suppose to learn from the sequence, negotiating the didactical contract and – not least – subsequently changing the evaluation method.

Lastly, a change from syllabus-oriented to competence-based teaching has profound consequences for the role of the teachers. When teaching forms are integrated in the bigger picture of competence development and the students are involved in project work, which also transcends some traditional subject borders, the requirement of openness and cooperation between teachers becomes clearer and more obvious. There may be a need for teacher teams that teach sequences together, for frequent exchanges of educational experiences between teachers, or for discussions of the state of affairs in terms of educational research and other initiatives that aims to make the teacher role less personalized and more professionalized.

In considering changes in the processes of teaching and learning at "ground level" – in this case, the transformation of a traditional lab course – which concentrates on the learning of factual knowledge and craftsmanship for its own sake, to a new course that aims at a competence-based learning of factual and procedural knowledge, I point in this chapter to some specific barriers to change and development, which of course can also be applied to other teaching situations. These barriers are the need for negotiation of the didactical contract, the need for well-pedagogically educated staff, and the need for changing all elements of a lab course, from the lab manual, over teacher roles, to evaluation methods.

This last observation leads me to my final point: changes in specific courses are not enough; higher education in science and mathematics as a whole must be transformed. In the particular lab course described in this chapter the relation analysis showed that the students did not obtain the intended skills and competencies from the course because of the mismatches between "the matter meant", "the matter taught" and "the matter learned". Similarly, the students must find it confusing and difficult to obtain the intended skills and competencies from their education when only few courses are organized as competence-based courses and others as traditional "broadcast" teaching. In order to educate citizens to function in a society where knowledge and competencies are central concepts, universities must offer education that is competence-based all the way through. All courses should correspond in their organization of learning and in their focus on first, competence development, and then choice of teaching form and content. Indeed, this may be the greatest challenge for transforming science and mathematics education at university level. Changes at ground level are not sufficient; the changes must happen all the way from ground level to penthouse level, at the same time and according to the same principles.

References

Anderson, O.R. (1976). *The Experience of Science. A New Perspective for Laboratory Teaching*. New York: Teachers College Press.

Arce, J. & Betancourt, R. (1997). Student-Designed Experiments in Scientific Lab. Instruction. *Journal of College Science Teaching*, 27, 114–118.

Argyris, C. (1992). *On Organisational Learning*. Cambridge: Blackwell.

Bauersfeld, H. (1979). Research Related to the Mathematical Learning Process. *New Trends in Mathematics Teaching*, 4, 204–206.

Bauman, Z. (1992). *Intimations of Postmodernity*. London: Routledge.

Beck, U. (1992). How Modern is Modern Society? *Theory, Culture & Society*, 9, 163–169.

Biessels, H.W.A. (2000). *From Mono-Ester to Poly-Ester, A Project in First Year Chemistry Curriculum at Utrecht University*. Personal communication.

Buning, J.B. & Thijs, G.D. (2001). *Comparing Teaching Assistants Guiding Students in Physics Undergraduate Laboratories*. Paper presented at the Third European Science Education Research Association (ESERA) Conference, August 21–25, Thessaloniki, Greece.

Giddens, A. (1990). *The Consequences of Modernity*. Cambridge: Polity.

Henderson, L. & Buising, C. (2001). A Research-Based Molecular Biology Laboratory. Turning Novice Researchers into Practicing Scientists. *Journal of College Science Teaching*, 30, 322–327.

Keulen, H.V. (1995). *Making Sense. Simulation-of-Research in Organic Chemistry Education*. Utrecht: CD-β Press.

Lewis, J. (1999). The Use of Mini-Projects in Preparing Students for Independent Open-Ended Investigative Lab Work. *Biochemical Education*, 27, 137–144.

Millar, R. (1991). A Means to an End: The Role of Processes in Science Education. In B. Woolnough (Ed.), *Practical Science* (pp. 43–52). Milton Keynes: Open University Press.

Parkinson, J. & Adendorff, R. (1997). Two Streams of Literacy in Science: A Look at First Year Laboratory Manuals. *Language and Education*, 11, 200–221.

Qvortrup, L. (2000). *Det hyperkomplekse samfund*. Copenhagen: Nordisk Forlag.

Ram, P. (1999). Problem-Based Learning in Undergraduate Education. A Sophomore Chemistry Laboratory. *Journal of Chemical Education*, 76, 1122–1126.

Simonsen, B. & Ulriksen, L. (1998). *Universitetsstudier i krise. Fag, projekter og moderne studenter*. Roskilde: Roskilde Universitetsforlag.

Tolman, C.A. & Parshall, G.W. (1999). Fifty-Year Trends in Chemical Industry: What do They Mean for Chemical Education? *Journal of Chemical Education*, 76, 178–189.

Troelsen, R.P. (2001). Kompetenceudvikling i kemiuddannelserne (Development of Competence in Chemical Education). In T.B. Hansen, K.N. Nielsen, R.P. Troelsen & E. Winther (Eds.), *Naturvidenskab, dannelse og kompetence* (pp. 61–132). Aalborg: Aalborg Universitet.

Troelsen, R.P. (2003). *Kemi og kompetencer – et forsøg værd? En undersøgelse af laboratorieundervisningens vilkår og muligheder i de tertiære uddannelser* [Chemistry and Competencies. An Investigation of Conditions and Possibilities of Laboratory Teaching in Tertiary Education]. Ph. D. thesis. Odense: Syddansk Universitet.

Wright, J.C. (1996). Authentic Learning Environment in Analytic Chemistry Using Co-Operative Methods and Open-Ended Laboratories in Large Lecture Courses. *Journal of Chemical Education*, 73, 827–832.

Ziehe, T. (1989). *Ambivalenser og mangfoldighed. En artikelsamling om ungdom, skole, æstetik og kultur*. Copenhagen: Politisk Revy.

Chapter 4
Assessment and Contract-Like Relationships in Undergraduate Mathematics Education

Niels Grønbæk, Morten Misfeldt, and Carl Winsløw

"Well, I let him pass, but I am pretty sure that if I had given my questions a twist, his lack of understanding would have been revealed." At traditional oral or written examinations, this kind of impression is not unusual among examiners. Knowing that the student has dutifully met the course requirements, the only reasonable verdict is "pass." Yet it is given with an uneasy feeling that the student has not really been put to the test. However, it seems that the circumstances and traditions governing university mathematics teaching make it difficult to assess more than the use of standard techniques or the passive knowledge of textbook material.

It is commonly acknowledged that assessment procedures influence university students' activities in many ways. At a very global level this influence can be seen as simultaneously inevitable, necessary and regrettable. *Inevitable* because universities need to grant a credible diploma and hence must assess and declare their holders' competencies in depth; the fact that this declaration may be decisive for the students' future career opportunities is, at a global level, an important reason for students to direct their study activity towards maximizing their "declared competency" (or, more modestly, to just get their diploma). It is *necessary* in the sense that (outside paradise) explicit reward is sometimes necessary to get people and hence society to work. This may be true not least for studies with many technical and difficult parts located more or less necessarily at the beginning of the curriculum: the necessity to do these parts (in order to achieve the target diploma) acts as a default motivation for students who do not themselves acknowledge their attraction or necessity. Finally it is *regrettable* because the existence of such a default incentive to study may seem to suspend the need for other rationalities for teaching and study, and hence reduce academic teaching and studies to something highly un-academic: work based solely on control and rule following. Obviously one would like to *minimize* the regrettable effects of assessment, while retaining a visible incitement for students to meet necessary work requirements as well as a credible declaration of the results of this work.

If we acknowledge the impact of student assessment on student activity during a course, and that students' activity is the root of their learning outcome, then we

N. Grønbæk, M. Misfeldt, and C. Winsløw
Copenhagen University, Denmark, and Danish School of Education, Aarhus, Denmark

O. Skovsmose et al. (eds.), *University Science and Mathematics Education in Transition*,

are led to conclude that the design of student activities and student assessment should be looked at as a whole. Very roughly one might think of the following pattern of influence, as shown in Fig. 4.1. The arrows in Fig. 4.1 are not all of the same nature. In particular the influence of assessment on student activity tends to be indirect, and to depend crucially on *students' images and expectation* of assessment.

In order to investigate the actual nature of the effects of assessment – and even more, to investigate how to control them – we must of course consider *contexts where these effects occur.* In fact, these effects are not easily observed. As we shall explain, they can be modelled as arising from a specific sort of "contract" between students and teachers, strictly conditioned by the context (in particular the institutional and academic context). However, in the absence of interventions this contract may be largely implicit and impossible to distinguish from *other* factors influencing the interaction of students and teachers, such as the conditions and possibilities offered by the subject to be learned and taught. As a consequence, beyond general remarks such as those professed in the first paragraphs of this introduction, we find it necessary to consider the effects of assessment from an *engineering point of view* (Artigue, 1994): in concrete contexts, with specific aims of control and perhaps change. It is a main point of this chapter to show how such an approach can be theoretically supported by the theory of didactical situations, with appropriate adaptations for the university level.

After introducing the relevant theory, we describe briefly the context from which our data come: a format for didactical engineering which is based on collaborative student work on written "thematic projects," subsequently assessed through an individual oral exam. We then explore two aspects of this context: the independent work of student groups and the learning outcomes as evidenced by the written projects. Both are analyzed with respect to the influence of the assessment format, including the perturbation of the "contract" that a new format leads to. Finally, we briefly discuss the potentials and limitations of local (single-course) changes in the contract underlying the mechanisms in Fig. 4.1.

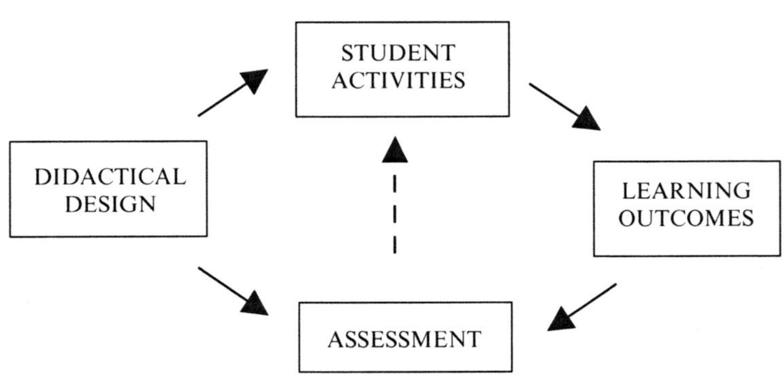

Fig. 4.1 A rough picture of the interplay between teaching, learning and assessment

4.1 Didactical Contracts in University Mathematics

The theory of didactical situations (hereafter abbreviated TDS), developed by Brousseau and his school since the 1970s, has chiefly been used in various contexts of school mathematics, such as the famous studies of the teaching of decimal numbers (Brousseau, 1997, Chaps. 3–4). However, the key features of TDS (Chaps. 1–2 and 5) seem highly relevant to university teaching as well, and indeed it is used in this context in some studies (for instance, Artigue, 1994). In this section, we shall try to outline some basic points of TDS that will allow us to talk meaningfully of didactical contracts in university teaching, and to understand their special features in this setting.

4.1.1 The Notions of Situation and Game

The basic assumption of TDS is that "learning is a modification of a student's knowing which she must produce herself and which the teacher must only instigate"(Brousseau, 1997, p. 227). This is, at face value, just the constructivist point of view: personal knowledge (termed "knowing" to translate the French word *connaissance*) results from various types of *action* in a situation, which somehow presents challenges to the learner. However, modern mathematics teaching – not least in the university context – implies that students must learn much more than is possible through spontaneous interaction with the world. Piagetian developmental psychology is mainly interested in showing the potential for the learner, at various ages, to make sense of quantities, motion, basic logic, and so on – using just elements of his immediate experience. Indeed, Piagetian theory is not concerned with teaching or other forms of institutionalised learning. From this background, TDS presents three original features, which are also crucial for the study of university teaching.

The first feature resides in insisting on the epistemological value of situations for teaching: they can and must be arranged by the teacher with didactical intentions in order to achieve learning of some specific knowledge. Such *adidactical situations* – prepared and "instigated" by the teacher – are achieved when the teacher withdraws to let the student(s) act. The term *adidactical* refers to the absence of explicit involvement of the teacher (with his intentions to teach). But the epistemological assumption is still stronger: "Each item of knowledge can be characterized by an (some) adidactical situation(s) which preserve(s) meaning: we shall call this a fundamental situation" (p. 30). A main objective for didactics is, in this perspective, to construct and investigate the use in teaching of situations corresponding to target knowledge such as the mastery of decimal numbers.

The second feature of TDS is to model the highly complex interplay between the teacher, the student(s) and the various forms of situations that allow the student to construct knowing and knowledge, and so to relate the epistemological and social dimensions of learning in a concrete as well as specific manner. The adidactical situation is the part of the teaching process where the teacher retires, in order to allow the direct *game* of the student with a *milieu* for learning. The milieu is arranged and, if necessary, rearranged by the teacher, in such a way that "winning strategies" successively approach the knowledge being aimed at. Of course, the teacher is not just conceiving of the milieu in order to subsequently step back and watch the students' failure or success in the game. The teacher *interacts with the students' game* while presenting it and modifying it in different phases. This interaction is called a *didactical situation* (p. 31). This is a larger form of game, involving the teacher as a crucial player. The process through which the teacher "hands over" a milieu to the students is called *devolution*; it is supposed to achieve the students' acceptance of the adidactical situation. The phases of the didactical situation may be roughly classified as follows (pp. 8–18, 65–71, 231–235):

- *Devolution of adidactical situations of action* (more or less "pure play") in which students explore parts of the milieu in a rather immediate way
- *Devolution of adidactical situations of formulation,* where the students are urged to articulate observations from their game with the milieu; the teacher may occasionally intervene in order to help students clarify their statements
- *Situations of validation*, where students and teacher consider the explicit statements (representing knowing) about the game with the aim of deciding their validity
- *Situations of institutionalization*, in which the teacher emphasises the validated knowing as common *knowledge*.

Notice that each of these phases may be considered as different sorts of games, with different objectives and rules, and with the players assuming different roles. A classical situation illustrating this is the so-called "race to twenty" (Ibid., Chap. 0), conceived for elementary school; the method (for university "lectures") of *scientific debate* developed by Marc Legrand (2001) is clearly conceived with a similar pattern of games in mind. Notice, however, that real time presence and interaction of teachers and students are clearly assumed in the didactical situations as described. It is a model of classroom teaching in a wide sense. But it is not a prescriptive model *per se*, as the case of university lectures show: in their usual form, these are mainly *situations of institutionalization* but without ensuring that the knowledge communicated is already established as validated knowing for the students.

The third feature of TDS is to relate the work of the teacher to that of the mathematician, which of course takes on a special meaning in the context of university teaching, where the two may be the same person (although research and teaching does not, typically, concern the same knowledge):

> Mathematicians don't communicate their results in the form in which they discover them; they re-organize them, they give them as general a form as possible. [...] The teacher first undertakes the opposite action: a recontextualization and a re-personalization of knowledge. She looks for situations, which can give meaning to the knowledge to be taught. (Brousseau, 1997, p. 227)

This means that we have two almost opposite *transpositions* of knowledge: from situation (of discovery) to formalized, official knowledge, produced by the mathematician; and from such official knowledge to didactical situations (including milieus and games) that enable students to acquire the knowledge. Of course the situations that are devolved to the students do not by any means need to be equal or even similar to the historical situation of discovery. But it is the close connection, if not similarity, of adidactical situations and mathematical discovery, as well as these complementary processes of knowledge transposition, which is the source of Brousseau's consistent insistence on the institutional and intellectual inseparability of mathematics and its didactics (Brousseau, 1999). In the most immediate sense, "all mathematicians are practitioners and consequently connoisseurs of didactics as applied to mathematics" (p. 44).

4.1.2 The Notion of Didactical Contract and Its Paradoxes

As was noticed above, the didactical situation in itself may be considered as a game, where the players are the teacher and the students. In order for it to succeed, the student must learn, and to do so, must accept her responsibility in the adidactical situation devolved by the teacher. On the other hand, the teacher is also responsible for the success of the students' game with the milieu, by designing it, by intervening or modifying it if necessary, and by evaluating it. In general, any teaching-learning situation implies – and requires – a set of mutual *obligations* between teacher and student. In a sense they *become* players in a didactical situation *through the establishment of these obligations*. Without the basic willingness of the teacher to help the student learn, and of the student to engage in the intellectual work proposed by the teacher, there is no didactical situation.

As we all know, good intentions are far from sufficient to teach and learn mathematics efficiently. And the obligations of students and teachers must be understood both with respect to the complexity of didactical situations, outlined above, and with respect to the high-stakes nature of mathematics learning in many contexts (such as university). Moreover, those obligations are mostly implicit and a considerable uncertainty about them may subsist for both parties. Nevertheless, "this system of mutual obligations resembles a contract" (p. 31) and indeed one does speak of a *didactical contract*. As with many other forms of regulation of social systems, it appears only – or at least most obviously – when it is broken. And, surprisingly, *breaking* the contract is also a condition for learning to take place (p. 32). While the contract is certainly necessary for devolution to succeed, it must not rule the students' game with the milieu: "the student's answer [to problems posed by the milieu] must not be motivated by obligations related to the didactical contract but by adidactical necessities of her relationships with the milieu" (p. 57). Thus, the didactical contract is the root of certain paradoxes linked to devolution and adidactical situations: after devolution and acceptance of the situation, "everything that [the teacher] undertakes in order to make the student produce the behaviours that she expects tends to deprive this student of the necessary conditions for the understanding and the learning of the target notion" (p. 41). It may be necessary for the teacher to intervene, on pain of leaving the student to fail (in which case both will have failed to honour the contract). But if this intervention reduces the students' task to little or nothing, the formal fulfilment of the contract leads to a meaningless "empty" milieu where learning cannot take place. Indeed, didactical situations – especially those devolving a very rich milieu – are "instable systems" in the sense that small perturbations, sometimes out of the teacher's control, may change success into failure.

4.1.3 University Mathematics Contracts

It is important to remember that learning achieved in adidactical situations has to be established and made common through situations of validation and institutionalization. So, teaching needs to both transform official mathematical knowledge into adidactical milieus *and* to transform the information obtained there *back* into shared knowledge compatible with the starting point. However, to achieve this personalized learning as a basis for institutionalization is costly and difficult, as suggested above. It is certainly simpler for the teacher to pass directly to institutionalization:

> There is a strong temptation for the teacher to short-circuit these two phases and to teach knowledge directly as if it were a cultural fact, thus saving the cost of this double manoeuvre. The knowledge is presented and students make it their own as best they can (p. 227).

Indeed, this is what happens in most university mathematics courses – at least, to some extent. More precisely, a very common format is to *first* institutionalize new knowledge in lectures (see Weber, 2004 for typical case) and *then* devolve milieus in the form of problems and exercises that require the knowledge and which are meant to help the students "make it their own." Subsequently, a kind of formulation and validation situation takes place during class sessions. We shall call this rough scheme of teaching the Lecture-Problems-Class Model (LPC for short). In some course contexts, the problems on which students get to work are all somewhat stereotypic applications of the theoretical material. These problems are meant to establish, for students, certain associated techniques – rather than a knowing of the full theory (see also Winsløw, 2006). In this case the "re-personalization" of knowledge, particularly the more theoretical parts exposed in lectures, will still largely be left to students.

Notice that we want to take TDS seriously as a descriptive rather than as a prescriptive model of teaching. We do believe that the four phases of a didactical situation, described above, are meaningful also in the university context; but there is no *a priori* reason to claim that they should necessarily appear in the order which seems to be the most natural and efficient for elementary school teaching (according to the massive experimental work done by Brousseau and his team). In particular, university students should not necessarily be fed with meticulously arranged situations of learning, but must also learn – as part of the academic trade – to access knowledge directly in its official, depersonalized form. But this is, indeed, something to learn, rather than a starting condition that can just be assumed.

As suggested by several authors (e.g., Weber, 2004, p. 131), there could be many rationales behind the use of LPC other than those linked to didactical finalities. But its massive use also shows that LPC offers a certain balance and stability with respect to the interests of institutions, students and teachers. Indeed, with respect to students and teachers, the LPC scheme demands – as any other didactical scheme – a didactical contract to establish and regulate the didactical situations involved. Parts of it are linked to assessment – which, in a university course, is rarely more than a few months away. Here is, tentatively, what the main clauses could look like, if made explicit:

> **Didactical contract for LPC-scheme (outline)**
> *Obligations of lecturer*: Clearly expose and explain the theory and examples required to do the problems assigned for homework. Make sure that doing the homework assigned will substantially help students towards passing the final exam (written or oral).
> *Obligations of students*: Follow the lectures attentively, read the corresponding texts. Do the assigned homework. At all times, ask questions if something appears to be unclear.
> *Obligations of classroom instructor*: Validate students' answers to problems. Institutionalise good solutions and answer questions of students.

Examination requirements do not necessarily dominate the contract. On the contrary, the tacit assumption that the situations devolved by the lecturer are "relevant" also with respect to these requirements, may help to put these in the background most of the time, and hence allow the student's work to be, some of the time, driven by "adidactical necessities of her relationships with the milieu." Of course a condition for such a silent consensus is that the contract above is actually observed, in particular that the work proposed by the teacher actually does help the students towards succeeding at exams. Now, the most common forms of examination – requiring, from the students, a few hours work on a collection of exercises, or a short oral presentation of textbook theory – may not motivate autonomous, in-depth work with mathematical theory according to the contract above.

Our basic analysis is therefore as follows: if we want to improve the quality and scope of students' adidactical work, we may need to modify *simultaneously* the exam requirements, the milieus worked on by students, and the role of the teacher with respect to devolution and institutionalization. All of this is likely to require amendments of the contract. A major challenge is therefore to avoid the logic of replacing the target adidactical work (searching) for a new contract. The establishment of the contract *and its disappearance* (in the adidactical situation) must be kept under strict control.

4.2 A Didactical Engineering Project: Thematic Projects in Real Analysis

In the following sections we describe and discuss the evidence of contractual relationships and their evolution during a didactical engineering project, carried out in the context of a second year course in real analysis at the University of Copenhagen. The necessity to consider assessment, the nature of student tasks, and the form of teaching simultaneously was at heart of the engineering, which we will briefly introduce in this section (it is described in much more detail in Grønbæk and Winsløw, 2007a,b).

4.2.1 A Difficult Course

The course in question, *Mathematics 2AN*, is a third semester course in real analysis crediting 10 ECTS (European Credit Transfer and Accumulation System). There are two weekly double lectures ($2 \times 2 \times 45$ min) and one weekly problem session of 3×45 min. It has an enrolment of about 175 students. In several respects the course is central to the overall study plan. It is the first course where students meet mathematics in full rigor; it is a prerequisite for many other courses; it addresses students with a wide spectrum of ambitions, talents, and time available for the course; and the syllabus is demanding (Carothers, 2000, Chaps. 1–11 and 15). The course is assessed by means of a written 3-h exam and an individual ½ h oral exam. All of the above are externally imposed boundary conditions on which a course designer has very little influence. Of course, this is a rigid system and a costly form of assessment but it is nevertheless common and traditional in major Danish universities. It is perceived to ensure a certain consistency among courses at different universities and with different teachers within the same university.

The first author had taught the course for 3 years. As in other courses of the programme, the didactical obligations were, roughly speaking, regulated by the LPC contract. The course was suffering from a number of functional problems, such as large dropout and low passing rates. It appeared that some of these shortcomings were rooted in students' amount and quality of independent work with course material and that the LPC contract was a part of the problem in the sense that it did not allow for different and more explicit demands on student activity. Some early attempts had been made previously to introduce project work and peer assessment in the course, but this had not resulted in significant improvements of student work. With hindsight, one could say that the LPC contract continued to be in force, mainly because the changes in assignment of student work were not reflected in accompanying changes of teaching and assessment, but also because the teachers did not succeed in making their motives and expectations clear to the students.

4.2.2 *Coherent Changes in Student Activity, Teaching and Assessment*

In order to address the above concerns, we developed a new format that we call *thematic projects* (see also Grønbæk and Winsløw, 2007a,b). A thematic project may be described as a written assignment consisting of mathematical tasks, some of which are quite open, and most of which are more complex and theoretical than standard "training exercises." The tasks are centred on a "theme" (e.g., *the Hilbert space* L^2) so that together the solutions represent a coherent piece of theory; in this sense, a thematic project is, ideally, a kind of fundamental situation. Indeed, it can be considered an adidactical milieu, its formulation being the devolution. The adidactical work implies, due to the open nature of the tasks, situations of action and formulation as well as situations of validation (integrated in teaching to some extent, cf. below). The students worked on the thematic projects in groups. The subjects of the thematic project were chosen in accordance with the progression of the course, so that the students could work with the projects successively and throughout the semester.

In order to support this work, parts of the class teaching (normally used for presenting solutions to training exercises) were devoted to work with the thematic projects, with the instructor available for questions. The lectures were partly changed in order to relate the presentation of theory and examples to the projects; and the lecturer was also available for questions related to the projects. For each project, a date was set where the students could hand in their work in order to get written feed back from the instructor, mainly of the form "this is good/OK/needs to be worked on."

The thematic projects of the course (six in total) replaced the traditional textbook presentation in the oral exam: instead of drawing a random "theorem" to present (as in the book), the students draw one of their thematic projects. Notice that the exam is individual, while the projects were worked on in groups. This had important consequences for the regulation of the students' work: each group member was required, in the end, to be able to explain and defend the common projects. It was explicitly said that the product of the work on thematic projects was not required or supposed to be an extensive report, but just 4–5 pages that could serve as background for the students' presentation and the examiner's interrogation (the written product had to be delivered to the evaluators at the time of examination). However, we shall in the following still refer to these written products as "reports."

4.2.3 *The Adidactical Contract of the Groups*

A key point to all of this was to change the work with theoretical parts of the course from merely acquiring the text book presentation to producing and formulating theory in projects, thus aiming deliberately at a *re-personalization-re-de-personalization* process (see Brousseau, 1997, p. 23). This organization of the work in groups was meant to amplify situations of action and formulation. An important factor turns out to be that the material goal of these situations is the production of a usable manuscript for the oral examination. Also, the thematic projects include tasks, which are optional and open, in particular tasks where the students must choose a level of ambition (for instance generality) in their interpretation of the task. The purpose of this is twofold. It gives the student the opportunity to demonstrate their competencies without the risk of "breaking their back" on sophistication. It also requires the students to be conscientious about the status of their knowledge; "knowing what I know" is not exclusively an issue of metacognition but is a crucial and advanced form of knowledge in mathematics. It becomes a part of the groups' work because this work is the basis for the oral examination. Here, the student simply draws a random project and presents selected parts of the solutions presented in his groups' report.

All these conditions for the groups' work, and in particular its relation to assessment, imply that the groups must *negotiate a common, internal contract for their work with the tasks*. We shall call these (implicit and explicit) agreements of mutual obligations within the group of students the *adidactical contract of the group*. It is clearly framed by the didactical contract and the didactical situation surrounding the adidactical work, but it is also, as we shall see later on, proper to the group.

4.2.4 Modifications of the Didactical Contract

The introduction of thematic projects, as described above, is based on a few explicit and official changes in the assessment procedure (mainly for the oral examination) and in the way the theoretical knowledge is taught. But its consequences for the didactical contract, regulating the mutual obligations of students and teachers, are more far-reaching. We shall exemplify and illustrate this in the following sections. There, we summarize the general tendencies and the relation to the LPC contract.

First of all, the written exam is still there, and so the LPC contract remains in force for the work on more elementary and technical tasks. The introduction of the thematic projects implies certain *amendments* of it, which address the more theoretical work that is assessed at the oral exam. These amendments concern potential conflicts in the new design, mainly coming from relating students' independent work and self-awareness so closely to the final assessment. On the one hand, the students are required to let go of the usual check marking from teachers, and rely on their own judgement (e.g., in choosing their level of ambition). On the other hand, students' work will be the basis of an exam, whose criteria is beyond their influence and experience. These conflicts are partly affective in nature, due to a fear of being "trapped" and an uncertainty about the requirements. This makes it very important that elements of the didactical contract are declared rather directly, in order to avoid uncertainty and misunderstandings. The students still assume, by default, the LPC contract. Accordingly the amendments have the nature of an allonge. In return for students' willingness to accept the proposed new format and the working mode, the course organiser declares:

- willingness to change the plan when students object (reasonably)
- intention to follow closely how things are developing at ground level
- intention to assess students according to the circumstances under which their results are developed (rather than comparing them to (fancier) textbook versions)
- detailed descriptions of formal requirements (at the beginning of the course).

Obviously, these declarations must be followed by practice. As an example of the willingness to change plan, in the first run of the course with thematic projects, it became apparent that the workload was too heavy. This led to allowing each student to discard one project for the oral exam (thus drawing from 5 instead of 6). This is (also) reasonable because it gives the students an opportunity to adapt the examination to their individual capacities, and represents yet another task of self-assessment.

The contract amendments concern the teaching directly. The lecturer must relate the course material with thematic projects in order for the course to become an integrated whole rather than a collection of isolated themes. Also, both class-room instructors and the lecturer offer guidance to students in their project work, and it is explicitly stated that they will not provide complete or partial solutions. Their task in this context is to help students in the phases of formulation and validation. The written feed back on the project notes must work as a situation of institutionalization – it turns out to be crucial for the students that they can really regard it as such.

Likewise, the final assessment must respect the genesis of thematic project outcomes, and ideally the winning strategies in the devolved games should correspond to winning strategies in the context of assessment. In most exams the grade is obtained as the result of a synthesis of two opposing principles: a subtraction process where one counts down from an ideal, thus penalizing the student for flaws an errors; and an addition process, where the student is credited for achievements. In the classical oral exam the evaluation is based on a comparison of the student's presentation with a "perfect" presentation from a textbook. This will almost automatically give priority to the subtraction principle. When the student is evaluated on the basis of her own product, the examiner must focus on the addition principle. This point requires deliberate attention.

Whereas the change in the lecture format and its relation with the contract are rather straightforward, the assessment aspect of the contract is much more complicated. It calls for a great deal of trust from the students and corresponding responsibility from the teacher. ("Can I trust that NN does give me credit for my rather weak version of theorem XX, knowing that there are much more powerful versions?"). Hence it is important that these elements of the contract are discussed during the semester. It can even help establish an ethos of the course.

4.3 Students at Work: Adidactical Situations or Contract Following

In this section we describe what goes on when the students work on the thematic projects between classes, without the presence of the teacher or lecturer. How do they organize their work? To what extent do the students follow a didactical contract and can their work be described as adidactical situations? These questions will be addressed by looking more deeply at the students' collaborative activities in connection to the thematic projects.

4.3.1 Investigation Method and Goals

The second author followed two groups of students, and investigated the students' work using a combination of informal interviews, video observations, and diaries kept by each of the students.

The purpose of the diaries was to obtain an overview of how different activities connected to the group work were situated during a specific week (Fig. 4.2). During this "diary" week, in the middle of the project period, all participants in the two groups were asked to keep a detailed record describing what project tasks he or she worked on, and when. Furthermore each student was asked to

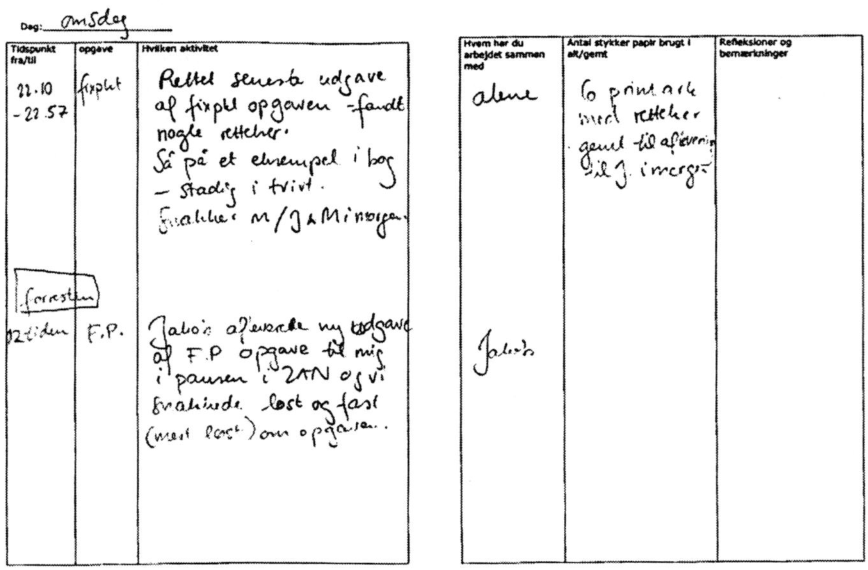

Fig. 4.2 A diary opening

describe the type of activity engaged in, explain if the work was done alone or in collaboration (if so, with whom), and describe how many pieces of paper had been used and kept (if they used a computer for writing they would write that). Furthermore the diary had a field that allowed for free reflections about the activity reported on.

The diaries were accompanied by video observations of working meetings. A total of seven meetings of various lengths (about 6 h of video in total) were recorded. The researcher was in frequent contact with the two groups of students, mainly by finding them in the area where they usually worked (a university cafeteria). This frequent contact served several purposes. The contact was necessary in order to gain access to the students working meetings, because the students typically saw each other several times a day, and hence were able to reschedule meetings with a very short notice; which they often did. The contact also allowed a better longitudinal picture of the activities by allowing frequent informal interviews with the group members. This proved to be a valuable source of information, and finally it helped to build trust between the researcher and the group of students.

The videos were summarized, and interesting parts identified and transcribed, using the Transana program for video analysis (Fassnacht and Woods, 2003). For more details about the methodology for collection and analysis of data see (Misfeldt, 2006).

4.3.2 Overall Impressions from the Meetings

The students in the two groups (Group One and Group Two) met in general once or twice a week to discuss their projects, with increasing intensity close to the deadlines for instructors' comments. The meetings were generally concerned with:

1. Dividing the labour
2. Discussing solving strategies
3. Actual problem solving activity
4. Reviewing previously written work.

In interviews the students point to (1) and (2) as the main reasons for having the meetings; but (3) and (4) also occurred during all the meetings observed.

The didactical contract appeared explicitly from time to time in these activities. When dividing the labour the students often referred to an explicit contract that divided the questions into kernel questions and other questions. This distinction plays at least two roles for the students. First and foremost it provides a prioritization of the tasks: it is clear that the students considered the kernel questions most important to solve. However, there was also an underlying assumption that the questions that were not kernel questions had to be very difficult, and hence perhaps above the students' abilities. The students seem to ignore them as a consequence of the didactical contract.

The contract was also present from time to time when the students were discussing solving strategies and when they were more deeply engaged in problem solving activities. The contract can pop up suddenly as we shall see in the following episode where three students (from Group Two) are working with accepting or rejecting a number of propositions and support their claim with either a proof or a counterexample. The first three propositions have been rejected with counterexamples, and we enter the conversation when they are brainstorming heuristics for the fourth. A possible counterexample is proposed when student B gets nervous:

18	Student 2b	Shouldn't any of this be true?
19	Student 2a	Yes
20	Student 2c	This can't be right, I think this is a little too easy
21	Student 2b	We were sure about this one, yes that was the argument from before
22	Student 2b	Can it be that these are the ones "*name of the instructor*" talks about?
23	Student 2c	Hmm I don't think so
24	Student 2b	Yes it must be them

There are several things in this quote that show that the students have the didactical contract in mind. Firstly, the concern that "some of the propositions should be true" is instantaneously supported by the two other students. We interpret this utterance as meaning "the teacher would not have asked us to prove or disprove if we only have to disprove." Secondly, once the contract is introduced in the conversation the students instantaneously reflect on what the instructor said in a previous session.

We have seen this type of contract pop-up many times when observing meetings amongst students. They tend to be short and do not in general guide the students' investigations, but as this quote shows the students are aware of the fact that their mathematical work is governed by a didactical contract and occasionally they make use of the logic of the contract as a part of the heuristics for their mathematical work. In this case it contributes to their solution of the problem.

4.3.3 An Example of an Adidactical Contract

Apart from the didactical contract the groups developed what we called an adidactical contract, regulating their collaborative work. Where as students learning fulfil the didactical contract, the product the students make together fulfils this adidactical contract. In this section we describe an example of such a contract.

In Group One there was a special division of labour implying that one of the students was responsible for writing an electronic version of the group work. The two other students were on the other hand more or less responsible for solving the mathematical problems. If we take a look at a schematic summary of their diaries, we see that the preparation of electronic documents is a very time consuming activity.

For each person and each day during the week the schema shows how much time he or she had spent at the university (above the line) and at home (below the line) and furthermore the letters P and G signify respectively whether the work was done personally or with the group (see Table 4.1).

The diaries and meetings showed that student 1a's personal work had to do with preparing an electronic version of the document they were working on. On Monday in this specific week, the group collaborated for an hour and 25 min on a task about fixed points and, in the evening, student 1a spent an hour and a half typing the results of their efforts. On Wednesday student 1a handed over his electronic version for review by the other students. On Thursday student 1a received a commented version of the manuscript and the same evening he finished the report on fixed points by doing the proposed corrections.

The two other students also worked at home on Thursday but they were preparing the next thematic project on homeomorphisms. On Friday they discuss this thematic project for 3 h and over the weekend student 1a works 6 h alone on making sense of their discussions and preparing a preliminary electronic version of the report.

If we compare the hours that each of the students put into the group work we see that in addition to the meetings student 1a uses 7 h and 50 min working on the thematic projects whereas student 1b uses 2 h and 20 min and student 1c uses 2 h and 30 min.

What kind of contract governs the students' collaboration here? And how does it relate to the didactical contract? The diaries and our observation and communication with the group suggest that the students' adidactical contract attributes the responsibility for the content oriented work in the thematic projects to students 1b and 1c, whereas the responsibility for writing it up electronically is given to student 1a. One can ask if it is desirable that one of the students only does the typing, and if the didactical contract should be changed in order to avoid such a division of labour.

It is worth noticing that this division of labour seems to be rooted in a shared wish to "make a nice report" rather than in a concern to ensure their learning outcome. In fact, looking at the reports collected during the oral examination, one notices a general tendency to take great care as to the professional design and layout of the report (in particular, the widespread use of TeX). If the work was dictated purely by the

Table 4.1 A schema of the diaries from group one

Person	Monday	Tuesday	Wednesday	Thursday	Friday	Saturday	Sunday
1a	G. 1:25 h P. 1:30 h		G. 0:15 h	G. 0:30 h P. 0:20 h	G. 3:00 h	P. 3:30 h	P. 2:30 h
1b	G. 1:25 h		G. 0:15 h P. 0:50 h	G. 0:30 h P. 0:30 h	G. 3:00 h		P. 1:00 h
1c	G. 1:25 h			P. 0:30 h G. 0:30 h P. 2:00 h	G. 3:00 h		P. 0:30 h

didactical contract it should have been a concern for the students to make explicitly sure that everyone could do all the tasks. Nevertheless this division of labour is, in our opinion, a nice example of the way that this new design supports adidactical situations in a very balanced way. The function of student 1a is not reduced to that of a typist. Indeed he does participate in the meetings where solutions and strategies for solutions are discussed and he does not receive a perfect manuscript for his preparation of electronic version; typically he brings home partially finished drafts of solutions developed by the other students. It is also well known that the situation of formulation is a significant aspect of the mathematical work.

Obviously the adidactical contract differs in details for different groups of students. For example Group Two took turns in writing up an electronic version of a document.

4.4 Evidence of Contractual Understandings in Students' Written Reports

A typical exam enrolment results in approximately 200 thematic project reports. This section is based on a perusal of 80 reports from the exam in January 2004.

4.4.1 Overall Characteristic Features

Two characteristics of the reports are prominent: they are quite *similar in form* but *very diverse in detailed contents*.

All of the reports (with the exception of one group's) are more extensive than required. In fact they are written as a full text rather than in the style of a synopsis. The mathematical argumentation is detailed and far more extensive than it is possible to present at the oral exam. In one case a group even included an appendix to incorporate material that was not accounted for in the body text. In a thematic project about "homeomorphisms," one needs to use a certain result from the textbook. About one third of the perused projects give a detailed account of this result, such that an uninformed reader in principle can start almost from scratch. In short, the students have adopted the genre of a textbook or lecture notes.

Also, as pointed out before, typesetting, graphics, etc. constituted a substantial workload. Almost all reports are carefully designed in terms of layout, in some cases even with a fancy graphical design on the front page. In a thematic project entitled "The Cantor Set," the students are asked to give at least two different representations of the Cantor Set. Most project groups choose "removing-middle-thirds-of-intervals" as one of these representations, and illustrate the process graphically in their chosen word processor (usually TeX). Such an illustration is time-consuming to type and potentially difficult to remember at an oral presentation, thus seemingly superfluous in an exam synopsis.

One may speculate as to why the groups dominantly chose to produce such elaborate reports. The most immediate hypothesis would be that the formal requirements (i.e., making parts of the didactical contract explicit) were somehow unclear to the students. However, these were discussed with the students at several "question lecture sessions," and it was repeatedly emphasized that the reports should just be a support for the students' oral presentation. A more important cause – expressed during focus group interviews with students in another run of the course – seems to be that the groups want to demonstrate, internally and externally, that their work is really "complete" mathematics of the type found in textbooks but elaborated by them. In fact, the internal functions may be more important. As suggested by the observations in the previous section, the written report is an important element in the adidactical contract: the work is organized around its production, and it is an important "mutual obligation" to ensure that everyone in the group will be able to use this product at the exam. Idiosyncratic "notes" or "memos" would not serve this purpose. The report becomes the means and end to fulfilling the adidactical contract.

The other striking feature is that the project reports are truly different from group to group. Of course, there are often only a few essentially different ways of attacking a problem, but these can be varied in the detail. For example, the uncountability of the Cantor Set is proved either:

1. By a direct diagonal argument on one of two representations of the Cantor Set as binary sequences (left/right intervals or 0–2 ternary expansions)
2. By using a binary sequence representation and referring to the cardinality of $\{0,1\}^{\mathrm{N}}$, which, being a power-set, the textbook informs is uncountable
3. By referring to the Cantor function as a surjection onto an uncountable set
4. By proving that a nonempty perfect set must be uncountable

Only approaches (2) and (3) can be derived directly from the textbook.

In fact, among the 80 reports perused, we found that no two reports overlapped significantly in their detailed formulation and approach. It seems fair to conclude from this that the groups have worked autonomously. While the homogeneity in form of presentation, discussed above, is related to similar adidactical contracts (and, perhaps, to misunderstandings among students about the formal requirements), the diversity of content can only be interpreted as a result of the adidactical work in the groups. This is amply confirmed by student interviews and observation of student work.

4.4.2 Specific Features

The thematic projects were designed to include several possibilities for the students to choose their own level of ambition and technicality. Different formulations of problems made it possible to respond at a level appropriate for the individual student. One of the ways to do this is to ask progressive questions: (a) give a conjecture, (b) substantiate your conjecture, (c) prove as much of your conjecture as you can. For example, in the thematic project "Homeomorphism," the students are given two (among other) statements about a bijection f between metric spaces M and N: (1) f is a homeomorphism, (2) K is a compact subset of M if and only if $f(K)$ is a compact subset of N. One group claims that (1) is equivalent to (2), prove that (1) implies (2) and sets out to prove that (2) implies (1), but concludes with the remark: "MISSING!!! Have not yet proven this in a satisfactory way." Apart from the optimist tone, this demonstrates confidence in the explicit part of the contract (partial answers will also be honoured). The collected answers to the questions (a)–(c) range from mere unsubstantiated guesses to full proofs of the statement.

Another mechanism is the application of a theorem or concept in various degrees of difficulty and complexity. In the thematic project "Fixed Points" the students may apply Banach's fixed point theorem in different settings. The "low-level" setting consists of finding an approximation to the fourth root of 2, and the "high-level" setting consists of showing the existence of an implicitly defined function. Both ways require full understanding of the statement of Banach's fixed point theorem and insight into the proof. In the thematic project "Interchanging limits" the students are asked to interpret known theorems as a statement about interchanging limits. Some are rather straightforward whereas others are somewhat obscure. In both of these projects the reports are distributed among all possible levels of ambition, thus demonstrating that they rely on the contract's stipulation that their reports are accounted for on a personal level and that this will be respected in the exam. As "adidactical situations are learning situations in which the teacher has successfully hidden her will and interventions as a determinant of what the student has to do" (Brousseau, 1997, p. 236), it is not surprising to find that sometimes the teacher's intentions with thematic project work can conflict with the effects of summative assessment. In the project "Homeomorphism" the students are asked to decide whether two given metric spaces are homeomorphic. Some groups respond to this by exhibiting explicit maps, which, as a very simple inspection shows, cannot possibly be homeomorphisms. However, the groups declare boldly that these maps are obviously of the desired type. In the project "Fourier Series" the students work with a version of Dini's test. Some technical calculations are necessary for the conclusion. There are several examples of "innovative" rules to achieve the goal. In other words, the logic of the contracts (demands from within and from without the group that it should produce definite results) supersedes, in these cases, the logic of the adidactical milieu.

4.4.3 Institutionalizing Meaning

When working on a thematic project, the institutionalization of knowledge is to some extent left to the groups, as the teacher is not present. Moreover, certain concepts are essential for the students' understanding but are not in themselves objects of knowledge. The student may have "constructed a meaning" but can it be institutionalized? Or at least depersonalized? Sometimes teachers and students must rely on convention established by common use. An example of this is the negotiation of the adidactical contract, concerning "how much detail should be given in our arguments" (in the report).

Construction of meaning may also depend on mental representations of mathematical objects. The need for institutionalization is perhaps clearest when such a representation is wrong or insufficient. One group trying to describe the Cantor Set by means of removing middle thirds writes:

> Even though one cannot continue to draw the corresponding picture, this procedure can be carried forward ad infinitum.

Later, when trying to prove that the Cantor Set is uncountable by means of *LR*-sequences (*L* for choice of left interval, *R* for choice of right) they write:

> By a descent in the diagram ("corresponding picture" above) we imagine that we choose, step by step, between *L* and *R*. Somewhere far down in the diagram we find the points of the Cantor Set.

Even though this is a misconception, the students refer (almost) correctly to the Nested Interval Theorem, which would be unnecessary if an argument such as the above was valid. One sees here a conflict between an incomplete situation of formulation and a misguided attempt to follow the contract ("the teacher expects us to quote the theorem at this point"). The institutionalization of knowledge has partially failed.

4.5 Conclusions

In this chapter, we have tried to theorize and exemplify the relations mapped out in Fig. 4.1. In the context of university teaching where a significant part of the construction and institutionalization of knowledge is delegated to project groups, we have shown that the "rules of the game" may be described by two sorts of contractual relationships: the usual didactical contract, concerning the relation between teacher and students, which is needed to achieve the devolution and acceptance of the assigned work; and also an adidactical contract among the students in a group, developed by them and regulating their different roles in achieving the aims of the group. The didactical contract is an arrangement between teacher and student that is successfully fulfilled by the student's learning. On the contrary the adidactical contract is between the students and in this case fulfilled by the product the students deliver, rather than their learning outcome. In this sense the contracts live side by side.

The didactical contract, as proposed by the lecturer to the students, is explicit on the demands for the written product not to be an extensive report but merely a synopsis to support the oral examination. Nevertheless the students developed an adidactical contract that implied investing a lot of energy into the elaboration of the report, sometimes by assigning the responsibility for this to one student. We have discussed the possible reasons for this result, but one can also question whether it is something one should try to avoid. A complete text rather than a synopsis may be a necessary part of depersonalization and shared validation for the students. A shared validation of the mathematical target knowledge for the thematic project will benefit from a fuller and less personal form of written product. A very short synopsis is essentially a personal tool and the more detail the report contains the easier it will be to share the report and gain shared confidence of its correctness.

The adidactical contract governing the collaborative work ensures that all students participate and it enforces a mutual commitment among them to accomplish the required work. The fulfilment of the adidactical contract is also needed because the students have to bring the reports to the final exam and give them to the evaluators. The strength of this specific kind of evaluation is that both these contractual aspects are evaluated by the final exam. The students have to achieve a mathematical product together, but they also have to show individually what they learned.

The development of contractual relationships within one course must be seen in relation to "normal" forms of contracts with which the students are familiar. Changing the form of assessment in a single course cannot be expected to result immediately in a retreat of the LPC contract and other well-established modes of conceiving the relationship between students and teachers. But as similar forms of work and assessment are increasingly introduced in other courses within the university (e.g., Rump and Winsløw, Chap. 11), one may expect that the associated renegotiation of the contract will no longer be just a local problem of a particular course such as the one we have considered.

References

Artigue, M. (1994). Didactical Engineering as a Framework for the Conception of Teaching Products. In R. Biehler, R.W. Scholz, R. Strässer and B. Winkelmann (Eds.), *Didactics of Mathematics as a Scientific Discipline* (pp. 27–39). Dordrecht: Kluwer.

Brousseau, G. (1997). *Theory of Didactical Situations in Mathematics*. Dordrecht: Kluwer.

Brousseau, G. (1999). Research in Mathematics Education: Observation and... Mathematics. In I. Schwank (Ed.), *Proceedings of the First Conference of the European Society for Research in Mathematics Education* (pp. 35–49). Osnabrück: Forschungsinst für Mathematikdidaktik.

Carothers, N.L. (2000). *Real Analysis*. Cambridge: Cambridge University Press.

Fassnacht, C. and Woods, D.K. (2003). *Transana* (Version 1.22). Madison: University of Wisconsin.

Grønbæk, N. and Winsløw, C. (2007a). Developing and Assessing Specific Competencies in a First Course on Real Analysis. In F. Hitt, G. Harel and A. Selden (Eds.), *Research in Collegiate Mathematics Education* VI (pp. 99–138). Providence, RI: American Mathematical Society.

Grønbæk, N. and Winsløw, C. (2007b). Thematic Projects: A Format to Further and Assess Advanced Student Work in Undergraduate Mathematics. *Recherches en Didactique des Mathématiques*, 27(2), 187–219.

Legrand, M. (2001). Scientific Debate in Mathematics Courses. In D. Holton (Ed.), *Teaching and Learning of Mathematics at University Level. An ICMI Study* (pp. 127–135). Dordrecht: Kluwer.

Misfeldt, M. (2006). Conversations in Undergraduate Students' Collaborative Work. In M. Bosch (Ed.), *Proceedings of the Fourth Congress of the European Society for Research in Mathematics Education* (pp. 852–860). Barcelona: Universitat Ramon Llull – ERME.

Weber, K. (2004). Traditional Instruction in Advanced Mathematics Courses: A Case Study of One Professor's Lectures and Proofs in an Introductory Real Analysis Course. *Journal of Mathematical Behavior,* 23, 115–133.

Winsløw, C. (2006). Research and Development of University Level Teaching: The Interaction of Didactical and Mathematical Organisations. In M. Bosch (Ed.). *Proceedings of the Fourth Congress of the European Society for Research in Mathematics Education* (pp. 1821–1830). Barcelona: Universitat Ramon Llull – ERME.

Part II
Changes in Academic Cultures

Chapter 5
Cultural Models of Physics

An Analysis of Historical Connections Between Hard Sciences, Humanities and Gender in Physics

Cathrine Hasse

Culture is generally understood to be the values, beliefs and practices that we inherit and transform over time. At a deeper level it can be argued that values, beliefs and practices stem from cultural learning. Science has downplayed the influence of culture in its internalist self-understanding. In this chapter I argue that science is formed through cultural processes, which influence the selection of who are able to perform as scientists. I use a cultural-psychological theoretical framework to analyze how selections of physicists' works through implicitly learned connections, which are only recognized as cultural, when they are contrasted with other ways of connecting. Cultural learning processes form conceptual connections over time, which are too self-evident to be questioned. They are only challenged when they are confronted with amazingly different connections. For many physicists in western European countries it has come as a surprise to learn that female physicists are found in larger numbers in the Southern and Eastern parts of Europe than in the Northern parts. Many possible sociological explanations have been proposed, but none have provided satisfactory answers. The cultural-psychological approach I propose offers a new understanding pointing to different historical formations of connections between gender, physics and the humanities. These connections can be understood as particular organizations of knowledge captured as different "cultural models" of physics. In Denmark, a Northern European country, we find a shockingly low representation of female physicists. In comparison, the southern European country Italy has a much higher representation of female physicists. It is argued that the difference is due to the work of historically formed implicit connections of gender and humanities in relation to physics, which are made explicit when the conceptions of physics in Italy is contrasted with conceptions of physics in Denmark.

C. Hasse
Danish University of Education, Denmark

O. Skovsmose et al. (eds.), *University Science and Mathematics Education in Transition*,

5.1 Challenging Internalist Science

In 1998 Sandra Harding posed the provoking question: Is Science multicultural? Internalist scientific epistemology rests on the central assumption that the internal features of science insure its success. Even when culture is recognized as an influencing factor on science this influence shall be weeded out. Society shall provide education and conditions for science, but not influence it.

> "When science progresses, at its very best, it shall produce nothing culturally distinctive to the representations of nature that appear in the results of the research. The aim is to produce scientific information in which one can find no culturally distinctive interests or discursive resources of the societies that have produced the research" (Harding, 1998, p. 3).

This would to a large extent be the internal tale of science, as it was told to Robert Merton back in 1942, where the ideal norms of science (the CUDOS-norms) explicitly state that anyone, regardless of gender, colour of the skin and family background has a right to do science.[1]

In the past 30 years Science and Technology Studies (STS) have discussed science *as* culture replacing the notion of a value-free objectivism with particular forms of reasoning which Karin Knorr-Cetina called "epistemic culture" (Knorr-Cetina, 1999). Many of these studies have undermined the internalist self-understanding, not least in physics.[2] In spite of these studies many scientists still want to believe that their world is basically culture-free. Though it is generally accepted that science plays an important part in forming society, it is much less accepted by scientific disciplines like physics that society forms science. In the words of the anthropologist Sharon Traweek, physicists belong to an "extreme culture of objectivity, a culture of no culture, which longs passionately for a world without loose ends, without temperament, gender, nationalism or other sources of disorder – for a world outside human space and time" (Traweek, 1988, p. 162).

As shown by STS-studies this is a false picture. Physics is not only a culture with internal cultural beliefs, values and feelings. Science *as* culture is imbedded in a wider national/western context: science *in* culture.[3] From postcolonial and cultural studies, we learn that science and society are co-constructing each other and they are

---

[1] CUDOS stands for the four most important ideals identified by Merton: Communism (meaning 'communality' —science is not connected to any specific personal, corporate or state interest— it is common and therefore has an obligation to be public accessible knowledge), Universalism (truth is evaluated through impersonal criteria and stands as universal regardless of the cultural background of the scientific practitioners), Disinterestedness (science proceeds without being influenced by any personal interests of individual scientists) and Organized Scepticism (Science understands all truths claims to be provisional and open to critique (Merton, 1942).

[2] See for example Knorr-Cetina (1999), Pickering (1984) and Galison (1997).

[3] It should be noted that many physicists have actually welcomed this debate with the STS-people and seen it as an opportunity to expand the understanding of the relation between technology and nature, but these studies rarely include the psychological understanding of man's understanding of nature. Other physicists have entered the dialogue with the STS-people and some have countered their claims in the fierce debate called 'the science war'.

also co-constricting each other in an array of different ways. Science is heterogeneous and complex. Even within the same discipline a glance through the magnifying glass can dissolve what appeared on the surface to be a coherent whole into a myriad of cultural practices, or, in the words of Joseph Rouse: "The practices of scientific investigation, its products, and its norms are historically variant. They also vary considerably both across and within scientific disciplines" (Rouse, 1992, p. 60).

The studies of science *as* culture have challenged notions of value-free scientific objectivism in the production of scientific facts (Knorr-Cetina, 1999; Latour, 1987). The studies of science *in* culture, including the present chapter, have tried to find explanations for variations in the way science is practiced and variations in who does the practicing in different national settings. It is still not fully understood why we find cultural differences, their nature and their influence, in the practice of physics. To understand the implications of cultural diversity in science we need a comprehensive theory of culture as well as informed empirical material. Neither is easily obtained. In the following parts of my argument, I shall first present a theory of culture, which connects historic variability and culture. Next I present empirical data, which supports the assumption that historical variability creates cultural diversity in relation to who are regarded as "intelligible" physicists. This analysis will finally be used to discuss the cultural diversity, which appears when we address who it is that actually make gendered careers within the "same" discipline in different nation states.

5.2 Cultural Models

The understanding that science is embedded in a culture, which is not confined to the scientific practice itself, demands a theoretical framework, which can render it probable that science *as* culture is connected with science *in* culture. The problem is that culture is not easily discerned as an empirical object. What one person sees as culture another may see as perfectly normal everyday life. Culture is therefore not directly what we *see*. It always appears as a foreground to an implicitly background carpet. When we decide something is cultural we need to explain the processes behind the creation of what is termed culture. Cultural psychology offers a comprehensive approach to the understanding of the processes behind the complex concept of culture. It is a fairly trans-disciplinary theoretical framework, which connects practice with psychological processes (Cole, 1996).[4] It is not *one* coherent theory, but a ramified network of theoretical approaches combining insights from psychology, philosophy and anthropology. Within this framework culture can be understood as inseparable from human activity (Engeström, 1987) and broadly

[4] Cultural psychology first appeared as a theoretical framework in connection with a meeting in San Diego in 1980, between representatives from psychology and anthropology where the declared focus was to study the relation between cultural and psychic processes in real life activities instead of laboratory arrangements with the aim of bridging the gap between practice and cognition (Shweder & LeVine, 1984).

understood as the "patterns of beliefs, values, and practices that we both inherit and transform over time". Individuals "never share the complete culture of the group to which they are said to belong. At the same time, cultural practices are open or responsive to their ever-changing environment" (Gutiérrez, 2002).

Within this wide framework the more specific theoretical approach of cultural models offers an analytical tool to capture both end-product (cultural models) and processes (the connections through which cultural models are formed). In these models, artefacts, discourse and activity combine to make a coherent whole (Holland and Cole, 1995). As an analytical tool cultural models aim to make implicit cultural organizations of knowledge explicit, exposing assumptions, tacit beliefs and connections (D'Andrade and Strauss, 1992). What we call cultural models are organizations of knowledge formed in practices and "doings", while they on their part give directive force to certain motivations, without making people "cultural dopes" (Holland and Quinn, 1987; Strauss, 1992). In this framework there is a strong focus on how different cultures seem to connect and organize local knowledge in particular ways, which implicates and directs certain practices (D'Andrade and Strauss, 1992; Holland and Quinn, 1987). It is through cultural models that our conceptions of the world are formed. Whenever we use a concept like "woman" or "physics" – or in one of the cases presented by Naomi Quinn a concept like "marriage" (Quinn, 1987) – we implicitly take a number of connections for granted.

Following a group of researchers working on parallel distributed processes, (McClelland et al., 1986) theories of cultural models use connectionism to explain how models are built up from everyday experiences. In connectionism, organization of cultural knowledge is learned implicitly in everyday practice. Contrary to classical theories in cognitive science the mind is not regarded as an information processor working on representations of the external world. Rather than making conscious association between otherwise discrete elements, the mind is considered a neural network. Whenever we experience the world around us, connections are made between neurons in the brain and when experiences are repeated the connections are reinforced (Strauss and Quinn, 1994). "Knowledge need not be explicitly learned or retained as explicit generalizations or formulae; instead regularities in behaviour reflect cognitive patterns extracted from repeated experience" (Strauss, 1992, pp. 11–12).[5] Cultural models organize the connections we make in relation to the categories we have learned mark the boundaries of the self-evident world around us.

The approach has, in my opinion, correctly been criticised for being overly mentalist and overlooking how actions and the body play a part in forming cultural organizations of knowledge. When the process is defined broader than connectionism, as a cultural learning process (Hasse, 2002), the body and its position in physical world become part of the process.

[5] The discussion of the formation of cultural models has often underestimated the importance of bodily presence for experiences, but in my understanding of cultural models they cannot be seen as inseparable from bodily presence in the world. It is our bodily movements in a physical world, which provide the experiences, which form the basis of connections.

What is relevant for my argument here though, is that culture in this theoretical framework is not explicitly learned as what we see, but what we see *with*. "Once learned, it [culture] becomes what one *sees with*, but seldom what one sees" (Hutchins, 1980, p. 12). In the process of forming cultural models, every kind of experience with discourse and artefacts takes part in the formation (Holland and Cole, 1995). We live in "naturalized" self-evident cultural worlds, where acts, artefacts and opinions around us are so well-known that we would not dream of questioning them. It is only when we meet with other cultural ways of life that we begin to question the naturalness of our own connections.

The problem with scrutinising this relation within the framework of cultural psychology is that when culture is what one sees *with*, how can a researcher, seeing with his or her own cultural connections, claim to detect what can be analyzed as the cultural models other people see with?

5.3 The Method of Culture Contrast

In the case of relations between gender and science, countries around the world seem to have surprisingly different cultural organizations of knowledge about how the female gender and physics are connected.[6] In countries influenced by Islam such as Turkey and Kuwait we find many female physicists (Megaw, 1991; Ebeid, 1998), where as we, with our western conceptions, would have expected the sharply divided gender roles to dictate that women stayed away from a "masculine" science such as physics. In countries such as Thailand girls are doing much better at physics and chemistry than boys (Klainin et al., 1989) and women seem to do much better in the "hard science" of physics in Eastern and Southern Europe (Barinaga, 1994).

When we regard this information as surprising we are using our implicit organization of knowledge. We perform an act of what Laura Nader has called "implicit comparisons". Cultural research is *always* comparative as the prerequisite for acknowledging culture (Nader, 1994). To find culture we need to select areas of contrast, which creates the effect that the culture we perceive is a precise inversion of our own self-evident culture. When in Western culture the Muslim culture is often perceived as repressive to women's rights, because women have to cover their hair and wear a chador, we in the same implicit comparison are seeing *with* a culture, which defines the western woman as free, because she can appear almost naked in public. Contrary to this what many Muslims see is a western culture that makes women into pornographic objects because they see *with* a cultural gaze, where a woman's sexuality should be covered in public.

When we find out girls are doing better than boys in physics in Thailand and that women do well in "hard sciences" in Eastern and Southern Europe, it is surprising

[6] Knowledge is in this theoretical framework more to be understood as embodied than propositional knowledge.

because in contexts such as the Nordic/Danish physics is associated with boys and careers in physics with male physicists.

If we combine the perspective of implicit comparison with cultural models, we can use the unexpected differences we find to tell us about the implicit connections we make ourselves in our own cultural models – and thus call forth explicitly the culture we see *with*. This is the aim in the method of culture contrast. The purpose is not to compare between two comparable objects, but to find connected to one object, what would be seen as the reverse in another context. This method opens up an analysis of surprising connections made between a science like physics and other culturally informed areas of life.

A starting point for the method of culture contrast can be for the researcher to compare statistical numbers connecting science and gender, but to get deeper layers of surprises it is necessary to combine with other fieldwork methods: participant observation and in-depth interviews. Both these approaches have the advantage that they open up the unexpected in research and thereby provide the sought after challenge to what we see *with* in the method of culture contrast. Below I first present studies, that connect gender with science and challenge our self-evident understandings of how physics and gender should be connected as seen from a Nordic European perspective. Next I present a number of semi-structured interviews I held with Italian physicists in 2002–2003, and physicists from other nationalities made in 2004–2005. "As semi-structured interviews the questions follow a general script, but are basically open ended" (Russell Bernard, 2002, p. 203). It is this open-endedness, which allows for the surprises, which can be analyzed through the method of culture contrast. In these interviews with 11 Italian women and 14 Italian men, a completely new connection to physics appears, which can reveal my own Nordic/Danish cultural model of physics in contrast to an Italian one.[7] This connection apparently has nothing to do with gender. It rather concerns the relationship between humanities and natural sciences in a local, cultural-historically shaped context. However, in the end this culture contrast can provide a new insight in the gender differences found.

5.4 Women in Science

In general, natural sciences are encountering increasing problems with recruitment, especially of female physics students – and notably in the industrial North Western world (Sjøberg, 2000, 2004; Mejding et al., 2004). The few women who choose to study physics tend to disappear after graduation.

[7] I have made 55 semi-structured interviews with physicists in all, mainly from Denmark and Italy (but also with physicists from the United Kingdom, Sweden, the Netherlands, Senegal and USA); and also a number of interviews and focus group interviews with Danish and Italian physics students. The interviews were gathered in two interconnected projects: *Gender-barriers in the Becoming of a Natural Scientist* (1996–1999) and *The Cultural Dimensions of Science* (2002–2005) both financed by the Danish Research Council.

This lack of women at the top of academia is not unique to the discipline of physics, though. Well-qualified female scientists seldom reach top-level positions to the same extent that their male counterparts do and often leave the research system prematurely. This has been well documented in a number of studies, notably the SHE-figures, the Helsinki Group Reports, and the ETAN- and ENWISE Reports.[8]

The difference between the women and men entering science studies and the women and men obtaining academic positions in academic science institutions are illustrated by the so called scissors diagrams (TERSTI, 2003, p. 263; Osborn et al., 2000). Even if women and men start their studies on an equal footing, the closer one gets to the top-academic positions the more men (the upper blade of the scissor) and the fewer women (the lower blade of the scissor) one finds. From this scissors diagram – formed from huge collections of data on women, men and science conducted by the European Commission (EU) – we can see that women are not moving up through the echelons of scientific careers as much as their male counterparts.[9] This lack of gender balance can, to a greater or lesser extent, be found in all areas of the European countries surveyed by the Helsinki Group, who commissioned the study of these subjects. Another apt metaphor for the same process is the leaky pipeline, introduced by Joe Alper in 1993.The background for using this metaphor was highlighted in the ETAN report on women and science (Osborn et al., 2000). Whatever country or discipline we're discussing, whatever the proportion of women among the undergraduates and whatever equality measures are put in place, we still see a disproportionate leakage of women from scientific careers at every stage in the academic hierarchy. These numbers are in themselves a challenge to internalist and CUDOS-driven conceptions of science.

Yet, in a number of EU-surveys cultural differences have appeared on a very general scale and have complicated the pattern of general exclusion of women from scientific careers; as well as troubling even more the notion of a coherent scientific enterprise. Theresa Rees, conductor of the Helsinki Report on national policies on women and science in Europe, summed it up in this way:

> So what have we learned about women and science from such a diverse range of countries? The first point to emerge is that there is a huge diversity in the approach to women and science among these countries. (Rees, 2002b, p. 53).

If we take a closer look at the general figures the numbers are puzzling, according to expectations. Many statistics deal with general sectors–as HES (Higher Education Sector) and GOV (Government Sector). If for example we take a closer look at the

---

[8] The reports can be found in Alper (1993), OECD (1996), Colosimo and Dewandre (1999), Osborn et al. (2000), Bebbington and Glover (2000), Laurila and Young (2001), Maxwell, Slavin and Young (2002), Rees (2002a), SHE-figures (2003), Blagojevic, Havelková, Sretenova, Tripsa and Velichová (2003).

[9] See for example the Helsinki report, the ETAN-report and the SHE-figures to mention a few (Rees, 2002a; Osborn et al., 2000; SHE-figures, 2003).

interesting pie charts presented by the Helsinki Group we find an almost similar number of women in HES in Denmark and Italy. In Denmark the percentage of women in HES is 27, 3%, whereas in Italy the number is 28, 4%. Rees (2002a, p. 38) says[10]:

> There is no significant difference between these numbers and that is a surprise in a Nordic context. You would have expected Danish women to have a higher share than Italian women. For one thing Denmark is a country with a long story of government supporting facilities easing women's work-life and women liberation. As it was stated in the Helsinki-report: "For more than 25 years there have been laws regulating equal rights between women and men within Danish society".

The first Danish laws on gender equality were passed back in the mid 1970s and even though in the past years we have experienced some setbacks (see Lykke, 2002, pp. 144–149), we are normally considered a paragon country when it comes to equality policies. We have in past years received much praise for the ability of the Danish State to supply assistance in childcare, kindergartens, maternity leave and so on. In politics, Denmark in 1999 had 37, 5% female members at the European Parliament, whereas Italy had only 10, 3%.

Italy, on the other hand, has only recently acknowledged the "women's problem". As it became clear at the European Conference on Gender and Research in Brussels in November 2001, Italy generally looked upon gender in this way: "as such, women's and gender studies were generally judged as an anomaly – which caused a delay in comparison with the situation of other European countries". (Cantú, 2002, p. 168)[11]. Furthermore, the female share of the workforce is 47% in Denmark and only 38% in Italy. In the "naturalized" connections we make in Nordic countries, we would expect to find that the percentage of women in HES-research in Denmark to be much higher than in Italy. But we find to our surprise, that even though women in Denmark have a much better foothold in the labour-marked in general, we are not performing better than Italian women in HES. When it comes to having gained a foothold in physics, Danish women perform worse. These kinds of puzzling differential patterns can be found not just in Europe between North and South – but also between East and West all over the globe.

[10] From the Helsinki report on National Policies we know that in Italy (p. 106) there are 50.501 HES-researchers of which 14.332 are women and 36.169 men. In my recalculation this is 28.4% women. In Denmark (p. 100) we have 9,685 HES-researchers of which 2,645 are women and 7,040 men, which in my recalculation equals 27.3% women (1999-figures) (Rees, 2002a).

[11] That Denmark concerning gender policy really has experienced setbacks since the 1970s has been apparent. Not only in the political and academic unwillingness to earmark professorships for women (see Lykke, 2002), but also in the fact that Italy already in 1996 appointed a woman as Minister for Equal Opportunities for the first time, and she was responsible for important mainstreaming functions. (Cantú, 2002) In Denmark the first Minister for Equal Opportunities was appointed in July 1999 three years after Italy.

5.5 Surprises across the Culture Divide

In March 1994 the special issue on *Women in Science* in the *Journal Science* presented a "world map" of women in physics showing the percentage of women working in university physics faculties for 31 countries based on a questionnaire sent out to 1,000 university physics departments throughout the world (Barinaga, 1994; Megaw, 1991).

This study showed that the participation of women in physics varied dramatically among countries. Japan had the lowest percentage of women physics faculty (1–2%), with Canada, West Germany, Switzerland, Norway and U.K., all having less than 5% women physics faculty. At the other end of the spectrum, Hungary had 47% women faculty in physics. Thailand, USSR, Italy, Philippines and Portugal all have more than 25% female physics faculty – some a lot more.

Even though such cultural comparative studies are still few – they have been followed by further studies carried out by the physicists themselves (e.g., Ivie, Czujko and Stowe, 2002). These studies, as Carlson (2000) notes, begin to give shape and form to the underlying causes and contexts for the low participation of women in statistics and the sciences. It will be seen that it is not a uniform condition and we have much to learn from the varying realities across countries.

In Megaw's data a kind of pattern seems to appear. If you were to take a card over Europe and put a little green pin for every female physicist in Europe (member states and associated countries of the European Union) an ever more lush and green landscape appears the further South and East we go, where as the Northern and Western countries remain rather white and arid.[12]

The results were unexpected, as noted in the special issue of Science, which was termed "surprises across the Culture Divide" (Barinaga, 1994). The new map showed that some of the most industrialized countries had the smallest percentage of women physics faculty, and seemed to contradict stereotypes about national cultures and how they treat women. The ten countries with the *largest* female physics faculty percentage included three European Mediterranean countries (Portugal, Italy and Turkey, with Spain and France in 11th and 12th place), apart from Asian and Eastern Countries. Countries with large physics establishments, high levels of industrial development, and strong women's rights movements provided six of ten countries with the *smallest* female physics faculty percentage: Canada, Germany, Norway, USA, UK and the Netherlands.

Though Denmark was not part of the Megaw study, I can from my own data supplement his work and confirm that in Denmark the percentage of women physics faculty is also below 5% and indeed when I started my study on physicists in Denmark in 1996, it was below 3.5%[13].

---

[12] Other surveys have found some differences from Megaw's very clear picture of the North/South-East differences, but the tendencies are more or less the same.

[13] As noted by Beverley Carlson, who finds the same puzzling dissimilarities within Women in Statistics, data like these clearly indicate that culture is a powerful influence on how well women do in science in general, as well as in statistics in particular. She notes that the Mediterranean phenomenon of much higher percentages of women in the scientific professions is also observed among the ISI members (Carlson, 2000).

These numbers are surprising and cannot be explained from an internalist view of physics. In their own self-understanding, physicists belong to a tolerant scientific community – only the explicit and official evaluation of scientific results matter. However, in practice their community is socially limited. Here some would argue that when we find more female physics students, we generally also find more female physicists. But as noted by Osborn (1994) there are places with up to 50% female university students studying science topics, but only 2–3% female full professors. Even if the numbers of physicist students followed the number of permanent staff more closely, it would still not explain why we find gender differences between the countries in relation to physics. In some countries it simply seems to be more feasible for females to study and make careers as physicists than in other national cultures.

According to internalist scientific epistemologies, the success of science is insured by its internal features (Harding, 1998, p. 2). However, the numbers suggest that physics, from the perspective of who gets access to study physics, is very far from the inherent ideals formulated by Robert Merton in the CUDOS-norms. According to these norms the evaluation of scientific results is based solely on impersonal and objective criteria. Nationality and gender are completely irrelevant when it comes to the evaluation of scientific expertise. If this was true we would expect either that male and female physicists or physics students were evaluated as equally competent no matter where they came from or we could expect a consistent pattern showing that male (or female) physicists or physics students were consistently better that their gender counterpart across national borders. Instead we find a complex pattern where girls can outsmart boys in physics in Thailand and the opposite is the case in Scandinavian countries. Female physicists apparently never outnumber male physicists when it comes to permanent positions. However, the share of female physicists fluctuates and we apparently find more of them the further South and East we move.

To understand the puzzling figures, we must first accept that knowledge about the natural world seems to be intertwined with knowledge about the social world (Harding, 1998, p. 11). When our knowledge about the social world changes, our knowledge about the natural world might also change. The very numbers in themselves dismiss the notion of an internalist understanding of science and cry out for more knowledge on consequences of cultural differences in gatekeepers (see Husu, 2003 for a discussion on gatekeepers) and recruitment policies to science and – from my point of view– most importantly: a better understanding of how cultural organizations of knowledge create cultural differences in practice.

5.6 Inconsistent Explanations

Several explanations of the numerical discrepancies have been put forward. In a special issue of *Science* in 1994, four discrepancies in particular were identified (Barinaga, 1994). It can be noted that these differences are on different levels of explanation:

1. Cultural differences in the economic development and the labour market
2. Cultural differences in perception of class
3. Cultural differences in perception of education
4. Cultural differences in state supported childcare.

At the most general level we can ascertain that in countries which seem in need of fast economic development (as the former communist Eastern European countries) women can make up to 50% of the scientific researchers, whereas in industrial countries with a long male dominated history of economic development (like USA and the Northern European countries) women's tenured participation in scientific activities can be as low as 3–5%. Connected to this argument is the point made about cultural differences in perception of class. In the Northern countries the pecking order has been considered by some to be organized by gender perception rather than class: rich men, poor men, rich women, poor women. Whereas in other countries, especially developing countries, the ordering principle has rather been economic wealth and the pecking order thus: rich men, rich women, poor men, and poor women.

These explanations can, from the point of view of cultural models, be discussed as examples of different implicit connections made between gender and physics. In some countries physics is not strongly connected with gender, so when the economy demands "more hands at work" being female is not seen as an obstacle. From the implicit comparative point of view we could then argue that in a Nordic cultural model of physics we regard gender as so important that even when more hands are wanted in the economy (and the lack of "hands" in physics has been strongly underlined in the Nordic countries the past ten years), we "think gender" before we think "hands". Any hand in this context is a gendered hand: and the connection between physics and women is apparently difficult to make. This could be connected to the second explanation that class overrules gender in some countries in so far Nordic countries are perceived as countries which have levelled out class differences. But this is contradicted by the fact that Nordic countries are also known to have levelled out gender differences. These two explanations can, in other words, not provide satisfactory answers and the pattern becomes even more complicated when we include the last two explanations given.

The third and the fourth explanations concern how the state is, or is not, providing structures for education and childcare. Some societies use gender as an ordering principle for education in the sense that boys and girls are sent to different schools. In other countries (as in the former Communist countries and in the Scandinavian countries) it is underlined that boys and girls must follow the same path. Again these explanations can be seen as connections made between gender and science. The division in boys' and girls' schools should explain why we, in some countries, find a better representation of female physicists. Education (and therefore also science education) is related to gender in so far as male and female students are thought to disrupt each other's education in mixed classes. The assumption therefore is that when boys and girls are kept separate (when gender is made the overruling principle) then female physicists have a greater opportunityto develop themselves as proficient physicists. But again the pattern of connections is not consistent. In some countries, like the former Eastern European countries, we find many female physicists and yet the schools are in no way gender divided. In the Nordic countries the schools are similarly not gender divided and again we find a low representation of female physicists.

The fourth explanation claims that women are having a hard time making a career in science, because the state does not provide childcare. Women are thus connected more with childcare than men, and childcare is seen as something that excludes women from science unless they are helped by the state. Yet we find the lowest representation of female physicists in some of the countries with the highest amount of state provided childcare, like Denmark, and a much higher representation in Italy, where very little state child care is provided. The four explanations given are all contradicted when we compare the conditions in Eastern, Southern and Northern Europe.

The question remains: why do we find more female physicists in a country like Italy without much explicit gender policy than in Denmark with all the government supported facilities and a long history of women's emancipation? If we take a closer look at the particular science of physics in Italy and contrast it with Denmark, we find even more puzzling data from the Nordic point of view.

5.7 The Hard Science

If we start looking at the everyday life of a Danish female physicist, it is very likely that her early every day experiences were somehow affected by the gender issue. Not least because she was considered odd when she chose to study physics. In Denmark, according to the OECD-survey *Education at a Glance* (OECD, 1996), boys consistently outperform girls in physics. In mathematics, the gender gap in achievement is moderate – with a slight advantage for boys. In science, however, it is considerable. Boys outperform girls in Denmark with more than 30 points – almost one grade-year equivalent (see OECD, 1996, Chart R10.2). Are the gender differences observed at this age predictive of later stages and future career choices? Although this question cannot be answered directly, when it comes to the percentage of students in higher education in Denmark female students comprise around 25% for mathematics and the "hard" natural sciences, whereas in Italy we find 50% of the students are female in the same natural science and the percentage is almost the same in mathematics (Osborn et al., 2000, Fig. 2.2).

In Denmark physics seems to be gendered. It is generally known to be a subject that does not attract women. They prefer health-studies or human studies (Henningsen, 1998). Physics is considered to be a very hard subject, and only students with a background in natural sciences at a certain level, generally obtained through enrolment in a high school with an emphasis on mathematical and physics skills. In a study I did at the Niels Bohr Institute for Physics at Copenhagen University, it was obvious that the students considered physics a very hard subject to study indeed (Hasse, 2002).[14] Most girls drop physics modules in high school and choose the language lines, thereby excluding themselves from studying physics at university level and pursuing a career in physics – or at least making it very hard for themselves to catch up at a later stage to meet university requirements in physics.[15]

Once students are able to study physics at university level, they constantly talk about how hard it is, and that becoming a physicist is equivalent to becoming an elite scientist. Female students in particular, seem to completely loose their self-confidence in this elitist physics culture. (Hasse, 1998a).[16]

There seems to be a conflation in connecting physics with "hard science" and the fact that hard science is mostly sought after by male students. Conceptualizations of physics in Denmark come very close to what Carolyn Merchant (1990) and others have identified as inherently sexist science. In Denmark there is a commonplace connection made between physics as a hard science dominated by males and the humanities as soft sciences dominated by females

The politicians as well as the physicists have for a long period tried to attract more women to a career in physics. Among the administrative staff and the elder established female physicists, there is a lot of focus on gender and even a special organization for women in physics. Physics is generally regarded as a male domain, in which it can be very difficult for women to find a place. Therefore, something extra needs to be done to help females succeed in physics (Hasse, 2000, 2002). The cultural-historical conceptualizations of physics connect physics with being a hard "male" science that creates highly elitist, sought after scientists and where the gendered nature of physics is highly enunciated and discussed.

5.8 The "Classical" Physicist

In many Southern and Eastern European countries the view of physics is different – t is not considered especially hard, difficult or elitist. Nor is physics conceptualized as particularly masculine. These views are reflected in the number of women

---

[14] In a survey done among first year students most considered physics to be a hard subject to study (Hasse, 1998a).

[15] Until a few years ago there was a sharp division between mathematical-physical studies, classical studies and general language studies in Denmark. The division line was also generally considered a gender division line between females choosing languages and males math-physics. A new reform has in 2005 softened these division lines but the high school choice of subjects is still decisive for further education. There is no free uptake at university level. In study physics at university level a student has to have passed physics and mathematics at A-level and chemistry at C-level. As not enough students in general apply for this study the students are not competing on who have obtained the highest marks in the gymnasium. If one fulfils the general obligatory demands, students who want can study physics at university level (in sharp contrast with a lot of humanistic studies where students compete for the seats). From 2008 the formal demands are changed so you now have to have Danish at A-level, English at B-level, and mathematics, physics and chemistry at either A-AB- or A-B-A-level.

[16] See also Hasse (1998b; 2001; 2002).

in physics, starting from the level of high school education. In Denmark female physicist students enrolling at university level comprise around 18–20% of a freshman group. In Italy female students comprise approximately 45%. During my fieldwork I asked Italian students questions about the oddity of a woman studying physics, but the students reacted with surprise. They expressed the opinion that they had never though of it this way. It was explained to me that "here in Italy physics is not a "gendered" subject" as it is in Denmark.[17]

On the other hand, in Italy physics is connected to something, which from a Danish point of view is very surprising, but rather self-evident for Italian students. In Italy, it is possible to study physics at university level with a background in the humanities. This fact was accidentally revealed through my interviews with Italian physicists. In these interviews I was surprised to find again and again that many of these physicists had apparently *not* entered physics from a background in a mathematical and physics oriented education. In a Danish context humanities and natural sciences are generally seen as mutually exclusive. The physics students I followed at the Niels Bohr Institute made this clear to me: the worse fate for a physicist was to be degraded to working with the soft humanities subjects (Hasse, 2002).[18] However, many of the very successful Italian professors in physics that I interviewed turned out to have entered their study with a high school background as classical studies. They were versed in philosophy and cultural history, studying Aristotle in Greek, reading Cicero in Latin and the like.[19] Through questions addressing why they began to study physics and general research biographies it became clear to me that the recruitment pattern differed substantially between Italy and Denmark. I was not aware of this when I planned the project in Italy and I did not select physicists to fit this pattern. This was something hat became apparent during my interviews. Apart from gender I selected the male and female physicists according to availability, geographic and topical interests. Even so 14% of the Italian men and 64% of the Italian female physicists I interviewed (or 7 out of 11 of the female physicists) turned out to have a background in classic language and philosophy.[20] From the perspective of someone familiar with the Danish culture of

[17] Barbara Mapelli (2002) has argued that Italian girls have a real desire to do science. However, they describe a female scientist different from a male scientist, and they perceive science differently from boys (Mapelli, 2002). This does not mean that physics in Italy is gendered in the same way, as it is in Denmark, though, where the girls have no desire for science. The girls Mapelli spoke to, and the Italian university students I spoke to, did not consider the discipline of physics to be a primarily 'male domain' as such —but considered that there can be gendered ways of doing physics.

[18] New reforms have tended to blur this distinction, but the political pressure is more directed at underlining that physics has a humanistic value, than the opposite found in Italy, that languages and philosophy have something to offer to physics.

[19] These surprising findings led me to start searching for information on criteria for the intake of students and the group interviews with students with and without 'classical' backgrounds.

[20] One of these women actually did not have a background in classical studies but entered physics science with a 'scuola magistrale', a background with even less connection to physics than the classical educated.

science this is quite strange indeed – though not at all odd from the Italian perspective. What is even more astonishing from a Danish cultural perspective is that neither the male nor the female physicists (who had all done very well in physics and were now employed as full time professors) had found that the classical background had been clogging them during their university studies.[21] In fact, on the contrary: the more technically trained students had even admired classical students. In Italy having a background in classical studies was seen as an asset. As one professor in physics, coming to physics with a scientific background, explained:

> I myself have a background equivalent to a mathematical-physicist high school student in Denmark. But here in Italy it is rather an advantage to have a classical linguistic background, when you start physics studies at university level. The 'classical' students are simply better at analyzing. What we learned in science high school was to think much more 'mechanically' – to think in the correct answers. I have always believed students with a classical background are the most advantaged.

An Italian teacher of philosophy expressed the same opinion in a slightly different manner, explaining that the "classical" students become especially apt physics students because they, through their knowledge of philosophical and classical subjects, learn to think in the abstract lines of thought of importance to both the natural sciences and the humanities. She underlines that until 15 years ago is was not uncommon that most of the students matriculated at the physics institutes in Italy originated from the classical language line in high school – and thus did not have a scientific background.

> The high school in classical languages was simply the best at training students in reasoning, because such a systematic and profound study of classical subjects had a formative influence on the students' intelligence – especially their ability to think in abstractions and make inferences.

Furthermore this teacher underlined that a non-specialized education system like the Italian one does not make too hard a division between humanistic and natural sciences. And we do not find the same kind of gender divisions that we find in countries like Denmark, where girls and boys have had to choose between a humanistic or a scientific line – and where we find that the girls in much greater numbers have chosen the humanistic line just as the boys in much greater numbers chose the scientific line. This also conflates somehow with a more relaxed attitude towards science in Italy.[22]

What the Italian female physicists brought with them when they entered physics at university level was a rather prestigious background, giving them self-confidence, as one of them explained to me. And this could be part of the explanation as to why we find more Italian and maybe also Southern European female physicists

[21] This was later confirmed in my study by young physics students from Rome with a classical background. They also claimed that they did not experience their classical background as particular cramming.

[22] Some might get the idea that Danes learn 'better' physics because it's considered a 'hard' subject in Denmark, but in the PISA survey Italy and Denmark score almost equally (Mejding et al., 2004).

than we do Nordic. Most of the Italian female physicists, just as their male colleagues, expressed satisfaction with their life as physicists. And even though many of them complained about problems with state support for childcare and maternity leave, they regarded these as problems belonging to the world outside of physics and many of them rejected even looking at themselves as "women in science". As an Italian physicist said:

> I do not want to speak of myself as a 'female-scientist' because I am a strange 'scientist-woman'. Why am I strange? We are alike. If I have a problem it' s because we do not have the kindergartens to bring the children to, but this is not a problem tied to the 'woman-scientist', this is a societal problem, where the Italian politics are failing.

This attitude is considered very problematic from the feminist point of view of making gender problems visible in Italy, but most Italian women I spoke to did not want to connect their career-paths in physics with gender.[23]

This classical connection to the humanities clearly influences female physicists' career paths. In 2000–01, 23080 students were enrolled in physics studies at Italian universities. Of these 43.9% were women and of that percentage more than a third had a humanistic background, whereas only less than a tenth of the men had such a background.[24] The requirements for physics studies at university level are very different from Denmark, and this seems to open up new possibilities especially for female students with a degree in the humanities – often in classical languages and philosophy. In interviews female students with a "classical" background also expressed very clearly that they do not experience this starting point as being set back, but rather an advantage.[25]

Whereas in Italy female physicists do not identify themselves as "female" and refuse to see their gender as connected to career problems, it is no problem in the Nordic countries to find female physicists who links career problems to being female. Even though I also spoke to some feminist oriented Italian physicists I heard none of the real horror stories about being a *female* physicist that I heard from their colleagues in the Northern part of Europe. My interviews with Danish, Swedish and English female physicists showed a much more problematic relationship between being female and being a physicist. I shall in the following refer to them as "Nordic" – because the patterns are more or less alike in the three countries – but also to protect the women telling sad stories about not being recognized. Many of the women belong to the same age group as the Italian tenured female professors, but have never had a permanent position. They were interviewed because we find so few women in permanent positions in the Nordic countries. When I interviewed these women a substantial number expressed real bitterness

---

[23] Here we find a curious continuation of the pattern found on large scale at the European map of gendered physics. In the Southern part of Italy female physicists are more satisfied than in the North.

[24] The numbers were collected from the Ufficio di Statistica, Rome, in the summer 2003 by my assistant Ketty Mazzara.

[25] I made group interviews with four groups of male and female students in Rome, La Sapienza and in Catania.

over their academic life stories. Their whole career they had felt like outsiders. One of the Nordic female physicists told a terrible story of being a male professor's assistant on his project for more than 25 years. When he died she had never had tenure, and proposed that the institute should finally employ her. Her colleagues looked at her in bewilderment: "But did you ever publish anything on your own?" they had asked. She was very hurt, she explained in the interview. They were simply not aware that even as the professor's assistant she had been able to make independent research and had published over 100 renowned articles within her field. Even so she never got tenure as a physicist researcher.

Another female physicist from a different physics institute in a Nordic country also expressed how she always felt like an outsider, though she had been connected to her physics institute on a more or less yearly renewal base for over than 30 years.

> I have never been taken seriously as a colleague, when they talked about opening up for new tenure positions at the institute[...] I have very strongly been made to feel I am not one of them [even though] I am specialized within the broad research profile of the institute.

This woman had attended a girls'-school before studying physics and she was regarded as being really bright. According to this woman it was in no way a help to her when she entered the physics labour market that she had had excellent marks and had always received praise for her work as a Ph.D. student. She had achieved her Ph.D. in another country at a university were four women dominated her rather small institute: one professor, one lecturer and two PhDs. They had all attended all-girls-schools and she was certain that this had something to do with the relaxed atmosphere around doing physics. She had simply not been aware that physics could be considered a "hard" subject until she returned to her native Nordic country and got a part time contract.

Even so, in her Nordic country, her colleagues passed her by again and again as an almost invisible person –until a certain episode. At this time she was a very renowned researcher within her field of particle physics, abroad. But among her own colleagues she was not regarded as an equal partner when it came to tenure. She explains how it really hurt her feelings when, after many attempts, she applied for a job, which everyone agreed that she was obviously qualified for.

> It was a very ugly story. Even though I was declared obviously qualified, one of my closest collaborators said I was unqualified [...] they had decided someone else should have the position. The head of the department became the Chairman for the re-evaluation committee. I could do nothing.

She did not get the job and she explains it this way: "I think he had a problem with me being a woman. He felt it as a threat, I think. And I have no mentors here".

5.9 Cultural Knowledge of Gender and Physics

Intelligibility is important for cultural studies of science (Rouse, 1992). What I argue is that the cultural organization of knowledge, the cultural model of physics, makes different connections between gender and physics intelligible. In Italy gender and physics are simply not strongly connected to the extent they are in Denmark (and the Nordic/Western countries).[26] In the Nordic cultures everybody has, through everyday experience, learned to reinforce connections between physics and "hard science". "Hard" connects to "male" and "elitist". Girls who want to perceive themselves as girlish, should, in other words, stay away from physics, which threatens their identity as female: thus connecting physics with terms such as "not-female" and "gender-segregation" in the educational system. With this cultural perception it is very surprising that someone could even think about considering a background in classical studies to be an asset for a physicist.

In Italy physics is easily connected to humanistic studies – studies of philosophy, Greek and Latin[27] – and far from seeing the "humanistic" background for physics studies as disadvantage it is seen as advantageous, because classical education is considered to be of a higher status than physics. Women in no way lose their female identity by studying physics, though all physicists might be considered to be on the boring side of the job-market. However, physics is generally considered a good education for earning money later in life – much more than an elitist education aimed at "lofty science". It is not the case that Italian physicists are less respected or considered less competent than their Nordic colleagues, who stress the "elitist" aspect of physics.[28]

There is a link back to the four reasons given in the special issue of Science (Barinaga, 1994), which make it possible to understand the conflicting explanations in a new manner. In Denmark and other Nordic countries physics is, from the outset, gendered (as physics has been connected with "hard science" and "hard science" with being male). Any woman, who chooses to study physics, has to overcome the inherent cultural knowledge that she is entering a masculine domain. She also enters with the culturally organized knowledge that this discipline is not only masculine, but "hard" and "difficult". In Italy, which is much more of a class society than Denmark is, being a classical student has always been connected to being "posh" and from the upper classes. This connection between classical studies and class makes classical students attractive to all kind of studies – and here physics is no exception.

Cultural conceptualizations and what can be analyzed as "cultural models" do not appear out of the blue. They are never simple, but rather complex. They are emerging in everyday life as historical processes.[29] Culture as cultural conceptualizations is never at rest (Hasse, 2002). Human science thinking has been almost

[26] This could also explain why much critique of natural science as androgynous comes from North-Western feminists.

[27] It can be noted here that the students Barbara Mapelli (2002) spoke to both connected physics with philosophy *and* to being a down-to-earth science that has practical implications for mankind (Mapelli, 2002). In my empirical material the women did not connect physics with more philosophical subjects, but I found the same connection with physics as a down-to-earth-science (Hasse, 1998a; 2002).

[28] Female PhD's in physics in Denmark often come from Italy, and Italian female physicists are highly respected in international collaborations in topics such as high-energy physics.

[29] In Denmark the first women graduated in physics in 1892. Italy has, on the other hand, a long history for women in science with the physicist Laura Bassi (1711–1778) as the most noteworthy. She was appointed to the Faculty of Physics in Bologna as early as in 1732.

completely excluded from physics in Denmark, but in both Italy and Denmark there are signs of new interrelations. In Denmark the Institute for Natural Science Philosophy has been placed at the Natural Science Faculty of Copenhagen University and in Italy the ministry of education is increasingly denouncing Italy's ongoing support of classical studies, which historically has been connected with a pride of the "Roman past". History and culture moves on, and physics *in* culture moves along with it.

5.10 Towards a New Humanistic Physics?

Cultural diversity challenges the notions of internalist physical science along with "ready made" stable categories of gender. Contrasting field-findings suggest that in the Nordic countries male gender and physics are linked whereas physics and humanity cannot be connected in general conceptualizations. Physics is considered masculine and incompatible with human science. The discipline is placed in an untouchable ivory tower of "hard elitist science". In the Southern part of Europe general cultural conceptualizations of physics seem much less connected to gender. Physics is more integrated with the general cultural history and is not seen as a particularly "hard" discipline.

In this analysis of cultural models we have to find and connect the pieces of the puzzle which seem to be the most important for the discussion at hand, but any cultural model is also an analytical construct constricted by the researcher's research and limited imagination. Even so from this perspective it seems that we can discuss the influence of the cultural models of physics in two ways. One points to an interesting discussion of how tacitly learned connections can influence women's (and men's) career possibilities and create gender segregated research fields, such as the field of physics, in a wider national cultural context. The other is a more speculative discussion concerning whether social epistemology in general can influence scientific epistemology *in* physics culture. In both cases questions are raised about an internalist notion of science.

In the first case it seems as if the very enunciation of gender, emancipation and gender policy in the Nordic countries could be followed by an engendering and segregation of science as a reaction to the strong focus on gender in society in general. Physics could, in this cultural context, be seen as the "last bastion" for men in a world where women seem to take over public space.[30] We find a pattern of women in human sciences and men in the natural science.

In Italy we find that women place a lot more engagement in areas like "hard" industries involved in physics, engineering and IT as well (Colclough, 2004). This can lead to the suspicion that gender segregation stems from a focus on gender and a substantial amount of women in the workforce. Gender barriers, in what is in Nordic

[30] The frequent newspaper statements underlining this point support this thesis.

countries typically considered "male domain", are not so explicit in Italy, where women's participation in the working market is more recent than in the Nordic countries. This supports the thesis that the higher the employment rate of women, the more gender segregated the labour market (Colclough, 2004). Our findings could further imply that the higher the educational level for women, the higher the disciplinary gender segregation. so for however we lack the reseach to confirm this thesis.[31]

In the present argument cultural models of science can be influenced by an integration of gender into the model. A Nordic cultural model of physics makes it difficult for women to be motivated to engage with "masculinized" domains, because it might mean a loss of identity as "female". The more women who enter the public arena, the more important gender as an ordering mechanism seems to become society – in education as well as work. Emancipation might historically have opened a lot of public doors to women in Denmark, but the price might have been a high degree of being connected as women the public arena. Some areas are easily connected to women (especially areas dealing with childcare, care of the elderly and the sick); others are more readily associated with men in this gender-segregated society. When a woman crosses a border into male "territory" she is risking her gender-identity. The result is that cultural models can force Nordic women to choose different career paths to Italian women if they do not want to be seen as "masculine women".

In Italy gender is not connected to physics in the same way, so female physicists do not perceive themselves as women-in-physics.

Contrary to Merton's ideals gender does seem to matter in the selection of those who get pass the intelligibility gates into the culture of physics. From a cultural-psychological perspective we can open the discussion of how differently connecting cultural models of physics and gender might influence scientific epistemology. For many scientists gender studies in physics are of no importance what so ever to internal science issues. In the so-called "science war" scientists such as Steven Weinberg, Paul R. Gross and Norman Levitt have attacked studies on the social practice of science. Feminist contributions to science studies from Sandra Harding and Donna Haraway have also been attacked (Parsons, 2003). For these opponents science/gender studies say nothing substantial about science even though they might say something on social relations. In the internalist science perspective a lack of women in science is a social problem, which does not concern science epistemology. On the other hand, even though it is not easy to determine exactly how different cultural models of science could have serious consequences for notions of internalist science, we also have some indications of how different cultural models of science with or without "masculinization" might influence science epistemology. We can clearly claim to find a pattern of a closer connection between humanities and conceptualizations of physics in the Italian case, but no research is available which could confirm or reject a direct link between the acceptance of "classics issues" in the practice of

[31] In the IT-area we find the same tendency as for women in physics. Women are severely under-represented in the IT-sector in Denmark and England, where they, according to 2001-figures, only comprise 26% and 24% of the employed. In Italy 38% of the employed in the sector are women (Colclough, 2004).

physics and the number of female physicists. It would be difficult to argue in any simple way that more women are equal to a different kind of physics, but it *has* been claimed that scientific knowledge is intertwined with social issues such as gender (Rolin, 2001). Gender ideologies can either restrict opportunities for scientific dialogue or distort the evaluation of scientific competence, as argued by the philosopher Rolin (2001). The strong opinions about women in physics in Denmark might influence evaluation of the work of female physicists.

In both models (the Italian and the Danish) connecting physics with either gender or classical studies could be of epistemic significance. This raises a number of questions about the relation between gender, science and the cultural recognition of excellence, following up on studies on gender bias and the mechanisms that appear to prevent women scientists from achieving excellence (Al-Khudhairy et al., 2004; Brouns, 2004).

It could be argued, that if natural sciences and their preoccupations in reporting on nature are embedded in and are complicitous with social projects, then a causal, scientific grasp of nature and how to study it must be embedded in (be a special area of) causal scientific studies of social relations and how to study them (Harding, 1991, p. ix). Not least because: adequate social studies of the sciences turn out to be the necessary foundations upon which more comprehensive and less distorted descriptions and explanations of nature can be built (p. 15).

Once it is acknowledged that there is no isolated internalist science, but a physics culture influenced by cultural-historical changes, we must start thinking about these studies of science and the practice of doing science in new ways. The physicist Karen Barad has, among others, encouraged readings of science and technology studies in physics classes and in general called for at more responsible natural science, which implies thinking about the nature of scientific practices and its relationship to other social practices (Barad, 2000, p. 246). Whether this approach will, in the future, show that there is a connection between many female physicists and a more responsible and more humanistic oriented science remains to be seen.

References

Al-Khudhairy, D., Dewandre, N., Wallace, H., Brouns, M. & Addis, E. (2004). *Gender and Excellence in the Making*. Brussels: European Commission. Retrieved May 2008. http://europa.eu.int/comm/research/rtdinfo_en.html.

Alper, J. (1993). The Pipeline is Leaking Women all the Way Along. *Science*, 260, 409–411.

Barad, K. (2000). Reconceiving Scientific Literacy as Agential Literacy, or Learning How to Intra-Act Responsibly Within the World. In R. Reid & S. Traweek (Eds.), *Doing Culture and Science*. New York: Routledge Press.

Barinaga, M. (1994). Surprises Across the Cultural Divide. *Science*, 263, 1468–1472.

Bebbington, D. & Glover, J. (2000) Women and Scientific Employment: Mapping the European Data. Brussels: Commission of the European Communities, Research Directorate General.

Blagojevic', M., Havelková, H., Sretenova, N., Tripsa, M.F. & Velichová, D. (2003). *ENWISE-Report. Waste of Talents: Turning Private Struggles into a Public Issue*. (Women and Science in the Enwise Countries). Luxembourg: OPOCE, European Communities.

Brouns, M. (2004). Gender and the Assessment of Scientific Quality. In D. Al-Khudhairy, N. Dewandre, H. Wallace, M. Brouns & E. Addis (Eds.), *Gender and Excellence in the Making*. Brussels: European Commission. Retrieved May 2008. http://europa.eu.int/comm/research/rtdinfo_en.html.

Cantú, F. (2002). Women in Science: The Italian Legislative and Institutional Framework for Gender Mainstreaming. In L. Maxwell, K. Slavin & K. Young (Eds.), *Gender and Research Brussels* (pp. 168–174). Luxembourg: Office for Official Publications of the European Communities.

Carlson, B.A. (2000). *Women in the Statistics Profession: A Status Report*. On-line Document. Retrieved May 2008., http://www.nass.usda.gov/cws/Status.pdf

Colclough, C. (2004). Passer kvinder ind på fremtidens IT arbejdsmarked? [Do women fit into the future labour marked?]. *Forskningscenter for Arbejdsmarkeds- og Organisationsstudier (FAO) Information*, Juni 2004, 6–8. Retrieved May 2008. http://www.sociology.ku.dk/faos/faosinf24.pdf

Cole, M. (1996). *Cultural Psychology: A Once and Future Discipline*. Massachusetts: The Belknap Press of Harvard University Press.

Colosimo, A. & Dewandre, N. (Eds.) (1999). *Women and Science: Mobilising Women to Enrich European Research*. COM (1999) 76 final, European Commission, Luxembourg. Retrieved May 2008. http://europa.eu.int/comm/research/science-society/documents_en.html.

Crenshaw, K. (1994). Mapping the Margins: Intersectionality, Identity Politics, and Violence Against Women of Color. In M. Fineman & R. Mykitiuk (Eds.), *The Public Nature of Private Violence* (pp. 93–118). New York: Routledge.

D'Andrade, R. & Strauss, C. (Eds.) (1992). *Human Motives and Cultural Models*. Cambridge: Cambridge University Press.

Ebeid, W. (1998). Enrolment in Mathematics: Problems and Aspirations in the Kuwait University Faculty of Education. In J.H. Jensen, M. Niss & T. Wedege (Eds.), *Justification and Enrolment Problems in Education Involving Mathematics or Physics* (pp. 259–267). Roskilde: Roskilde University Press.

Engeström, Y. (1987). *Learning by Expanding. An Activity-Theoretical Approach to Developmental Research*. Helsinki: Orienta-Konsultit.

Galison, P. (1997). *Image and Logic: A Material Culture of Microphysics*. Chicago: University of Chicago Press.

Gutiérrez, K.D. (2002). Studying Cultural Practices in Urban Learning Communities. *Human Development*, 45, 312–321.

Harding, S. (1991). *Whose Science? Whose Knowledge? Thinking from Women's Lives*. New York: Cornell University Press.

Harding, S. (1998). *Is Science Multicultural? Postcolonialisms, Feminists, and Epistemologies*. Bloomington: Indiana University Press.

Hasse, C. (1998a). Kulturelle forestillinger og køn i videnskabens samfund [Cultural conceptions and gender in the scientific society]. In I. Henningsen (Ed.), *Køn i den akademiske organisation* (Working paper nr. 4). Copenhagen: Københavns Universitet.

Hasse, C. (1998b). Learning to Pattern Physicist Virtues: Male and Female Dissimilarities. In J.H. Jensen, M. Niss & T. Wedege (Eds.), *Justification and Enrolment Problems In Education Involving Mathematics Or Physics* (pp. 106–122). Roskilde: Roskilde University Press.

Hasse, C. (2000). Feedback-Loop Among Physicists – Towards a Theory of Relational Analysis in the Field. *Anthropology in Action*, 3, 5–12.

Hasse, C. (2001). Institutional Creativity – The Relational Zone of Proximal Development. *Culture & Psychology*, 7(2), 199–221.

Hasse, C. (2002). Kultur i bevægelse. Fra deltagerobservation til kulturanalyse – i det fysiske rum, [Moving culture – from participant observation to cultural analysis in 'physical' space]. Copenhagen: Forlaget Samfundslitteratur.

Henningsen, I. (1998). Kønsdifferentierede adgangskrav på overgangen til de lange videregående uddannelser. In I. Henningsen (Ed.), *Køn i den akademiske organisation* (Working paper nr. 3). Copenhagen: Københavns Universitet.

Holland, D. & Cole, M. (1995). Between Discourse and Schema: Reformulating a Cultural-Historical Approach to Culture and Mind. *Anthropology & Education Quarterly*, 26(4), 475–489.

Holland, D. & Quinn, N. (Eds.) (1987). *Cultural Models in Language and Thought*. Cambridge: Cambridge University Press.

Husu, L. (2003). *Exploring Gender and Power in Academia: Gate-keepers in Research Funding in Focus*. Paper Presented at the Workshop 'Har vi kjønn i akademia? at the Nordic Conference Könsmakt i Norden. Oslo, June 12 and 13, 2003.

Hutchins, E. (1980). *Culture and Inference*. Cambridge: Harvard University Press.

Ivie, R., Czujko, K. & Stowe, K. (2002). *Women Physicists Speak: The 2001 International Study of Women in Physics*. College Park, MD: American Institute of Physics – IUPAP. Retrieved May 2008. Http://www.aip.org/statistics/trends/reports/iupap.pdf.

Klainin, S., Fensham, P.J. & West, L.H.T. (1989) The Superior Achievement of Girls in Chemistry and Physics in Upper Secondary Schools in Thailand. *Research in Science and Technological Education* 7(1), 5–14.

Knorr-Cetina, K.D. (1999). *Epistemic Cultures: How the Sciences Make Knowledge*. Cambridge: Harvard University Press.

Latour, B. (1987). *Science in Action: How to Follow Scientists and Engineers Through Society*. Cambridge: Harvard University Press.

Laurila, P. & Young, K. (2001). *Synthesis Report. Gender in Research. Gender Impact Assessment of the Specific Programmes of the Fifth Framework Programme: an Overview*. Brussels: European Commission – Directorate-General for Research.

Lykke, N. (2002). Response to the Gender Impact Assessment of "Improving Human Research Potential and the Socio-Economic Knowledge Base". In L. Maxwell, K. Slavin & K. Young (Eds.), *Gender and Research Brussels* (pp. 144–149). Luxembourg: Office for Official Publications of the European Communities.

McClelland, J.L., Rumelhart, D.E. & the PDP Research Group (1986). *Parallel Distributed Processing, Volume 2: Psychological and Biological Models*. Cambridge: MIT Press.

Mapelli, B. (2002). Girls' View of Science and Choices in Education. In L. Maxwell, K. Slavin, & K. Young (Eds.), *Gender and Research Brussels* (pp. 235–240). Luxembourg: Office for Official Publications of the European Communities.

Maxwell, L., Slavin, K. & Young, K. (Eds.) (2002). *Gender and Research Brussels*. Luxembourg: Office for Official Publications of the European Communities.

Megaw, J. (1991). *Science and Engineering: Increasing their Numbers*. Washington: National Academies Press.

Mejding, J., Lindenskov, L., Egelund, N., Weng, P., Sørensen, H., Andersen, A.M., Rangvid, B.S., Krone, M. & Andersen, T.Y. (2004). *PISA 2003. Sammenfatning*. Copenhagen: Danmarks Pædagogiske Universitets Forlag.

Merchant, C. (1990). *The Death of Nature: Women, Ecology and the Scientific Revolution*. San Francisco: Harper.

Nader, L. (1994). Comparative Consciousness. In R. Borofsky (Ed.), *Assessing Cultural Anthropology* (pp. 84–96). New York: McGraw-Hill.

OECD (1996). Education at a Glance. Paris: OECD.

Osborn, M. (1994). Status and Prospects of Women in Science in Europe. *Science*, 263, 1389–1391.

Osborn, M., Rees, T., Bosch, M., Hermann, C., Hilden, J., McLaren, A., Palomba, R., Peltonen, L., Vela, C., Weis, D., Wold, A. & Wennerås, C. (2000). *Science Policies in the European Union: Promoting Excellence Through Mainstreaming Gender Equality. A Report from the ETAN Network on Women and Science*. Luxembourg: Office for Official Publications of the European Communities.

Parsons, K. (2003). *The Science Wars. Debating Scientific Knowledge and Technology*. New York: Prometheus Books.

Pickering, A. (1984). *Constructing Quarks: A Sociological History of Particle Physics*. Chicago: University of Chicago Press.

Quinn, N. (1987). Convergent Evidence for a Cultural Model of American Marriage. In D. Holland & N. Quinn (Eds.), *Cultural Models in Language and Thought* (pp. 173–192). Cambridge: Cambridge University Press.

Rees, T. (2002a). *National Policies on Women and Science in Europe*. Luxembourg: Office for Official Publications of the European Communities.

Rees, T. (2002b). First Results from the Helsinki Group on Women and Science Conference: Policy review. In L. Maxwell, K. Slavin & K. Young (Eds.), *Gender & Research Brussels* (pp. 53–56). Luxembourg: Office for Official Publications of the European Communities.

Rolin, K. (2001) Gender and Physics: A Theoretical Analysis. *Journal of Women and Minorities in Science and Engeneering*, 7, 1–8.

Rouse, J. (1992). What are Cultural Studies of Scientific Knowledge? *Configuration*, 1(1), 57–94.

Russell Bernard, H. (2002). *Research Methods in Anthropology. Qualitative and Quantitative Approaches*. New York: Altamira.

SHE Figures (2003). SHE-Figures. Women and Science, Statistics and Indicators. Luxembourg: Office for Official Publications of the European Communities.

Shweder, R.A. & LeVine, R. (1984) (Eds.). *Culture Theory: Essays on Mind, Self and Emotion*. New York: Cambridge University Press.

Sjøberg, S. (2000). *Science and Scientists: The SAS-Study. Cross-Cultural Evidence and Perspectives on Pupils' Interests, Experiences and Perceptions. Background, Development and Selected Results*. Oslo: Department of Teacher Education and School Development, University of Oslo. Retrieved May 2008. http://www.uio.no/~sveinsj/

Sjøberg, S. (2004). *Naturfag som allmenndannelse. En kritisk fagdidaktikk*. Oslo: Gyldendal Akademisk.

Strauss, C. (1992). Models and Motives. In R. D'Andrade & C. Strauss (Eds.), *Human Motives and Cultural Models* (pp. 1–20). Cambridge: Cambridge University Press.

Strauss, C. & Quinn, N. (1994). A cognitive/cultural anthropology. In R. Borofsky (Ed.), *Assessing Cultural Anthropology* (pp. 284–300). New York: McGraw-Hill.

TERSTI (2003). *Third European Report on Science & Technology Indicators 2003. Towards a Knowledge-Based Economy*. Brussels: European Commission, Directorate-General for Research. Retrieved February 2008. Retrieved May 2008. http://www.dife.de/~mristow/2003EU_3rd_report.pdf

Traweek, S. (1988). Beamtimes and Lifetimes. The World of High Energy Physicists. Cambridge: Harvard University Press.

Chapter 6
The Gap Between University and the Workplace
Examples from Graphing in Science

Wolff-Michael Roth

The orthodox view in the teaching of science and mathematics at the university level is that during lecture courses, knowledge and information are transmitted (as if "piped") from the heads' of the professors to those of the students. The latter then (fail to) apply what they supposedly "learned" during the lectures to world problems or "real-world contexts." Even those who adopt a constructivist stance to learning appear to assume that students transfer to the workplaces that they enter after graduation whatever they have learned in their university lectures. The reality shows that this is not the case. My experience and research shows that university science and mathematics professors complain that their undergraduate students come with little knowledge; those who employ university graduates, in turn, also deplore the substantial lack of graduates' mathematical and scientific knowledge required on the job. This can be interpreted in at least two ways. First, we may infer that both high school and university students have cognitive deficits so that they either or both (a) do not learn and (b) do not transfer what they have learned to a new setting. Second, we may infer that very little relevant knowledge has actually been transferred from textbooks and teachers' or professors' minds to the students. In any case, there appear to be knowledge gaps fittrst between high school and university, then between university and workplace. Being successful in the former institution does not guarantee success – at least initially – in the latter. How then should university science and mathematics educators approach this problem? What good does it do to teach if little of what has been taught is of actual use in the places that the university intends to prepare students for?

In this chapter, I track the problem of the knowledge gap between university and workplace. I begin by describing and exemplifying the results of nearly a decade of research involving both think-aloud protocols among science students and professional scientists and long-term ethnographic studies among scientists and technicians. My paradigm case comes from graphing, that is, a "skill" or practice that lies at the very heart of and defines the nature of science (Roth, 2003). I briefly articulate the problem in terms of a theoretical framework that is centrally concerned with *what people do* rather than with what they might carry around in their brain case.

W.-M. Roth
University of Victoria, Canada

O. Skovsmose et al. (eds.), *University Science and Mathematics Education in Transition*,

This theoretical approach not only explains the gap but also allows us to articulate constraints on the redesign of university education intended to do a better job in preparing science and mathematics students for their future workplaces.

6.1 Graphs and Graphing

In the history of science, visual representations other than text in general and graphs more specifically are essential components of the increasingly rapid development of science and scientific knowledge (Edgerton, 1985). During the twentieth century, for example, the knowledge in theoretical ecology exploded with the arrival of a particular form of graph that embodies models of and theories about populations (Kingsland, 1995). It is therefore not surprising to find many such representations in scientific journals: surveys of journals in biology (Roth et al., 1999) and physics (Lemke, 1998) revealed that there are, on average, 14.8 and 12 visual representations, respectively, per 10 pages of scientific text, of which 4.2 and 10, respectively, are histograms, scatter plots, and line graphs. Learning how to use graphs ought to be an important aspect of formal schooling so that by the time that students graduate from some university science program, they know how to read them or how to make use of them in oral communications (e.g., in conferences) and in written form (e.g., journals, reports).

To find out more about graph interpretation practices, I studied how individuals from different populations – including university science students, university graduates, scientists, and science teachers – interpreted a series of graphs all culled from a first university course in ecology. The graphs used in my research constitute a representative selection of those found in a popular 800-page introductory ecology textbook. One of these graphs displays a plate figuring birth rate and death rate as functions of population size (density); the two curves intersect twice (Fig. 6.1). The graph is paradigmatic for the discipline and constitutes a model of the type that

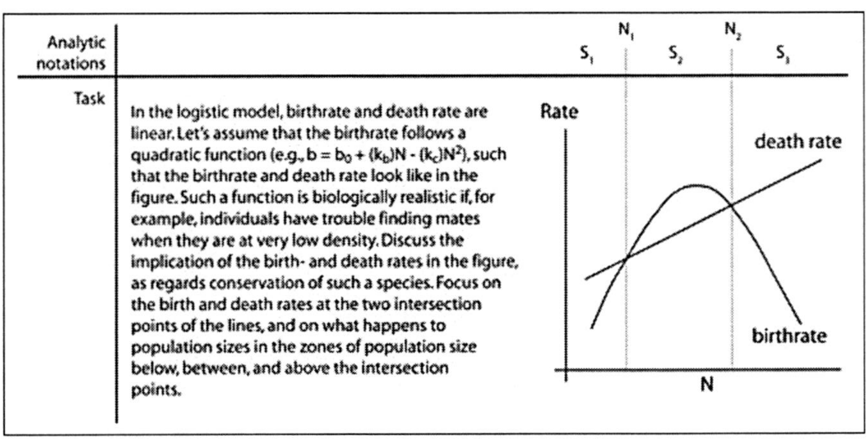

Fig. 6.1 A paradigmatic graph from a first course in ecology depicts birth rates and death rates of some population as a function of its size N

ecologists began using during the 1950s and 1960s. Although this is one of the first graphs university students encounter in their studies, our research shows that representatives of all populations – including practicing scientists – encounter difficulties in providing an answer that the professor of an ecology course would accept as correct. In the following two sections, I articulate some of the similarities in the difficulties that showed up in our controlled studies of expertise.

6.2 University Graduates' Interpretation of Graphs

Nearly a decade and a half ago, I began a research program into the use and development of representation practices in science. Since then, I have conducted think-aloud protocol and ethnographic studies in the disciplines of biology and physics: the participants in my research ranged from eighth-grade students to university students, graduates from university science programs, professional scientists and professors, and technicians. One of my studies had shown that eighth-grade students developed tremendous competence in analyzing data they had collected themselves, representing them using statistics and graphs, and defending their use of graphs and other mathematical representations (Roth, 1996). These results stood in stark contrast to a review of the literature, which showed that students in general do not do well when it comes to graphs and graphing (Leinhardt et al., 1990). Pursuing the question of whether universities prepared science teachers to teach inquiry and the use of mathematical representations, we were surprised to find that there was a statistical difference in the performance of college graduates with B.Sc. and M.Sc. degrees and our eighth-grade students, who, working in pairs, outperformed the future teachers (Roth et al., 1998). We conducted several more studies to track down the source of the fact that college students had such tremendous difficulties in using and interpreting mathematical representations in science generally and graphs more specifically. The following two episodes – featuring two university students in the process of interpreting a population graph (Fig. 6.1) – exhibit some of the main, characteristic features that I found.

The two students were among a small select group of students in an advanced fifth-year course preparing them to become science and mathematics teachers. They stood out among the cohort of future elementary teachers because of a strong preparation in the subject matter and because of their high level of success in their coursework. Despite this preparation, they, as others in the advanced course, failed to identify the two equilibrium points as *unstable* (N_1) and *stable* (N_2) and the point in the graph where the population should exhibit the largest increase in individual organisms[1] – 44, 22, and 0% in the total (pre-service) teacher group provided correct answers with respect to these three features. The episode picks up about 1 min into the 6-min session with respect to this graph. Tara and Karin are still working their

[1] The point where the increase in individual organisms is largest is of special significance in population ecology from a management perspective, for it is the point that permits the *maximum sustainable yield.*

way through the text next to the graph (Fig. 6.1), which they read while going back and forth to the graph depicted next to it.[2]

6.2.1 Episode 1

01 Tara: such a function is biologically realistic if, for example, individuals have trouble finding mates when they are at very low density. (0.47) yea, because that's (*) mo:re people die, (0.68)

02 Karin: (*)uh um.

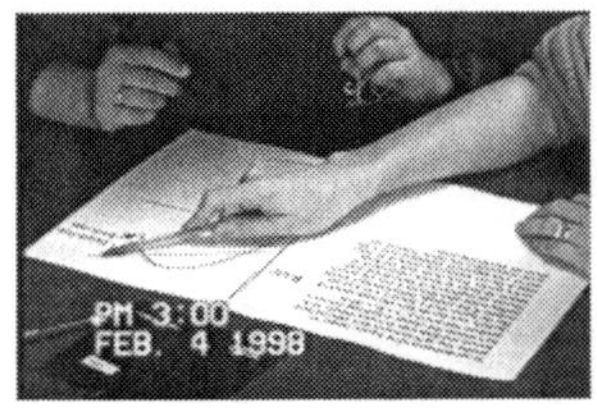

03 Tara: is less people, so there's less births. (*)

---

[2] The following standard transcription conventions have been employed.

A: ϒi'm/ looking at	Brackets indicate the extent to which speech or actions overlap.
D: ≤oh ƒ (0.56)	Pause measured in tenth of a second are enclosed in single parentheses.
(.)	A dot in parentheses indicates a slight pause, less than 0.10 seconds long.
Faster	Underline indicates emphasis or stress in delivery.
YEA	Capital letters are used when a syllable, word, or phrase is louder than the surrounding talk.
A: S:o::	When a sound is longer than normal, each colon indicates approximately 0.1 of a second of lengthening.
A: Envi–	The n-dash marks a sudden stop in the utterance.
.,?	Punctuation marks are used to capture characteristics of speech rather than grammatical features.
(stay out?)	The question mark following items enclosed in single parentheses denotes an uncertain hearing.
((circles))	Double parentheses enclose transcriber comments.
°Or?°	Degree signs enclose utterances produced in a low voice, that is, with low speech intensity.
(*)	Denotes the exact place in the transcript that corresponds to the image presented above.

As Tara (sitting to the right from the observer perspective) reads the text about the biological realism of the birth rate curve at low densities, she actually moves to look at the death rate curve, follows it with her pencil from low to high population densities (see off prints in turns 01–02). She then establishes a causal relationship between the higher death rates to the right of the graph and the lower birth rates in that same region (S_3, Fig. 6.1). It is important to note that her gestures exhibit the same conceptual relations as her words, which – because gestures and words have been reported to express different conceptions (Church and Goldin-Meadow, 1986) – is evidence for the rather consistent and stable understanding expressed in her presentation. In this instance, the two students agree about the fact that the graph depicts a causal relationship between death rate and birth rate. That is, their talk establishes a relation $b = b(d)$ rather than the relations $b = b(N)$ and $d = d(N)$ that are salient to a theoretical ecologist. Furthermore, the two students provide a static description – higher death rates (01) means "less people" (turn 03) despite the fact that the population (density) is actually higher to the right of the graph than to the left.

From the perspective of an ecologist, the graph constitutes a model for the *dynamic* of a population given its birth rates and death rates as functions of population size. Thus, the two intersections clearly identify two equilibrium points, one unstable (N_1) and one stable (N_2). (Not articulated here but evident to a theoretical ecologist is a third equilibrium point at $N = 0$.) To get to this (standard, correct) interpretation of the graph, one has to perceive it as representing a dynamic, iterative relation: The population size determines rates, which change the population size leading to new rates, and so on until the population has settled in a stable equilibrium. Or, in more mathematical terms,

$$N_{t+1} = N_t + (b_t - d_t)N_t \qquad \text{(i)}$$
$$b_{t+1} = b(N_{t+1}) \qquad \text{(ii)}$$
$$d_{t+1} = d(N_{t+1}) \qquad \text{(iii)}$$

The following episode exemplifies the approach of many university students: they tend to view the graph as depicting a set of stable rather than functional relations. The episode begins when Karin articulates that birth rates larger than death rates mean population growth. She first gestures a positive growth rate (right hand moves upward with respect to top of left hand), then points to the middle section on the graph (S_2, Fig. 6.1). She then articulates the inverse relation as meaning "non-growth," simultaneously pointing to the left (S_1, Fig. 6.1) and right-most areas on the graph (S_3, Fig. 6.1).

6.2.2 *Episode 2*

01 Karin: cause any time there's (*) more people being

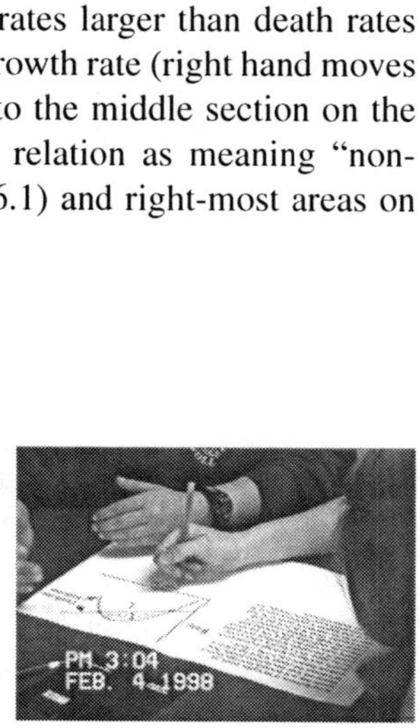

(*) born than

there is (*) dying, (0.40) you're having a population growth. (0.74) at anytime there is less (points to S_1) people being (points to S_3) born than dying (points to S_1, S_3) you have a population (.) like (0.20) non-growth what are you gonna call it ϒand here/

02 Tara: ≤°yea in° *f*

03 Karin: (points to P_2, P_1) there is no population growth at all (0.19)

04 Tara: in those two points (0.47) okay, (0.43) yea, (0.29) i guess what I was thinking about this (*) top (0.25) ϒsection (0.34) right here (0.31)/

05 Karin: ≤this's all i remember of this one now*f*

06 Tara: is that this is the point where (0.56) the, (0.64) oi: i don't know where but, (0.58) u:m (0.93) there is a certain relationship between (0.23) how (*) fast (.) this is going,

07 Karin: ϒuh hn /

08 Tara: ≤and how*f* (*) fast this is going, (0.31) and this is a point where it changes (0.33) in the reverse (0.32) an

09 Karin: ϒuh hn/

10 Tara: ≤this *f* starts to slow down (.) in respect to this, (0.22) not really speeding up but (0.22)

11 Karin: ϒuh hn /

12 Tara: ≤in rel-*f* in relative terms, (0.90) it would like this is speeding up,

Karin ends her summary of their findings with the statement that at the two intersection points: "there is no population growth" (turn 03). Whereas her description is certainly correct if taken to be the description of the population parameters at a particular point in time, it does not express the dynamic relations that are salient and important to (many) ecologists. More importantly, Tara then articulates not the comparison between the two rates but the steepness of the curves as the salient feature. First, she points to and talks about the maximum of the birth rate curve (turns 04, 06) and then articulates the relationship between the slopes of the curves ("relationship between how fast this is going, and how fast this is going" [turns 06, 08]) while pointing to the two rates. She then describes the maximum of the birth rate as a special point, where "it" (where the nature of the *it* remains under-determined at the moment) "changes in the reverse" (turn 08). In relative terms, the difference between the slopes of the death rates and birth rates get larger the further one moves to the right.

In this Episode 2, therefore, two features that prevented university students reaching the correct interpretation are exemplified. First and visible in the transcript of Karin's talk, the graph is perceived as depicting a series of static comparisons between birth rates and death rates. This relationship then determined what would happen to the population independent of the fact that any changes in the population size would entail a change in the birth rates and death rates, which are interpreted to have specific consequences for the population dynamics (see equations i–iii). Thus, Karin and Tara, like the predominant number of university students in several studies, inferred that in the right-most section of the graph, the population would crash. In a number of instances, students talked about the graphs as if they depicted functions of time (i.e., $b = b(t)$; $d = d(t)$) rather than functions of population size N.

Second and exhibited in the transcript of Tara's talk, the slopes of the curves often stood out for the university students rather than the relative heights of the curve. This generally led to confusion between *rate* used to denote the derivatives of mathematical functions (e.g., $b'(t) = \mathrm{d}b(t)/\mathrm{d}t$) and the rates depicted here (e.g., $b = b(N)$), which are already functions of the population size or density N. This frequently caused confusion, for the students were no longer aligned with respect to their understanding; and, unbeknown to them, some students within the same groups discussing the graph talked about the slopes whereas others spoke about the values of the two curves.

In both instances, the *perceptual* aspects of the graph became mediating elements that interfered with arriving at those interpretations that the professor of the introductory ecology course from which we culled the graph would have accepted as correct. *Perceptual salience* played an important role on another point as well. When asked which population there would be the largest number of individuals added to, all university students and science graduates responded either by pointing to the maximum of the birth rate curve ($\mathrm{d}b/\mathrm{d}N = 0$) or to the point where the distance between the two curves is largest ($\mathrm{d}(b–d)/\mathrm{d}N = 0$). The number of individual specimens is maximized, however, when the function $(b(N) - d(N))N$ is maximized.

A final point to be mentioned here about characteristic features in science students' interpretation pertains to the role of concrete examples used to contextualize the graph and to make its implications concrete. The science students often described the curves literally (i.e., "birth rate goes up" and "birth rate decreases") but did not talk about actual or possible populations that it might be used to describe or model (explain). Some students and student groups, such as Karin and Tara, talked about the population in terms of "people." This sometimes led to confusion, especially when they talked about birth rates being high in countries (Ethiopia and India were used as examples) where the death rates are high.

6.3 Scientists' Interpretation of Graphs

The research concerning graphs and graphing among scientists began with my interest in identifying some answers scientists would provide to a variety of tasks, which I wanted to use as a standard in my work with middle and high school and university students. Our (researchers') preconceived notion was that scientists would be experts – I write "preconceived" because graphing has long been touted to be one of the core scientific process skills. One of our investigations, framed as an expert/expert study using a standard think-aloud protocol, showed that scientists were far from perfect (Roth and Bowen, 2003). Although they were most successful on the population graph – only 56 and 50% correctly interpreted a correlation and isocline graph, respectively – 25 and 37% failed to characterize N_1 and N_2 (Fig. 6.1) as unstable and stable equilibrium, respectively. Only one of the sixteen scientists in the first expert/expert study correctly identified the point in the graph where there is a "maximum sustainable yield." Our study showed a statistically reliable difference between biology professors, who exhibited higher success rates, and scientists working in the public sector, who did not teach undergraduate biology courses. A close analysis of the think-aloud protocols revealed many of the same problematic features that I had detected among high school and university students. The following Episode 3 – which has been excerpted from the protocol involving Annemarie, a physics professor and highly successful (award-winning) teacher with more than 30 years of experience – exemplifies some of the features I noted. A cursory look at the episode shows that there are many pauses, some very long (5.06, 20.76, 6.24, and 4.98 s), which are indicative of the troubles she experienced and expressed with this task.

This episode begins when Annemarie articulated an interpretation for the right-most section of the graph (S_3), which she takes as indication of a decreasing population. Although not explicit in the first few lines, it is evident from other parts of the transcript that she means the population crashes. Much like the students discussed in the previous section, she takes a static rather than dynamic perspective, leading her to make incorrect inferences. She then points to the maximum of the birth rate graph (second offprint) and talks about the change between two situations, the decreasing population just articulated and an increasing one inferred a little earlier. Annemarie then talks about rates of change.

6.3.1 Episode 3

so in this region (*) the death rate is higher than the birth rate (1.20) (*) so the population is decreasing again. (2.55) um – so somewhere in the middle (0.57)

(*) there was a change between these two. (0.84) ↓A::hh. (5.06) the birth rate (.22) it's a – (1.18) it means rate of change? (0.47) °But the rate of° (0.95) yea:: I see. (20.76) (pencil moves from S_3 to S_1, to word "death rate") Conser – an conservation of such a species° (6.24)

Well it just – (1.14) feels that as long as the (1.20) n birth rate (4.98) is increasing (*) (1.20) faster than the death rate is increasing

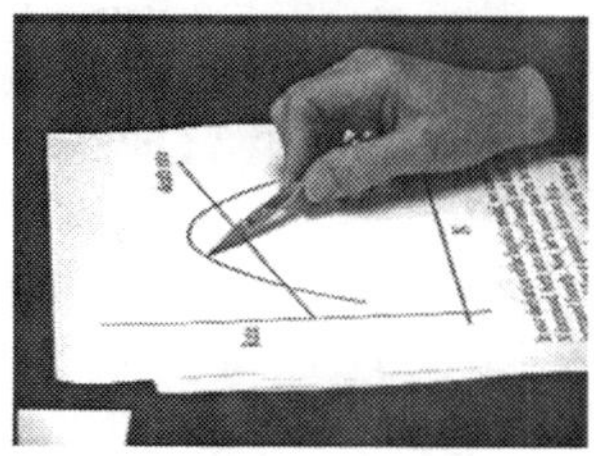

(1.20) (*) then the population is::: in good shape. (0.90) but when the birth rate begins to decline, (2.03) and the death rate stayed the same (pencil aligned with the death rate curve) (3.52) (pause, pencil comes to left of intersection 2)

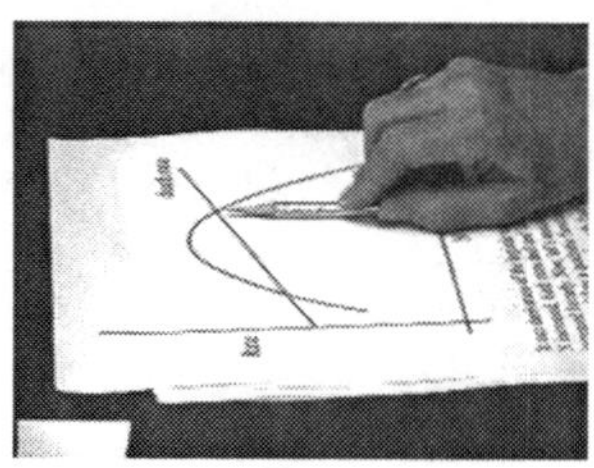

it's okay for a while (*) but eventually – (0.86) clearly the ah (0.82) population is going to diminish. am I on the right track?

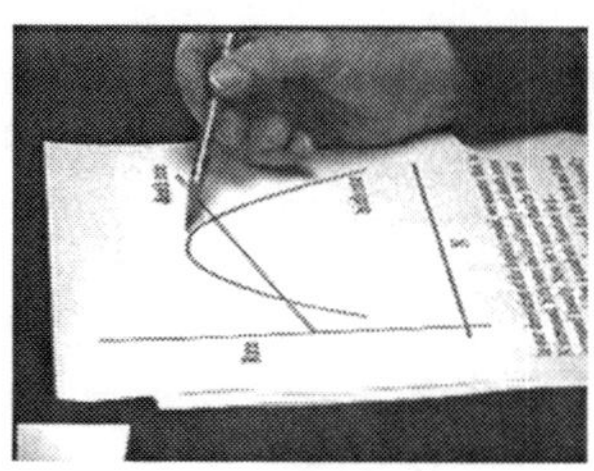

The transcript shows that Annemarie not only talks about rates of change in general but also about the derivatives of these particular curves ("birth rate is increasing faster than the death rate is increasing"). Other parts of the think-aloud protocol confirm that Annemarie is talking about the slopes as salient features; she even uses the term *derivative* to name what she is looking at. This is made even further evident when she compares the birth rate and death rate as declining and staying the same, respectively. It is evident that the slope of the death rate stays the same when one goes from left to right – as she does with her pencil. That is, at this point, she envisions trouble for the population as soon as the slope of the birth rate becomes negative with a constant slope of the death rate. Toward the end of the episode, with her pencil moving repeatedly from the maximum of the birth rate curve toward the intersection, Annemarie suggests again that the population is going to diminish – in fact, as she says elsewhere, is going to disappear.

This episode exhibits three problematic aspects identified even in the scientists' protocols:

a) taking the graph as a representation of static birth rates and death rates
b) failing to account for the underlying dynamic, the salience of the slopes
c) a potential confusion between the functions (birth *rate* $b(N)$ and death *rate* $d(N)$), their derivatives (rates = slopes: $\mathrm{d}b/\mathrm{d}N$ and $\mathrm{d}d/\mathrm{d}N$), and the temporal changes in the population and the curves depicted (rates = speed: $\mathrm{d}N\mathrm{d}t$, $\mathrm{d}b/\mathrm{d}t$, and $\mathrm{d}d/\mathrm{d}t$).

These results (including those pertaining to the inappropriate attribution of the maximum sustainable yield to the maximum difference between the two curves) underscore the fact that scientists also commit the errors previously ascribed to children and older students, especially when unfamiliar with a (type of) graph. They interpret the perceptual aspects of a graph rather than some "deep" feature that those who know attribute to representations.

Scientists not only exhibited difficulties but also experienced them as something they were talking about as they went along. Thus, depending on the task, between 13 and 50% of the scientists suggested that the graphs made little or no sense, that they were lost, or that they were not doing well because the graphs featured bad practices (one scientist even sent me an article about how to construct better graphs). The scientists attributed their experience of difficulties to these reasons.

In contrast to the difficulties scientists experienced with the graphs from the ecology course and textbook, they exhibited tremendous competencies when it came to talking about graphs that issued from their own research or from others working in the same domain. There was a gap in their practices between the two situations, interpreting an undergraduate-level but unfamiliar graph and interpreting a work-related graph. The most important difference between talking about the tasks versus talking about their own graphs were the copious amounts of detailed knowledge scientists articulated with respect to the experiments conducted, the tools and instruments used, the transformations that led from the natural phenomenon to the graphical representation, the contextual details pertaining to origin and history of the samples studied, and so on (Roth, 2003). I came to understand that all of these contextual details and knowledge about what did or conceivably might have happened was a central element in competent graphing and graph use. That is, competent graphing seemed to be not a *general* but in fact a highly *contextual* skill, requiring tremendous knowledgeability about the phenomenon, experimentation, data generation and transformation, and theoretical background. The graphs and categories these employed turned out to be highly indexical. To be able to interpret the graphs and use the categories, even the scientists needed to identify a relevant context within which graphs and categories became resources for marking and making sense.

These and similar results initially were perplexing to me. Researchers on graphing among high school students generally suggested cognitive deficits for explaining why the students did not arrive at standard answers. For example, drawing on certain perceptual features of graphs rather than on some deep structures was said to constitute perception-based misconceptions; reading the height of graphs rather than their slopes was attributed to slope/height confusion (Leinhardt et al., 1990). My problem was that my population involved scientists, most of whom had completed Ph.D.s, who had been very successful in their work as measured by their publication rates and the amount of external funding they attracted to their institutions. It would not have washed well had I tried to argue that their problem was due to a cognitive deficit. This, then, became a defining moment in my research. First, I was cured forever of attributing students' learning problems to cognitive deficits and of accepting such explanations – if I ever did. Second, it led me to a double-pronged research approach. On the one hand, I wanted to see how scientists used graphs in their workplace and how they developed the tremendous competencies with respect to graphs that issued from their own research or from the research of others working in the same domain. I decided to conduct ethnographic studies both among scientists (experimental biology, field ecology) and among technicians (water technician, fish culturists). On the other hand, I was changing the way in which I looked at knowing and learning, turning to praxis-oriented theories, including cultural-historical activity theory. In the following two sections, I articulate my findings and developments in these two domains, respectively.

6.4 Graphs and Graphing in the Workplace

The results from the research on graph scientists' interpretations pointed me in a different direction with respect to how to theorize knowing and learning. Rather than thinking about how people know and learn in terms of stuff ("knowledge," "skills," "conceptions," "cognitive structure") in their heads, *knowledgeability* appeared to be closely related to what they do and how they do it in everyday praxis. To find out required more than thinking or philosophizing about knowledgeability: I decided to conduct ethnographic studies. Initially, these studies were confined to scientific research situations – an ecological fieldwork station, a forest engineering research group, and an advanced laboratory studying fish vision. But then the research in these situations was extended to include an environmentalist organization and a fish hatchery. In each of these I studied the use of graphs and graphing practices for periods covering 3–5 years.

It turned out that not only scientists but also technicians and individuals with no more than high school education were highly competent with respect to graphs and graphing that were the result of, and appeared in, their work. To exemplify the particular nature of work-related graphing competencies, I draw on the ethnographic study in the fish hatchery, where fish culturists kept detailed *brood records*, that is, records concerning the particular salmon species they raised from egg collection to smolt size before releasing them into the neighbouring river. In the following Episode 4, Erica, one of the fish culturists currently responsible for raising coho, has taken me to her office and leaves through her brood records. Erica attended 2 years of a business program at a local college but then dropped out to first work as a temporary worker in a fish hatchery and then, when a position opened up, became a fulltime employee.

At the moment where the episode begins, Erica has already talked about a graph that she uses to track fish growth over an 18-month period, and has already suggested that the graph really only represents average fish weights and that the population is actually better represented in the histograms she creates after each sampling episode. The episode picks up when Erica explains the nature of the bimodal length distribution, which is associated with a mono-modal weight distribution (Fig. 6.2): The first mode means that "there's a few runties" and the second mode suggests the presence of "a few really big fish" (turn 01). That is, without hesitation Erica matches the graphical representation and a state in the world, here the size (length) of her fish.

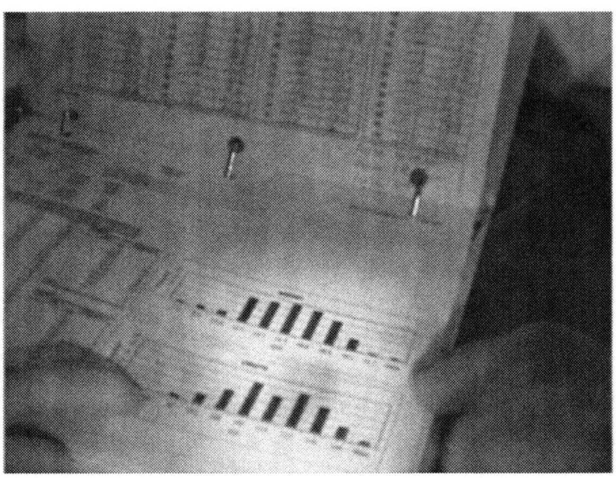

Fig. 6.2 As part of her database tracking the average sizes and weights of the coho salmon in her care, Erica plots the two distributions for each sampling episode

6.4.1 Episode 4

01 Erica: there's, there's a few runties down here (*) and there's, there's a few really big fish there. This (data sheet) is my last one.

02 WMR: which is much more um–

03 Erica: so this is yeah this is like 3 days ago

04 WMR: oh because here it's (*) again it's

05 Erica: it's looking nice isn't it except for here (points to bottom histogram in Fig. 6.2) the lengths are all different and in the same breath I really don't, well I don't want all my fish to look the same. people come in different shapes and, and forms so if they're all the same then yeah I would be concerned see that–

06 WMR: then your sampling is not that high that's why you get, probably, you know you only have what a hundred fish?

07 Erica: hundred fish, if I had a larger sample it would probably–

08 WMR: yea, more, be more even.

09 Erica: reflective, it would be a better reflection of the population.

10 WMR: because this one here (points to same spot as previous offprint) looks more like a long tail rather than–

11 Erica: um um so I've got a few really big fish so this tells me I've got oh there's my weights but look it like I've got some, I've got some really short fat ones (*)

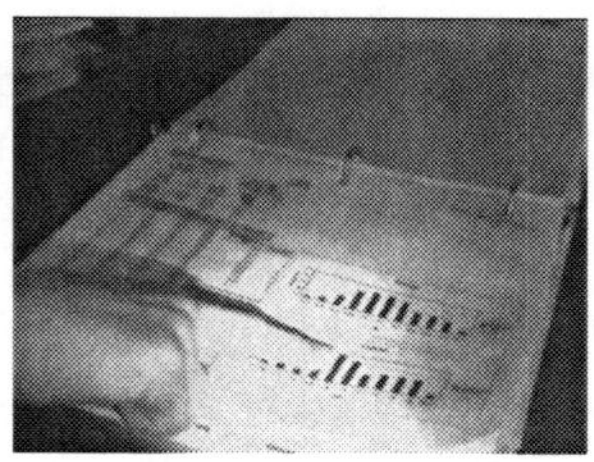

and I've got some really skinny long, long ones as well (*)

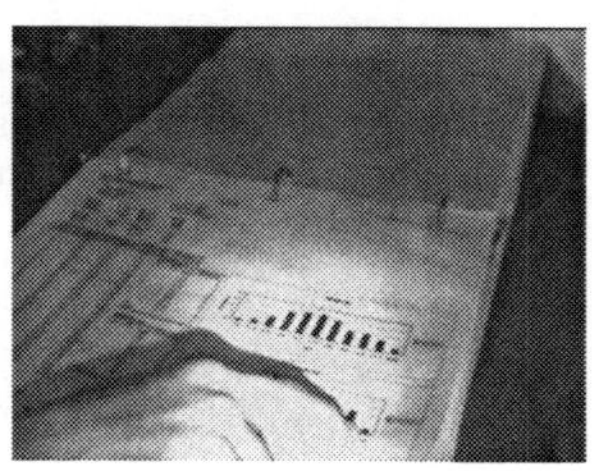

Erica then explains that the considerable distribution in fish length is something desirable and she would be concerned if there were no variation. Variation She wants to guarantee variation in the same way that it occurs in nature; she thereby exhibits an intuitive understanding that variation is necessary rather than an evil. She raises fish to live in the wild, able to return to the hatchery to spawn again, and thereby maintain a healthy population rather than producing fish for the market. In this latter case, having more consistent size would be a reasonable target, but it might not matter at all whether the fish come in various sizes.

In this episode, Erica also exhibits an understanding of the influence that sampling size has on the shape of the histogram. She knows that if she were to sample more fish, the distributions would better represent the population in the pond (turns 07, 09) – each of which holds about 350,000 coho salmon in the fry stage of the life cycle. The latter part of the episode provides evidence that Erica understands not only the relationship between the graph and what her fish actually look like along a simple dimension but also along two dimensions simultaneously. On the page, she has plotted weights and lengths in two histograms, and now makes a comparison across the two (turn 11). Here, the tail is longer in the length histogram than in the weight histogram, which means that there are longer fish at constant weight, and the higher peak or rather double peak with the first one below means that some fish are shorter and therefore have to be heavier.

My ethnographic research shows that Erica not only understands the histograms but also the graphs, formulas, ratios, and other mathematical representations. More so, she easily talks about some physical state for each of the representations she talked me through – as if she were talking about geographical maps while taking me through the corresponding terrain. Ultimately, therefore, Erica does not merely talk about and explain mathematical representations: her interpretive practices exhibit a deep understanding of the relationship that the representation has to the physical world of her hatchery generally, and the fish that she is raising to compete in the wild more specifically.

In the light of the earlier presented difficulties of university science students, graduates, and research scientists, the tremendous knowledgeability that Erica and her co-workers display raises many questions. An important question for me was how Erica could be so knowledgeable about the mathematical representations and the relationship between representation and the physical world. Here, my ethnographic research provided at least part of an answer.

As I am following Erica around and sometimes filling in for her, I come to understand that she continuously oscillates between the two worlds – working with the fish and producing records (Fig. 6.3). While she is feeding, for example, she does not merely throw food into a pond but closely looks at fish behaviour, how they feed, how fast they swim, how vigorously they break the water surface, and so on (Fig. 6.3a). Over time, she has evolved an intuitive understanding about how all of these behaviours are mediated by weather and other factors – lower temperatures mean more sluggish fish and lower feeding rates. When she enters data into her computer (Fig. 6.3b), she then finds out that fish growth rates are small and sometimes, for example in the winter months, fall to zero. Erica understands that physiology underlies all of this and, as a consequence, she has developed a much better

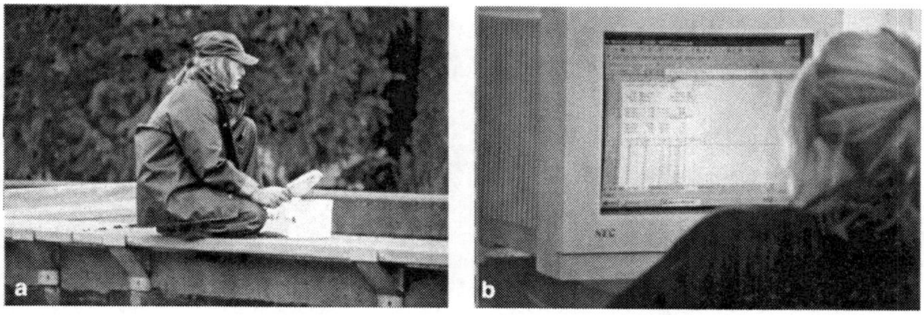

Fig. 6.3 Erica constantly moves back and forth between working in the physical plant of the hatchery and her database. (a) While feeding the fish, Erica observes their every move, responses, and behavioural patterns. (b) While working with the database, Erica inquires and reflects upon mathematical patterns, including the graphs, histograms, tables, and summary statistics

growth trajectory for the fish than the one that scientists have previously proposed. Her feeding schedule is better because, as a comparison of the scientists' and her own graphs show, the fish of the former lost weight during the winter months, which means, they were starving. Erica's fish never starve but, with few exceptions, always grow even if only a little during the coldest month of the year (January).

The constant back and forth between working with the fish and the computerized database – including its many plotting capabilities – allows Erica to develop her tremendous knowledgeability. In fact, the mathematical competencies are not independent from the tremendous competencies related to the fish. For Erica, the mathematics does not exist independently of her hatchery work but is just another tool that she uses on a daily basis. That is, Erica's mathematics is *for the sake of* getting the job done rather than *about* the hatchery world; it is part of her everyday world without her making "mathematics" thematic. In fact, asking her whether and what mathematics she uses, Erica might be stymied for a moment, because she does not at all think of herself as doing mathematics. She is a fish culturist, and she concretely practices fish culture. There are two different relations between mathematical representation and the world: in the former situation, the mathematical representations are *primary artefacts* whereas they function as *secondary artefacts* in the second situation. Primary artefacts are those directly used in everyday praxis; secondary artefacts are those employed in talk about praxis and therefore "in the preservation and transmission of the acquired skills or modes of action or praxis by which this production is carried out" (Wartofsky, 1979, p. 202). Erica's mathematical knowledgeability is associated with their function as primary artefacts, whereas the scientists I asked about the graphs were facing these as secondary artefacts, as representations of modes of actions and objects that they were unfamiliar with.

The notion of primary artefact points us to actual use and the relationship between artefact and objects (fish, environment) as being one of constant association. Thus, pertaining to the histogram, I gathered a lot of evidence showing how Erica's knowledgeability is constantly arising in and through the concurrent handling of the fish

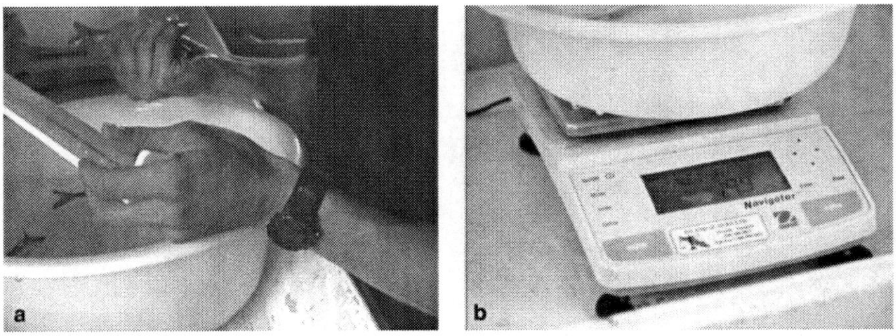

Fig. 6.4 Erica's tremendous understanding of the statistical distributions literally goes hand in hand with the handling of the fish. (a) Erica closely inspects each fish and then lays it on a ruler to measure its length. (b) She then places the fish in a bowl of water on a tarred scale to weigh it. Both measures are entered immediately into the database and plotted when the sampling is complete

and the database. For example, during the monthly sampling episodes, Erica visually inspects each of the 100 fish she collects. In this, she does not just look from a distance, but holds each (anaesthetized) fish in her hand, turns it about, lays it onto a ruler to measure its length (Fig. 6.4), and then places it into a bowl of water on a tarred scale so that she can read off the weight. Both measures are then entered into the computer immediately to her right. She often greets particularly long and large fish with expressions of joy, and exhibits concern when fish are small, have eye or body abrasions, or appear affected in some other ways.

When all 100 fish have been measured, Erica immediately plots the results of the sampling episode, providing her with a graphical representation of the lengths and weights of the fish that she has just physically handled and inspected, which yields graphs such as those featured in Fig. 6.2. She has a good sense of the number of really long fish or really short fish she has handled and is aware of the distribution of weights. Now she sees the lengths and weights plotted, providing her with a different access to these measures than when she physically handles them. But the two different ways of accessing weight and length are so close in time and space that it allows her to develop a direct correlation. That is, her tremendous knowledgeability is the direct result of *handwork*; in fact, knowing graphs *is* a form of handwork.

Some (academic) readers may want to argue that Erica is a technician, and technicians, in contrast to scientists, are less capable and therefore only learn much of what they know on the job. It turns out that my ethnographic work among university and public sector scientists yielded exactly the same results (e.g., Roth, 2004, 2005a). Even in their own laboratories or fieldwork situations, scientists often learn to read graphs as part of their evolving understanding of the object they study. This knowledgeability with respect to graphs is inseparably tied to their familiarity not only with the object but also with the measuring instruments, the preparation and transformation that the object undergoes, and, of course, with their evolving understanding of scientific principles and theories governing the research object and measurement. Any representation that scientists produce and use has an indexical nature such that the context is required for specifying their sense but the representations also constitute the context. The upshot is that graphs and context comprise reflexively constituted relational configurations. My work amply shows that scientists *learn* to read graphs at work, through a protracted process of becoming familiar with and correlating the various features of the research situation, rather than being able to read them the first time they see it. Here, too, knowledgeability appears to be related to *extended embodied praxis* rather than to an *ability* hardwired in the scientists' brains. Before drawing some implications, I need to present a theoretical framework that accounts for these research results better than the information processing theories that underlie the general presuppositions about learning among science faculty.

6.5 Explaining Graphing: Toward a Cultural-Historical Activity Theoretic Perspective

In the beginning of this chapter, I note that graphs and graphing constitute some of the fundamental objects and actions of science. Furthermore, not only science students but also science teachers and scientists experience difficulties reading rather simple graphs that are part of entry-level university courses in ecology. Yet scientists and even technicians with relatively little or no postsecondary education are highly competent and demonstrate deep understanding when it comes to graphs related to their work. There therefore exists a gap between the kinds of knowledgeability that make people successful in formal learning environments (school, college, and university) and the kinds of knowledgeability required and exhibited in the workplace. Such results challenge the common epistemology underlying much of university teaching – information processing.

Information processing metaphors and analogies are frequent among professors. Expressions such as "In my lectures, I am trying to *get across*..." "This student didn't *get it*..." and "Students *regurgitate what* they memorise but cannot apply concepts..." are common at the university. They depict knowledge as something that can be transferred ("gotten across") from professor to students who do or do not "get it" and who can or cannot translate and therefore apply "*it*." Process skills are thought of as hardwired or ability-dependent and context independent so that they can be applied irrespective of the particular context. Research results such as those I present here show that these information-processing analogies are inappropriate. Why would biology and physics professors experience difficulties interpreting really basic graphs presented in introductory textbooks of their own discipline? Why does their competence to read graphs increase with their familiarity concerning research object, tools, instrumentation, and concepts? Such results show that we need to think about knowing and learning, that is, about knowledgeability, in different ways. Research results such as that presented above already point us into a useful direction– knowledgeability is relative to the current activity.

Cultural-historical activity theory has been developed to explain real, concrete, and observable human behaviour rather than ephemeral things and structures in people's heads (Leont'ev, 1978). Here, activity denotes events or processes that are part of a societally defined division of labour – *doing research* to create scientific knowledge, *studying* to get a degree, or *hatching salmon* to mitigate over fishing. Each of these activities is not only directed toward different collective-need-related motives but also involves different tools, rules, and divisions of labour. Cultural-historical activity theory takes activity systems as their fundamental unit of analysis so that participation, knowing, learning, and even identity are inherently mediated by the system as a whole (Roth et al., 2005). Thus, consistent with my research in the hatchery and the scientific laboratories interacting with it, knowledgeability pertaining to the same graph will be different when it is used or created *for the purpose* of generating scientific knowledge or *for the purpose* of successfully raising a brood of coho.

To understand this systemic approach to activity, let us look at an analogy: we cannot understand a sentence such as "the fish culturist monitors fish growth" by only considering any one of its elements, subject, verb, or object. The transitive verb *monitoring* takes its sense from the object and subject of the action, a person doing something on or to something. (The adjective "transitive" means that the verb denotes an action that passes from the subject [agent] to the intended object.) Conversely, the subject and object of the sentence not only presuppose the verb but also presuppose each other. A person is a fish culturist because he or she monitors fish growth, and fish growth is an object of actions in hatcheries. More so, to push this analogy, the available tools mediate actions such as monitoring. Thus, Erica monitors fish growth using a ruler and scale to measure their lengths and weights; and she uses a computer to monitor fish growth over time. The available ready-to-hand tools shape the action of *monitoring*. Thus, the graphs allow Erica to compare her own growth trajectories with those that scientists have proposed as ideal growth curves; if she did not have the graphs available, her monitoring action would look different.

Activities are general level events; they are realized by means of concrete, goal-directed actions. These two levels of processes stand in a dialectical relationship: actions both realize activities, thereby bringing them into existence, and presuppose them. As a result of this dialectic, the same action has a very different sense when it realizes a different activity. The sense of mathematical equations and the associated mathematical actions may substantially vary between scientists and engineers or technicians (Brown et al., 1989); and hatchery mathematics is very different from mathematicians' or scientists' mathematics (Roth, 2005b). Thus, the research scientists become highly competent at the kinds of graphs they use and need at work, but are much less competent in dealing with the graphs that are useful in the context of, and for doing well in, an entry-level university course. We can therefore understand why professors do significantly better than public sector scientists on the graphing tasks – professors lecture and teach seminars in which graphs of *this type* are standard objects and where interpreting *this type* of graph are standard actions.

My research in the fish hatchery shows that associated with motive of an activity are e*moti*on and *motiv*ation, which in turn mediate mathematical actions such as modelling

(Roth, 2007a). Even ethics and ethical behaviour enter our considerations of knowledgeability when we take a cultural-historic activity theoretic approach. Thus, when someone suggested to Erica during one sampling episode that she omit the measurements of some very short and light fish, therefore creating a database that gives a better impression of the current status of the fish population, she became very upset (Roth, 2007b). For Erica, ethics and care for the fish and the way in which they are mathematically represented in the database mediate her every action in this hatchery.

It is important to note that the activity systems do not *determine* but rather *mediate* actions; they offer resources to the human subjects who realize the possibilities of these resources in different ways. Although they work in the same hatchery, the practices of the five fulltime fish culturists differ, each individual realizing the available possibilities in different but equally legitimate ways. The subjects of and in the activity systems produce outcomes that are useful either in the same – the actions of the maintenance persons in the hatchery benefit the hatchery first – or in other activity systems – commercial, sports, and indigenous fishermen enjoy large salmon runs made possible by the hatchery activities.

The subjects in the activity systems not only produce outcomes but also produce and reproduce themselves. In each action members of the hatchery also provide evidence to whoever watches, including hatchery managers, research scientist, and ethnographer, of their knowledgeable practice. Erica is widely known as a highly competent fish culturist because of what she does with the fish and the database. More so, with each action she also develops her knowledgeability so that over time she becomes more knowledgeable and develops a deeper understanding between the shape of a histogram and the weight or size distribution of the fish that she handles and inspects. That is, each action not only produces outcomes and reproduces her as a competent person, but also makes her a more knowledgeable fish culturist.

6.6 Producing the Gap

Currently there is a gap between learning and knowing in formal education – e.g., university science and mathematics – and the workplace. More so, the results of my research program exemplified in the four episodes shows that knowledgeability is highly contextual. Knowledgeability is contextual to such an extent that even the interpretation of very simple graphs requires familiarity with the phenomena represented, data collection procedures, instrumentation, and so on. This has substantive implications. If knowledgeability is tied to the motive of activity, emotion, motivation, ethics, and identity, we need to raise questions: why do we make students attend lectures if knowledge and knowledgeability cannot be transferred to any meaningful activity outside the class context? Why do we make students attend lectures if knowledge cannot be transferred from the professor to the student? Why have students do laboratory exercises that have no motive other than to subject students to more hurdles toward a degree?

My research results imply that rather than sitting in university lectures and doing fake laboratory exercises – the purpose of which is merely to create a rank order to be used in the distribution of more and less desirable places in graduate studies and jobs (Roth and McGinn, 1998) – students need to be involved in the real stuff, learning while doing what we are training them for. From an activity-theoretic perspective, by going to university students first of all become good at doing "going to university," realizing the possibilities it embodies. Some even develop identities of successful students embodied in their grades, which they can trade in for subsequent career opportunities in graduate school or the job market. Some readers might turn around and charge me with extremism. You may ask, how can we teach otherwise? and explicate, students need to have a knowledge base before doing what we train them for. To this I would respond that I already provided an existence proof for the possibility of teaching science by involving students – in my case seventh-graders (age 11–12) – in forms of environmentalism that pre-exist as activity system in their community (Roth, 2002). These students, despite their tender age, designed research for the purpose of contributing to a community-wide database concerning the health of a watershed generally and one creek in particular. Their research results were later featured on a website, in the local newspaper, and during an open-house event in the community. In the process of doing their research, they became knowledgeable not only in science and mathematics but also other school subjects as well.

Problem-based learning and internships (e.g., in medicine and dentistry) are some of the alternative learning environments created in the attempt to bridge the gap between knowing at the university and knowing in the workplace (see Kolmos, Chap. 12). While it is the case that these learning environments constitute improvements over lectures, they do not entirely mitigate some inherent distinguishing features that are of structural nature. Thus, an ongoing investigation in my research laboratory shows that even the highly practical and applied internship experiences of dentists only allow development of formal concepts and raw skill, failing to develop the knowledgeability needed for *right action* (Ardenghi et al., 2005). Whereas the dentistry students do work with real patients in real dental clinics, the existing division of labour (hierarchical professor/supervisor–student relationship) and the associated aspect of performing for assessment mitigate the development of ethical practices. As long as the really important goal of doing "going to university" is the production of grades and certificates, very little will change because students will get good at producing high grades and getting certificates. Really knowing and understanding such things at graphing is neither required nor does it lead to an expansion of students' power to act so that when it occurs it in fact constitutes a by-product.

In the earlier discussion, I suggest that emotion, motivation, and ethics are all implied and theorized when we take the perspective of cultural-historical activity theory. What Erica does and how well she does it is mediated by emotion, motivation, and concerns for the implications of her actions; conversely, her actions influence emotion, motivation, and concerns. She wants the fish to be healthy and do well, which motivates what she does and influences her short and long-term emotional valence. It mediates her desire to have the mathematical representation reflect the fish population rather than making her look better in the eyes of hatchery management. But receiving a lay-off notice influenced her emotional valence to such an extent that she forgot to do what she had been doing under *normal* circumstances; her actions provided her co-workers with evidence that she was not being her normal self (attribution of identity, emotion) and that she was not doing her job as knowledgeably as she normally would. This linkage of emotion, motivation, and ethics in activity theory also provides us with an important inroad to understanding the frequently reported practice of "cheating."

At the university, professors complain about students who "cheat" on exams or, more frequently, on course assignments completed at home. The theoretical perspective developed here allows us to understand. The motive of the activity system is not to know – everybody is aware of the knowledge gap between university and workplace. One therefore goes to university to get a degree and one develops relevant knowledge in the workplace. Students do not consider harming themselves if they "cheat," that is, if they make use of all *available* resources in the university culture to get assignments handed in. They simply realize one of the possibilities inherent in the system; assignments can be completed by copying the work of someone else. The universities (as institutions) and each individual professor are as much responsible for cheating as the students to whom they confer degrees and whom they teach.

Ultimately, then, the activity-theoretic approach allows us to understand that universities, by conferring diplomas and grades, contribute to producing and shaping a workforce. The types of knowledgeability required for being successful in the formal learning environment and workplace are generally different, leading to the observed and observable knowledge gap between "ivory tower" and the "real world." The gap therefore is not simply out there; universities contribute to producing it. By being professors, we produce and reproduce universities in the way they are, and we thereby produce and reproduce the knowledge gap. Most importantly, our students do not contribute to society while at university; all their productions are produced so that they can be evaluated and converted into grades, which students accumulate in their grade point average. Their efforts correspond to energy lost – given that they could learn while contributing to science and society in useful ways.

6.7 Coda: So What? and Where to Next?

Some readers may think of my perspective as overly pessimistic and ask me questions such as: "so what?" and "Where do we go from here?" I neither have *an* answer nor *a* panacea. I do, however, have experiences and have conducted research that constitutes an alternative to the predominant way of *doing* "going to university." It comes from a large experimental project in the training of science teachers, preparing to teach in the most difficult inner-city schools existing in the USA.

At the University of Pennsylvania, my colleague Ken Tobin and I created a program in which those with aspirations of becoming science teachers participate in teaching from the very moment they enter the program. But rather than letting them teach on their own, we create opportunities for them to teach at the elbow of one or more teachers, usually a more senior person, but sometimes also a peer (Roth and Tobin,

2002). During breaks, after school, and in the evening, they meet with other teacher aspirants and university personnel to discuss their experiences and learn about theories that might be useful in understanding the events of the day.

The program has been tremendously successful to a large extent because students no longer experience a gap between university and workplace: they participate in both at the same time. More so, rather than being mere interns, they are recognized as legitimate contributors to the activity system as a whole. That is, they learn *while* contributing in a useful way to their school in particular and to society more generally. Rather than producing assignments for evaluation, they first and primarily contribute to the teaching of needy youths. Any evaluation only occurs in a second instance. More so, in and through their presence at the schools, they also contribute to the regeneration of the resident teachers, who, normally isolated to teach on their own, in the past have frequently left the schools as a consequence of stress and burn out. Their presence also leads to the developmental transformation of the teaching practices.

This example shows that it is possible to rethink university teaching in such a way that students become knowledgeable in ways that are useful and applicable after they have left their institution of formal learning. Whether this model is transferable or transposable into the undergraduate programs of the natural sciences and mathematics remains to be tried and documented. An old saying states, "Where there is a will, there is a way." From the perspective of cultural-historical activity theory, we (professors, secretaries, deans, presidents, students) are all a constitutive part of university education. We are not merely cultural dopes who, like cogwheels, do as they are told; the university only exists in and through our actions. When we change, the university changes; and changes in the university provide new resources for us to do what we do best. From my perspective, the main aspect to be changed in university education is its current inability to teach in ways that allows our students to develop *useful* practices. It is not useful for one person to say that we need to change, just as it is not useful for one professor to lecture how to interpret the population graph. All of us involved in university education, including the students, have to work together to bring about the necessary changes that at least narrow the gap between the different forms of knowledgeability. I strongly believe that in such a grassroots movement, we can transform university science and mathematics teaching and decrease – if not eliminate altogether – the gap between knowing at the university ("ivory tower") and knowing in the workplace ("real world").

References

Ardenghi, D., Roth, W.-M. & Pozzer-Ardenghi, L. (2005). *Learning Ethics in Dentistry Practice.* Paper presented at the annual meeting of the American Educational Research Association, Montreal, QC.

Brown, J.S., Collins, A. & Duguid, P. (1989). Situated Cognition and the Culture of Learning. *Educational Researcher,* 18(1), 32–42.

Church, R.B. & Goldin-Meadow, S. (1986). The Mismatch Between Gesture and Speech as an Index of Transitional Knowledge. *Cognition,* 23, 43–71.

Edgerton, S. (1985). The Renaissance Development of the Scientific Illustration. In J. Shirley & D. Hoeniger (Eds.), *Science and the Arts in the Renaissance* (pp. 168–197). Washington, DC: Folger Shakespeare Library.

Kingsland, S.E. (1995). *Modeling Nature: Episodes in the History of Population Ecology* (2nd ed.). Chicago: University of Chicago.

Leinhardt, G., Zaslavsky, O. & Stein, M.K. (1990). Functions, Graphs, and Graphing: Tasks, Learning, and Teaching. *Review of Educational Research,* 60, 1–64.

Lemke, J.L. (1998). Multiplying Meaning: Visual and Verbal Semiotics in Scientific Text. In J.R. Martin & R. Veel (Eds.), *Reading Science* (pp. 87–113). London: Routledge.

Leont'ev, A.N. (1978). *Activity, Consciousness and Personality*. Englewood Cliffs, NJ: Prentice Hall.

Roth, W.-M. (1996). Where is the Context in Contextual Word Problems?: Mathematical Practices and Products in Grade 8 Students' Answers to Story Problems. *Cognition and Instruction,* 14, 487–527.

Roth, W.-M. (2002). Henderson Creek. In L.M. Richter & R. Engelhart (Eds.), *Life of Science: Whitebook on Educational Initiatives in Natural Sciences and Technology* (pp. 155–166). Copenhagen: Learning Lab Denmark.

Roth, W.-M. (2003). *Toward an Anthropology of Graphing*. Dordrecht: Kluwer.

Roth, W.-M. (2004). Emergence of Graphing Practices in Scientific Research. *Journal of Cognition and Culture,* 4, 595–627.

Roth, W.-M. (2005a). Making Classifications (At) Work: Ordering Practices in Science. *Social Studies of Science,* 35, 581–621.

Roth, W.-M. (2005b). Mathematical Inscriptions and the Reflexive Elaboration of Understanding: An Ethnography of Graphing and Numeracy in a Fish Hatchery. *Mathematical Thinking and Learning,* 7, 75–109.

Roth, W.-M. (2007a). Emotions, Motivation, and Identity in Mathematics and Activity Theory. *Mind, Culture, and Activity*, 14, 40–63.

Roth, W.-M. (2007b). Mathematical Modeling 'in the Wild': A Case of Hot Cognition. In R. Lesh, J.J. Kaput, E. Hamilton & J. Zawojewski (Eds.), *Users of Mathematics: Foundations for the Future* (pp. 77–97). Mahwah, NJ: Lawrence Erlbaum.

Roth, W.-M. & Bowen, G.M. (2003). When are Graphs Ten Thousand Words Worth? An Expert/ Expert Study. *Cognition and Instruction*, 21(4), 429–473.

Roth, W.-M. & McGinn, M.K. (1998). Science Education:/Lives/Work/Voices. *Journal of Research in Science Teaching,* 35, 399–421.

Roth, W.-M. & Tobin, K.G. (2002). *At the Elbow of Another: Learning to Teach by Coteaching*. New York: Peter Lang.

Roth, W.-M., Bowen, G.M. & McGinn, M.K. (1999). Differences in Graph-Related Practices Between High School Biology Textbooks and Scientific Ecology Journals. *Journal of Research in Science Teaching,* 36, 977–1019.

Roth, W.-M., Hwang, S., Lee, Y.-J. & Goulart, M. (2005). *Participation, Learning, and Identity: Dialectical Perspectives*. Berlin: Lehmanns Media.

Roth, W.-M., McGinn, M.K. & Bowen, G.M. (1998). How Prepared are Preservice Teachers to Teach Scientific Inquiry? Levels of Performance in Scientific Representation Practices. *Journal of Science Teacher Education,* 9, 25–48.

Wartofsky, M. (1979). *Models: Representations and Scientific Understanding*. Dordrecht, The Netherlands: Reidel.

Chapter 7
The Culture of Mathematics and the Mathematical Culture

Leone Burton†

In a recent study of the epistemologies of practising research mathematicians and how these related to and affected their disciplinary practices, I identified differences between the culture of mathematics, those aspects of mathematics which are recognizably discipline-related (such as the particular attitudes towards beauty, rigour, succinctness, etc.) and the mathematical culture, the socio-political attitudes, values and behaviours that constitute how mathematicians, and their students, experience mathematics in the settings of conferences, classrooms, tutorials, etc. In this chapter, I provide empirical justification for drawing this distinction and then explore the ways in which the culture of mathematics and the mathematical culture influence attitudes, behaviours and values within the discipline. Whereas aspects of the culture of mathematics have, historically, been defined as integral to mathematics and are seen as part of what students are expected to acquire in the process of becoming mathematicians, the mathematical culture is a product of stereotypes and biases that control who can enter the discipline and how they do so. However, it is not so easy to differentiate the two. I argue that it is the mathematical culture that creates barriers to entry by members of certain groups and facilitates others. Consequently, it is the mathematical culture that exercises power over how the culture of mathematics is understood. I conclude, therefore, that it is the mathematical culture that must be addressed if mathematics is to achieve widespread accessibility.

7.1 Setting the Scene

In 1997, I undertook a study (Burton, 2004a) with thirty-five women and thirty-five men in career positions as research mathematicians in universities in England, Scotland, Northern Ireland and the Republic of Ireland. The purpose of the study was to elicit their descriptions of how they come to know mathematics when they are researching. These descriptions I compared with an epistemological model

L. Burton†
Professor Emerita, University of Birmingham, United Kingdom

O. Skovsmose et al. (eds.), *University Science and Mathematics Education in Transition*,

(Burton, 1995) that used the critical literature in the philosophy, history and sociology of scientific ideas (for example, Bloor, 1991; Davis and Hersh, 1983; Ernest, 1991; Harding, 1991; Lakatos, 1976; Lerman, 1994; Restivo et al., 1993; Rose, 1994; Skovsmose, 1994). Although authors such as those above have critiqued positivist science and mathematics, and they, as well as others such as Evelyn Fox Keller (1985) have specifically addressed the epistemology of science, there was surprisingly little to be found which critically assessed the epistemology of mathematics. I hoped, therefore, to be able to make a contribution to this field. In mathematics, discussion of the learner's experience of mathematics rarely includes an epistemological critique. Indeed, it is a Western European perspective that "objective" knowledge of mathematics is independent of the knower and/or their culture but accessible through the power of clear reason (for a critique see Thayer-Bacon, 2000). One consequence of this is that learning is understood, mistakenly in my view, as a product only of pedagogical transmission.

The model I generated has five categories: person and cultural-social relatedness; aesthetics; intuition and insight; styles of thinking; and connectivities. I wished to see if these categories, taken together, provided a comprehensive picture, a model, of how these research mathematicians understood the epistemological practices through which they came to know mathematics. In exploring the categories, I was also interested in establishing in what ways they functioned for the mathematicians. For example, the category of "styles of thinking" was derived from a literature that identified two such, visual and analytic thinking. Were these all there were and would research mathematicians move freely between the two?

I was interested in how research mathematicians spoke about coming to know, in part to see how well the model, as an epistemological framework, described the processes of their coming to know research mathematics. But, because I believe that coming to know is learning, I was also concerned with exploring the wider relevance of the model to mathematics learning by less "sophisticated" learners. Would I find evidence to support Ernest (1998) that the learning of mathematics by mathematicians has both similarities with and differences from that of less sophisticated learners, and in what ways?

It must be emphasized that the focus of this study was not mathematics as a discipline. That is, I was not concentrating on the acquisition of knowledge objects that constitute the content of the subject and are to be found on the curricula of schools and universities. I wanted to know how those inside mathematics learn something new to them, that is the changing state of their knowing of mathematics. For me, this is the difference between epistemology as a theory of knowledge, and epistemology as a framework (a model) for explaining knowing.

The study was based upon data collected both by tape-recording and taking detailed notes of 64 face-to-face, and 6 telephone interviews, which averaged an hour and a half in length. Since I was interested in interviewing equal numbers of females and males, I first approached female mathematicians. When they expressed a willingness to join the study, I asked them to find a male "pair" preferably in their own institution. I did not dictate what constituted a "pair" but asked that they would indicate to me what governed their choice. In no case did this appear to present a

Table 7.1 Distribution of participants in the study

	Post-doc	Lecturer	Senior lecturer	Reader	Professor	Senior research officer	Research fellow
Females	1	19	7	3	3	1	1
Males	1	17	9	2	6		

problem to the research. The majority chose someone at a similar level of the hierarchy to themselves and/or in the same mathematical speciality. Three were marital pairs. Table 7.1 shows the distribution of the participants, by status and sex:

I interviewed in 22 universities in England, Scotland and Ireland, North and South. Prior to the interview, participants were provided with a one-page outline of the topics of interest. These related to their "history", their current research practices, and to how they came to know mathematics through solving research problems. They were offered the option, which none took, of deleting anything they did not wish to discuss. Notes of the interview were returned to them so that they could agree their contents, amend, change or delete. They were guaranteed anonymity. The data were entered into *Nud.ist*, a qualitative analytical computer-based tool and quantitative data were collected, computed and tabulated on *Excel*.

The interviews were discursive in style and the participants were free to introduce and explore issues of their choice. Although the subjects of the interviews were not teaching and learning, in the course of describing their experiences participants inevitably talked of themselves as learners and teachers. In this chapter, I am entirely concerned with a distinction that became apparent to me from the data, between how the mathematicians spoke of the *culture of mathematics* and how they spoke of the *mathematical culture*. By the culture of mathematics, I mean those aspects of mathematics that are recognizably discipline-related (such as the particular attitudes towards beauty, rigour, structure, etc.). Learning the importance of these aspects of the culture of mathematics is part of induction into the mathematics community of practice (Wenger, 1998) and therefore seen as a necessary part of learning mathematics even though strictly cultural. For example, beauty and rigour are two such highly valued aspects yet the mathematicians made clear that not only are they mal-defined but they are also contested. In part, they depend upon the area of the discipline in which the mathematician is working. So, it cannot be asserted that aspects of the culture of mathematics are intrinsic to the distcipline. But acquiring that culture, entering that community of practice, is a necessary part of learning to be a mathematician. The *mathematical culture*, on the other hand, is constituted through the socio-political attitudes, values and behaviours that dictate how mathematicians, and their students, experience mathematics in the settings of conferences, classrooms, tutorials, etc. So the mathematical culture is the environment in which the mathematics is encountered and learned and inevitably influences the culture of mathematics. I use data from the study to justify this distinction and to explore how the culture of mathematics and the mathematical culture operate to shape and influence the values, attitudes and beliefs of mathematics learners.

7.2 The Model and Its Relationship to the Cultural Distinction

The epistemological model at the focus of the study proved to be remarkably robust. The mathematicians spoke freely of its five interacting categories as they described how they came to know mathematics. Of the five categories, their beliefs about the personal and cultural/social-relatedness of mathematics not only spanned the conventional distinction between discovery/invention but, to my surprise, almost 10% of the mathematicians maintained that the discipline is a cultural artefact (Burton, 2004a, particularly Chap. 2; 1998, p. xi). Like Reuben Hersh (1998) they believed that mathematics must be understood as a human activity, a social phenomenon, part of human culture, historically evolved, and intelligible only in a social context. This belief has not, unfortunately, permeated the culture of mathematics where the somewhat unproductive debate between discovery and invention continues, as do beliefs in the objectivity, homogeneity and impersonality of mathematics. However, as researching mathematicians, whichever stance they took on the nature of mathematics as a product, they embraced heterogeneity in their research practices. So, while describing researching mathematics in culturally rich and heterogeneous terms, the mathematicians also positioned themselves within a cultural story about the discipline itself, which promoted the kinds of approaches within the mathematical culture that confirmed exclusivity, hierarchy and competition.

The discussion around aesthetics and intuition underlined the heterogeneity of approach of the mathematicians. While many of them subscribed to the importance of both aesthetics (61%) and intuition (83%), there were those who asserted either that there was no such thing, or that it was unimportant. There were others who adopted the diametrically opposite position of saying that it was the only thing about mathematics that was important! Efraim Fischbein, the only mathematics educator to make a major study of the role of intuition in mathematics education, similarly drew attention to this contradiction amongst mathematicians:

> According to Poincaré, no genuine creative activity is possible in science and in mathematics without intuition, while for Hahn (1956) intuition is mainly a source of misconceptions and should be eliminated from a serious scientific endeavour (Fischbein, 1987, p. 4).

Here are some examples from the interview data, first statements with respect to aesthetics and then to intuition[1]:

> There is the beauty of results if they are surprising or simple but give you an awful lot of information and turn out to be a key piece of the jigsaw.
> Beauty lifts it from how to why.
> Beauty doesn't matter. I have never seen a beautiful mathematical paper in my life.
> I don't think you would ever start anything without intuition.
> I don't think intuition plays a part.
> Intuition is not a word I would use.

[1] All quotes are taken directly from the data.

As can be seen, personal interpretation was very important in influencing how the mathematicians talked about aesthetics and intuition, as was whether they were pure or applied mathematicians or statisticians. But other factors, that I came to recognize as emanating from the mathematical culture, also played a part. For example, the put-down in the next quote, was not unusual:

> Beauty is in the eye of the beholder. What is beautiful is something that is simple and clear [...] Sadly, for some pure mathematicians, beauty is to put things in a symbolic language that only they understand.

These kinds of inter-community competitive statements reflect aspects of the mathematical culture that disrupt productive community relations. However, I also uncovered a resistance to using the word "intuition" to label something that some of them recognized but preferred to call either "insight" or "instinct". This resistance constituted, I believe, a reaction to the social stereotype female/intuitive, male/cognitive that perceived the word "intuition" as inappropriate to mathematics, which is valued as supremely cognitive, ergo male. Such gender-influenced positions are imported into the mathematical culture and help to explain the sense of isolation and "stranger-ness" reported by the female mathematicians (see below and Burton, 2004a, Chap. 9).

Until this work was done, only two mathematical thinking styles were recorded in the literature, the spatial and the analytic. I found a third, the classificatory. I also found a surprisingly high proportion (roughly 36%) of the mathematicians who used only one of the three exclusively. While this does not directly influence either the culture of mathematics, or the mathematical culture, it does have implications for the mathematical culture as it appeared, from the ways in which the mathematicians spoke, that those who could *not* think spatially saw themselves as slightly deficient. In other words, yet another hierarchy was at play.

The only category from the model on which all the mathematicians agreed, without exception, was on the importance of connectivities; this was whether they were inwardly-oriented within the different fields and methods of mathematics itself, or outwardly-oriented between mathematics and other disciplinary areas through real-world problems or multi-disciplinary activities. Although many of the mathematicians were extremely territorial about their discipline area and, as I have said, occasionally sarcastic about others, they all agreed that, "making connections is what you are trying to do" or expressed their interest "in connecting up areas". Whether important to them, personally, or not, they asserted that "ideas from one area often do have relevance to another area". The sadness about this agreement, for less sophisticated learners, is that mathematics is more frequently encountered as a disconnected and fragmentary set of facts and skills, rather than as the coherent whole that underlay how the mathematicians spoke. Additionally, the heterogeneity that became so apparent amongst the mathematicians is rarely recognized in the practices in either school or university teaching.

So, aspects of my epistemological model, particularly aesthetics, and intuition, contributed to the construction and maintenance of the mathematical culture, and other aspects, thinking styles and connectivities, had relevance to the ways in which the mathematical culture is performed. Although beliefs about the nature of the discipline, which I categorized as personal- and cultural/social-relatedness of mathematics, are particular to the individual, they help to define and/or reinforce the culture of mathematics and they do not arise without influence and experience, both performed within the mathematical culture.

7.3 The Mathematical Culture

The mathematicians, particularly the women but also some of the men, were very unhappy about attitudes, behaviours and values that I am calling, together, the mathematical culture. In particular, they drew attention to three: hierarchy, competition and isolation. In their discourse they demonstrated discomfort with many of the ways in which these three descriptors constrained and influenced their participation in the community. Etienne Wenger (1998, p. 4) describes knowing as "participating in the pursuit of [valued] enterprises", so any cultural constructions that inhibit or influence such participation must have implications for the well being of those community members so influenced. Indeed, we know from research done with postgraduate students (Herzig, 2002, 2004), that many abandon or express disaffection for their postgraduate studies for precisely these reasons. But even those who do not, still express their discomfort with the kinds of attitudes, behaviours and values being discussed here.

7.3.1 *Hierarchy*

Hierarchies are common in academe, often role-legitimated (in the UK, these roles are doctoral student, post-doc., research associate, lecturer, senior lecturer, reader, professor). But mathematicians appear to be skilled at developing many other hierarchies:

> The biggest hierarchy is the one that mathematicians are always putting themselves in. People rank each other.

A female mentioned this ranking with respect to a distinction drawn between "real" and others:

> I consider myself a mathematician because I lecture and research in mathematics and I suppose that by definition makes me a mathematician. But I have met people whom I consider to be 'real' mathematicians and I am not one of those people.

And Margaret Murray (2001) found much the same in her study of female mathematicians:

> She is careful to distinguish her work from 'real research' [...], which is the work of 'top mathematicians' such as 'Gauss and Archimedes and Newton', who reside 'somewhere near heaven'. (2001, p. 214)

However, even a senior mathematician with an established international reputation managed a similar differentiation:

> There is a level above the standard level at which we work. The top mathematicians can transcend these different fields and know enough. Often the dramatically new things are accepted at that level and then the smaller communities have to adapt themselves and, for example, learn the new language. Recently in complex dynamical systems very original ideas from probability theory have come in and have been accepted now by the world's top mathematicians and the complex dynamicists have to get on with it and accept it.

However, it was not only the labelling of people into hierarchies that was noticeable in the interviews. The mathematicians were also quick to invoke a hierarchy of judgment when speaking of the quality of published work. Work of quality was "significant"; work could also be "important" or "interesting".

> Interesting is a polite way of saying it hasn't done very much. Significant means that it has made a major contribution whereas important indicates a contribution.

The kiss of death was to declare something "trivial":

> Trivial –they aren't really saying anything different; it is a re-hash that might have been said before, and possibly better.

Apprenticeship, as a model for learning, is not discussed from a gender perspective despite the fact that many studies point to the gender implications of its use (Etzkowitz et al., 1992; Herzig, 2002). One particularly poignant way in which apprenticeship operates is to institutionalize hierarchies, ensuring that the learners are at the bottom of the hierarchy. Elena Nardi and Susan Steward identified some effects of this in schools:

> The hierarchy inherent in the above outlined elitist situation alters the nature of the classroom experience from one that focuses on catering for the individual learner's needs to one that focuses on establishing and assessing each learner's position in this hierarchy. The students express their alienation from this depersonalized, deterministic mathematical experience. (2003, p. 359)

So, the making of hierarchies is part of the mathematical culture. From school into university and then into the community of practice of mathematicians, one learns of the importance of hierarchies, not only within the actual discipline, but within the practices as well. But as an aspiring mathematician, one rapidly comes to appreciate the sacrosanct nature of the hierarchy and one's position within it. Establishing such hierarchies, invokes competitive practices since moving up and down the hierarchy is subject to the judgments of your peers and those in positions of power.

7.3.2 Competition

Pat Rogers pointed out: "lack of competition is not usually associated with the mathematics classroom" (1995, p. 184). A female mathematician said: "The competitiveness of mathematics is institutionalized from the beginning. So people are made to feel stupid if they don't achieve in mathematics". A male mathematician reflected: "There can be tremendously savage competition in our field and that is not something I feel particularly comfortable with". Thinking back to her undergraduate studies, a female said: "The thing I hated about doing mathematics was that the boys were so competitive". It is the institutionalization of this competition that has been observed as starting in school classrooms and continuing throughout undergraduate and postgraduate studies into the practices of the community of mathematicians.

I am labelling such competition as constituting part of the mathematical culture. It can be found in practices at both school and university level. Competitive practices emanate from many different assumptions but are implicitly part of institutionalized hierarchies. Pointing to a dominant view of mathematics as cognitively difficult, held by those within the discipline, Jo Boaler and Jim Greeno challenged the effects of this stance as "unusually narrow and ritualistic [...producing] environments in which students must surrender agency and thought in order to follow predetermined routines" (2000, p. 171). The result, observed in their research, was that "by emphasizing drill and practice of procedures, they [mathematicians] create a rite of passage that is attractive only for received knowers" (p. 190). Not only does this exclude the very type of learner who might be able to benefit from learning mathematics and, possibly, contribute to the development of the discipline, a learner interested in generating questions, ideas, and pursuing challenges, it also establishes a climate of discrimination and consequent competition. Furthermore, as Candia Morgan has shown, because assessment is also a social practice, it should be understood as "a process by which a student may gain or be denied access to particular forms of privilege or power" (2000, p. 231). Discriminatory outcomes of school assessment have been well charted by Barry Cooper and Máiréad Dunne (2000).

As a teacher, once one has adopted an assumption about some students having "ability", the next step is to classify those in the class into a hierarchy which is helped by a cultural assumption that mathematicians are "born not made". This was certainly not substantiated by the mathematicians in my study and was refuted by Margaret Murray (2001) who referred to "the myth of the mathematical life course" (p. 16). Of the 70 mathematicians in the study, only 18 (approximately 26%) spoke of early influences, 26 (approximately 37%) chose mathematics during their secondary school years and for 5 (approximately 7%), their choice was only confirmed at university. One said: "I certainly cannot say that I ever set out to be a mathematician. It was more a drifting into mathematics". Nonetheless, mathematics students at one of the UK's most prestigious universities commented in a study on the ways in which "the mathmo" was identified and nurtured compared with "the also-rans". One student said:

> I think the way in which [this university] carefully selects the best candidates for each subject and then fails to fully exploit them is bordering on the criminal [...] all staff should be reminded that the aim of higher education is to make students able to use their subject.
>
> Quoted in Burton, 2004a, p. 175.

I continued by noting that:

> 54% of the females and 25% of the males in that study responded to the question of what they would be doing next, now they had graduated, by saying something along the lines of "I intend to have nothing further to do with mathematics. I have found my experience here very demoralizing."
>
> Burton, 2004a, pp. 175–176

This, of course, was not the perspective of a "mathmo", one of whom (a male) identified "the opportunity to discourse with like-minded and intelligent people (both students and dons)" as the best thing about being at this particular prestigious university and said: "I intend doing part III, followed by a [university specified] Ph.D., in some area of applied maths to be determined". I would conjecture that his socialization into the mathematical culture had been extremely successfully achieved but at great cost to those not so nurtured and with little evidence that such identification can be achieved without invoking stereotypes.

My research data show that competition features within the community, even though many mathematicians express dislike of it. Research of mine (for example, Burton, 2004b), and that of others, has also recorded the dislike of competition expressed by school students. There can be little doubt that competition flourishes inside mathematics and that its effects are not welcome to many. One of these effects is to isolate, particularly women, so that they feel alienated from the discipline and from the mathematics community of which, as mathematicians, they are entitled to be a member.

7.3.3 Isolation

Many female mathematicians spoke about the sense of isolation that they experienced within the world of mathematics. Sometimes this was because fewer women study the discipline. This was the case for a female mathematician who said:

> One thing I have learnt from my own experience of undergraduate and postgraduate work was the isolating experience of being the only one.

But this kind of personal isolation was not the only kind to which they referred. One female mathematician explained her movement into statistics because:

> A statistical problem is never as isolated, it has all sorts of things impinging on it; it is always in a context. Whilst mathematics cannot be given a context, it can be very isolated.

A graduating female student who had decided to leave the discipline explained: "This is partly why I'm leaving – I love the maths but hate the isolating way of learning". And a number of mathematicians cited collaborative work as being a good way to overcome isolation.

> If you are collaborating with mathematicians, you feel much less isolated. Because mathematics is an isolating experience, you often feel burnt out, dried up, you have no skills and everyone else does. When you start working with someone else, you discover that there are things that you can do, perhaps better than they can, and things that they can do that you can't.

This is, of course, one of the reasons for collaborative group work given by those mathematics educators who advocate it in mathematics classrooms. But in the above quotation, I would like to draw attention to the intensity of the feelings (*burnt out, dried up, no skills*) expressed by this female mathematician and the degree to which they coincide with descriptions to be found in the literature from research with school pupils as well as with university staff and students. In one of my own studies with 16–18 year old pupils specializing in Advanced Level mathematics, I noted:

> The students wanted their teachers to facilitate discussion, teamwork, a light-hearted approach, a relaxed classroom environment where you are not afraid of making errors. They told anecdotes to support the fact that they did not want to be put down, persistently asked the same questions, made to look a fool or feel patronized, be put into a position where others laugh at you, be thrown 'in at the deep end'. (Burton, 2001, p. 67)

These are among the features of the mathematical culture to which I have been drawing attention. But I would dispute that they are an inevitable part of the mathematical culture. They are constructs that have come about for particular sociocultural reasons and should be changed because they no longer match or contribute anything positive to the conditions of practice of mathematics. On the contrary, they create an environment which is only conducive to some and yet which is reproduced year by year as past students carry the practices with them out into the world of work, for example teaching in schools. While students at every level persistently call for collaboration, a discursive environment in the classroom, freedom to make errors without criticism or loss of respect, they equally persistently report experiencing competition, being laughed at or made to feel a fool, being put down or feeling isolated. These features of the mathematical culture are both widespread and destructive. I believe that they help to explain the loss of interest in mathematics as a discipline of study in the university and the composition of the student community who enter mathematics, predominantly male and of the dominant culture. Not only do such negative behaviours affect those who experience them, however, but they also impact upon the discipline itself. I now discuss the culture of mathematics, as perceived by mathematicians.

7.4 The Culture of Mathematics

In searching the database for the ways in which the mathematicians used signifiers of the culture of mathematics, I found the following: structure (56% of the mathematicians discussed this feature of mathematics), rigour (41%), beauty (37%), pattern (31%). Other aspects that were raised were power (19%), simplicity (12%), symmetry (10%) and conciseness (6%). Beauty was a component of the epistemological model and, as such, was discussed above. The latter four were each introduced by fewer than 20% of the mathematicians in the study, so I am not regarding them as major to this discussion of the culture of mathematics; I focus on structure, rigour and pattern. Of course it is important to recognize that these terms are not necessarily used in identical ways and I certainly found differences between pure and applied mathematicians in their approach, for example, to rigour. However, for the purpose of the argument that I am making here, that is that the mathematical culture frequently dictates the ways in which the culture of mathematics is described and used, I think it is sufficient to demonstrate with structure, rigour and pattern. Both "structure" and "pattern" were discussed in language that was free of judgmental, hierarchical or competitive statements but that can, nonetheless, be traced back to the mathematical culture. "Rigour" is heavily imbued with judgmental statements.

7.4.1 Structure

The participants in the study appeared to agree on the centrality of structure to mathematics. That is, from what was said, structure was a way of understanding what makes something mathematical. Structure, therefore, can be said to constitute part of the culture of mathematics and is a part which is apparently unaffected by the mathematical culture. None of the statements that were made about structure were contentious in the ways already described with respect to "intuition" and "aesthetics". The descriptions that the mathematicians offered, therefore, appeared to be located in the culture of mathematics without the influence of the mathematical culture. They could be used to provide an entry, for learners, into the meaning and function of structure. Here are four:

> You can take a variety of apparently different problems and abstract out the features that are the same, common across them, rather than the features which differentiate them, so that you can see that, in general, there is a structure which can usefully be seen to apply in different contexts.
>
> I also might have a structural view, putting things in relation to one another, so that I can think of these things in relation to one another, this definition, these properties. So the answers must be able to be put into a bigger global structure.
>
> I would like to be able to say that certain things are true because these structures are underlying them. It is not just an application, it is also an explanation.
>
> Mathematics is, for me, understanding structures, how to solve problems that have been modelled mathematically.

Despite being contentious-free, it is important to recognize that such uniformity of approach to structure is deeply rooted in the practices of the mathematical community. It is not surprising therefore, that the mathematicians agree about structure since such agreement is part of the pattern of their membership in the mathematics community of practice. "In perceiving themselves as mathematicians, they use a similar set of symbols to give meaning to their experiences" (Meaney, 2005, p. 115). Nonetheless, when the mathematicians came to discuss "rigour" within the culture of mathematics, their descriptions were more transparently connected to the mathematical culture.

7.4.2 Rigour

Philip Davis and Reuben Hersh described the *Ideal Mathematician* pointing to "the discrepancy between the actual work and activity of the mathematician and his own perception of his work and activity" (1983, p. 34). They continued: "He rests his faith on rigorous proof ... Yet he is able to give no coherent explanation of what is meant by rigor, or what is required to make a proof rigorous" (p. 34). Of course I found similar discrepancies in the discourse of the mathematicians that I interviewed, and have written about them elsewhere (Burton, 1999, 2004a, 2005). Here I draw attention to the linguistic formulations used by the mathematicians about 'rigour' and their implications.

> Of course some proofs are not rigorous because the writers don't have the necessary expertise but, more interesting, sometimes we just don't have the mathematical structures needed to turn the heuristic arguments into rigorous proof.

This mathematician is demonstrating not only that the colleague about whose work they write is deficient, but that they themselves are not. In other words, power play through the making of judgments, which was shown above, is at work in this quotation. Another mathematician was also judgmental about the person when speaking of "rigour":

> I am a little ambivalent towards rigour in that it is certainly necessary to attend to the quality of maths but people tend to concentrate on rigour at the expense of explanation.

Two mathematicians understood the social nature of "rigour" but one, again, focussed on the person:

> There is a social thing, a very clear idea of what a correct proof is and that is rigour as I understand it. Something isn't part of mathematics until proved that way.
>
> Rigour is the ability to re-construct the argument. But of course with different people, different amounts of detail are adequate.

Traces of the jockeying for position between pure and applied mathematicians and their consequent positioning also showed themselves:

> That is what rigour is – the ability to convince someone else of truth. I am demanding that a paper convinces me that it is true.
>
> But I have never experienced writing mathematics using a rigorous approach. I have used mathematics but always in the context of physics where the emphasis was not put on the minutiae of rigour but on demonstrating that the maths was right as far as possible, it made sense – the application was very important.

Finally, one mathematician gave a form of definition but one that is so indeterminate as not to be very helpful in elucidating "rigour":

> Rigour is avoiding hand-waving, proving every argument in detail; "it's obvious" – that is not acceptable [...] Rigour is adding understanding because of the detail of the proof and also when you prove something non-rigorously you have the impression that you have proved it but you haven't really.

Not only can it be seen that the mathematicians could not give a clear statement of what constitutes "rigour" but their discourse displayed other features of the mathematical culture that are unacceptable, the implicit use of power, the making of judgments on the basis of personal preference, and the institutionalizing of hierarchies.

7.4.3 *Pattern*

Although pattern searching was agreed to be a central activity by the mathematicians, to find a pattern provides a warrant for proof and proof is cultural. As Imre Lakatos put it:

> …informal, quasi-empirical, mathematics does not grow through a monotonous increase of the number of indubitably established theorems but through the incessant improvement of guesses by speculation and criticism. (1976, p. 5)

Or, in the words of David Bloor:

> At any given time mathematics proceeds by, and is grounded in, what its practitioners take for granted. There are no foundations other than social ones. (1991, p. 153)

One mathematician explained the process:

> One is finding what the pattern is, why things are happening. By the time you have understood that, it is simply a case of translating the picture into words and symbols. The final written proof is a lot harder to understand than the proof that I originally had which let me know that it worked which was essentially the picture.

Lying behind these words is the gap between what is described pejoratively in the discipline as "proof by demonstration" and a formal, abstract proof. This gap is frequently the butt of comments by pure mathematicians for whom the formal proof is *de rigueur*. The mathematician above is alluding to the gap as well as to the behaviours that often accompany its identification. Of course "simply a case" is not simple at all and many of the mathematicians made reference to the difficulties that present themselves when this "translation" is being attempted. So while pattern searching is accepted in the discipline as a natural, and an important mathematical activity, it has cultural repercussions, which are influenced by those aspects of the mathematical culture, which I discussed above.

7.5 Relationship with Epistemology and Consequent Pedagogy

It has been the argument in this chapter that cultural influences have created and continue to maintain the mathematical culture which is inhabited by mathematicians and, consequently, by the students that they teach. Further, this mathematical culture is not benign. On the contrary, it has elements, which cause intense discomfort both to students and to staff and, in some cases, contribute to individuals abandoning the discipline. That such elements are neither necessary nor acceptable is clear. But they will not be changed without a very serious consideration of how mathematics is experienced in schools and universities. As is so often the case, those in positions of power, and consequent influence, need to acknowledge and reflect upon the negative effects of the mathematical culture on the discipline itself, that is, on the culture of mathematics, as well as on how learners experience the discipline; then we need, actively, to set out to make changes.

However, it is not all negative. A most important finding of my study was that how the mathematicians, themselves, came to learn in the course of their research had very close links to the pedagogical style, which has, for quite some time, been recommended by mathematics educators. Mathematicians themselves, in their research practices, were engaged in personal and socio-cultural activities that acknowledged the interplay of emotion and cognition, depended upon a discourse community, valued heterogeneity, and reflected upon the complex ways in which mathematics inter-connects, both internally and externally. (See Roth's definition of knowledge, Chap. 6). Far from the stereotypic image that portrays the mathematician as a loner, working in isolation, the mathematicians I interviewed were working collaboratively, were agents of their own learning and were engaged in the messy process of authoring their work through the complicated passage from early thoughts to ultimate acceptance by the community of peers. It is precisely this process for which students call when they are critical of the transmissive, "banking" concept of education (Freire, 1971), that they most frequently encounter in classrooms. For example, a number of students, in the study to which I have already referred, remarked upon poor experiences of teaching:

> The lectures are in my opinion very bad and are generally an exercise in dictation.
>
> I would remove/penalize bad lecturers. Currently it seems to be that all faculty members must lecture irrespective of how bad they are.
>
> Techniques of algebra/DEs are being taught with no context for why these are there or what the fundamental mathematical/physical objects are like.
>
> There are different thinking styles but unfortunately if you cannot understand that of the lecturer then they generally will not give an alternative perspective; if the supervisor does likewise then you have absolutely no hope with that course.
>
> My impression is that the average person in the average lecture is totally lost within ten minutes of it starting and spends much of the rest of it copying down the symbols with no idea of what they mean. Arguably we'd lose little if we were given a book of notes in advance, and [...] I'd much prefer to be given such a book and then for the lecture time to be spent getting the meaning, not the technical manipulations, across.

A female student had been convinced of a sex difference in approach, which left her blaming herself rather than the style of teaching and the epistemological view of the subject matter:

> I think the way Maths is taught is for the way males think – they give you facts and you go away, chew on it and come up with the missing bits and the solution. Female minds do not work like that. If I'm presented with the bare outline of something and it is not explained well, there is no way I can fill it out, and more time will make no difference.

The sadness is that the open, questing, negotiated enquiry style that the mathematicians use when collaborating on their research represents the opposite of the closed, tired, dry ways in which they teach. Their embrace of heterogeneity in styles, approaches and methods was in stark contrast to their teaching of one "right" way, a single route and a closed answer. From many research studies done in schools, we know that the enquiry style is one which is successful and motivating with learners (for example, Alrø and Skovsmose, 2002; Boaler, 2002). But, unfortunately, it has failed to permeate into many university settings (although Aalborg and Roskilde University in Denmark are outstanding counter-examples. See Kolmos', Chap. 12). Again, part of the explanation for this is the heavy influence of the mathematical culture together with the widespread attitude that, in universities, teaching is not important. It is a major irony that, while the mathematicians learn in an enquiry style, they revert to traditional teaching practices with their students. By this means, they carry with them into their classrooms some of the worst features of the mathematical culture, in particular the reification of hierarchies in learning, competitive practices and isolation.

7.6 Drawing Conclusions

While the mathematicians that I interviewed spoke very positively of the excitement, satisfaction and, indeed, euphoria that they experienced from their engagement with mathematics, as has been seen from the discussion above, this was often associated with some very negative feelings. And we know that students reflect these negative feelings to such a degree that many who could study mathematics, are failing to choose it or, like some of the students I interviewed, having chosen it, then reject it. Whether we are concerned about the loss of these potential mathematics students, or of the mathematicians they might have become, or we are more simply concerned that the social experience and reputation of mathematics is of a discipline which is inaccessible, unattractive and difficult, much negativity can be explained, I believe, by focussing, as I have done, on the mathematical culture. It is to the mathematical culture that I think we should turn when trying to explain and change rejection of the discipline and failure to succeed at it. I cannot feel at ease within a discipline that is described in terms of the misapplication of power through the maintenance of hierarchies, competitive practices and isolation of those who do not appear to match some mythical prescription of a mathematician. Nor do I feel at ease with the unthinking acceptance of those aspects of the culture of mathematics that have themselves been affected by the mathematical culture.

The evidence of heterogeneity was, for me, one of the most positive features to emerge from the research. Building in acceptance, even celebration, of heterogeneity would not only change the experiences of those within the discipline, student, researcher and mathematician, but would open the discipline itself to the promotion of variation and to ways of encouraging learning that valued differences and sought the means to compare and contrast them. These are the means that many of the mathematicians I interviewed are pursuing in their research practices but, unfortunately, are not to be found in classrooms. The next major question is how to shift an acceptance of these research practices into the classroom so that learners can feel, behave and be accepted as researchers.

References

Alrø, H. & Skovsmose, O. (2002). *Dialogue and Learning in Mathematics Education: Intention, Reflection, Critique*. Dordrecht: Kluwer.

Bloor, D. (1991). *Knowledge and Social Imagery* (2nd Ed.). London: University of Chicago Press.

Boaler, J. (2002). *Experiencing School Mathematics: Traditional and Reform Approaches to Teaching and Their Impact on Student Learning*. (Revised and Expanded Ed.). Mahwah, NJ: Lawrence Erlbaum.

Boaler, J. & Greeno, J.G. (2000). Identity, Agency, and Knowing in Mathematics Worlds. In J. Boaler (Ed.), *Multiple Perspectives on Mathematics Teaching and Learning* (pp. 171–200). Westport CT: Ablex.

Burton, L. (1995). Moving Towards a Feminist Epistemology of Mathematics. *Educational Studies in Mathematics*, 28, 275–291.

Burton, L. (1999). Epistemology, Philosophy and Pedagogy of Mathematics. *Philosophy of Mathematics Education Journal*, November. Retrieved May 2008. http://www.ex.ac.uk/Pernest/pome12/default.htm

Burton, L. (2001). Mathematics? No Thanks – Choosing and then Rejecting Mathematics. In S. Johnston-Wilder (Ed.), *Proceedings of a National Day Conference on Key Stage 3 Mathematics Teachers: The Current Situation, Initiatives and Visions* (pp. 58–71). Milton Keynes, UK: The Open University.

Burton, L. (2004a). *Mathematicians as Enquirers – Learning About Learning Mathematics*. Dordrecht: Kluwer.

Burton, L. (2004b). "Confidence is Everything" – Perspectives of Teachers and Students on Learning Mathematics. *Journal of Mathematics Teacher Education*, 7, 357–381.

Burton, L. (2005). Mathematicians' Narratives About Mathematics – and Their Relationship to its Learning. In B. van Kerkhove & J.P. van Bendegem (Eds.), *Perspectives on Mathematical Practices. Bringing Together Philosophy of Mathematics, Sociology of Mathematics, and Mathematics Education*. (pp. 165–174). Dordrecht: Kluwer.

Cooper, B. & Dunne, M. (2000). *Assessing Children's Mathematical Knowledge: Social Class, Sex and Problem-Solving*. Buckingham, UK: Open University Press.

Davis, P.J. & Hersh, R. (1983). *The Mathematical Experience*. Harmondsworth, Mx.: Penguin Books.

Ernest, P. (1991). *The Philosophy of Mathematics Education*. London: McGraw-Hill.

Ernest, P. (1998). The Culture of the Mathematics Classroom and the Relations Between Personal and Public Knowledge: An Epistemological Perspective. In F. Seeger, J. Voigt & U. Waschescio (Eds.), *The Culture of the Mathematics Classroom*. Cambridge, UK: Cambridge University Press.

Etzkowitz, H., Kennelgor, C., Neuschartz, M. & Uzzi, B. (1992). Athena Unbound: Barriers to Women in Academic Science and Engineering. *Science and Public Policy*, 19(3), 157–179.

Fischbein, E. (1987). *Intuition and Science and Mathematics – An Educational Approach*. Dordrecht: D. Reidel.

Freire, P. (1971). *Pedagogy of the Oppressed*. New York: Herder & Herder.

Harding, S. (1991). *Whose Science? Whose Knowledge*? Milton Keynes: Open University Press.

Hersh, R. (1998). *What is Mathematics, Really?* London: Vintage Books.

Herzig, A.H. (2002). Where Have all the Students Gone? Participation of Doctoral Students in Authentic Mathematical Activity as a Necessary Condition for Persistence Toward the Ph.D. *Educational Studies in Mathematics*, 50(2), 177–212.

Herzig, A.H. (2004). "Slaughtering this Beautiful Math": Graduate Women Choosing and Leaving Mathematics. *Gender and Education*, 16(3), 379–395.

Keller, E.F. (1985). *Reflections on Gender and Science*. New Haven, CT: Yale University Press.

Lakatos, I. (1976). *Proofs and Refutations*. Cambridge, UK: Cambridge University Press.

Lerman, S. (Ed.) (1994). *Cultural Perspectives on the Mathematics Classroom*. Dordrecht: Kluwer.

Meaney, T. (2005). Mathematics as Text. In A. Chronaki & I.M. Christiansen (Eds.), *Challenging Perspectives on Mathematics Classroom Communication*. Westport Conn: Information Age.

Morgan, C. (2000). Better Assessment in Mathematics Education? A social Perspective. In J. Boaler (Ed.), *Multiple Perspectives on Mathematics Teaching and Learning* (pp. 225–242). Westport, CT: Ablex.

Murray, M.A.M. (2001). *Women Becoming Mathematicians: Creating a Professional Identity in Post-World War II America*. Cambridge, MA: MIT Press.

Nardi, E. & Steward, S. (2003). Is Mathematics T.I.R.E.D.? A Profile of Quiet Disaffection in the Secondary Mathematics Classroom. *British Educational Research Journal*, 29(3), 345–367.

Restivo, S., van Bendegem, J.P. & Fischer, R. (Eds.) (1993). *Math Worlds*. New York: SUNY Press.

Rogers, P. (1995). Putting Theory into Practice. In P. Rogers & G. Kaiser (Eds.), *Equity in Mathematics Education: Influences of Feminism and Culture* (pp. 175–185). London: Falmer Press.

Rose, H. (1994). *Love, Power and Knowledge*, Cambridge, UK: Polity Press.

Skovsmose, O. (1994). *Towards a Philosophy of Critical Mathematics Education*. Dordrecht: Kluwer.

Thayer-Bacon, B.J. (2000). *Transforming Critical Thinking*. New York: Teachers College Press.

Wenger, E. (1998). *Communities of Practice: Learning, Meaning and Identity*. Cambridge, UK: Cambridge University Press

Chapter 8
Becoming a Teaching Scholar

Concepts of "Good" Teaching Among Science Teachers Participating in Training Programmes

Søren Kruse, Kirsten Nielsen, and Rie Troelsen

University teaching is going through a professionalization process as part of a change or reform of the teaching and learning culture. This process builds on the notion that it is no longer sufficient for a university employee to be an excellent researcher; she must also be an excellent teacher. The relation and interaction between research and teaching is of great importance, but it is documented that being a good researcher does not necessarily correlate with being a good teacher (Feldman, 1987; Hattie and Marsh, 1996, 2002).

In higher education there is a reinterpretation of academic scholarship including educational/teaching scholarship (Boyer, 1990; Entwistle, 2003; Entwistle et al., 2000; Trigwell et al., 2000). The primary argument for scholarship is that we share common knowledge about good teaching, which is promoted by the community of scholars communicating their concepts, findings, methods, and principles. The idea is that educational scholarship promotes high quality teaching (Shulmann, 1993). So if the aim of teaching is "to make student learning possible" (Ramsden, 1992), the aim of scholarly teaching is "to make it transparent how we have made learning possible" (Healey, 2000, p. 171). In other words, we need theories about teaching. Andresen (2000) points out that educational scholarship is not just about describing *what, how and why*, but also a term of recommendation or challenge. We are engaged in promoting a set of intellectual values, so teaching scholarship is also a moral discourse. Scholars' communication is also a negotiation of status and power in the establishment of teaching. Boyer (1990) points out that the scholarship of education cannot be isolated from academic scholarship in general. The aim of professionalism is to change the academic culture towards including knowledge about teaching and learning in academic scholarship. Staff development programmes aim to promote this educational scholarship or professionalism. Thus today, university teachers are being educated as educators. In an international context, this professionalization has become one of the most important parameters in the further development of research-based education (Felten and Pingree, 2003; Lauersen, 2003). But why is this professionalization process taking place now?

The what, how and why questions of teaching are crucial internal parameters with regard to both the total curriculum and the actual teaching situation.

S. Kruse, K. Nielsen, and C. Winsløw
Danish School of Education, Aarhus, and University of Southern Denmark

O. Skovsmose et al. (eds.), *University Science and Mathematics Education in Transition*,

Increasingly, every teacher, every departmental teacher team, and every institutional authority with responsibility for and influence on teaching faces substantial didactical[1] tasks. They have to choose what (content) and how (form) they are going to teach, and they must have good reasons for their choices (aim).

In a publication like this, which aims at understanding some of the multiple aspects and dimensions of the transition of science and mathematics education in the current information society, it is important that teachers are also taken into account. It is essential that the teacher recognizes the importance and relevance of implementing the development of competencies, ethics, theory of science or meta-reflection in their teaching, and finds it realistic to actually teach in that particular way. In that way, the challenges and possibilities for university science and mathe matics education depend on teachers acknowledging the challenges and seeing the possibilities. Changing the culture of teaching and learning is by all means a complex process that depends on both the individual teacher's ability to reflect upon their beliefs, concepts and approach, but also on the patterns of social relations and the academic culture related to teaching and learning (Trowler and Cooper, 2002). Science teachers especially face the challenge of developing professionalism or educational scholarship (Cope and Prosser, 2005; Healey, 2000).

For 6 years we have been involved in the development of training programmes in Learning and Teaching in Higher Education (LTHE programmes) for new staff members at the Royal Veterinarian and Agricultural University in Copenhagen. In the period 2001–2003 we conducted an investigation among science teachers attending such training programmes for novice academic staff at science and technical faculties at eight universities in Denmark.

Based on this investigation and our experience as teacher trainers we focus, in this chapter, on what the teachers themselves consider to be "good teaching." After presenting the results of our investigation of the teachers' views on good teaching, two interrelated discussions follow:

- The complexity of good teaching. How can teachers' views on good teaching be interpreted compared with relevant research on teachers' beliefs or concepts of teaching and learning?
- Implications for teacher training. What kind of consequences and demands do these views make on the organization of future LTHE programmes? How do we promote educational professionalism and scholarship?

8.1 Lessons in Good Teaching: Inspiration from the Literature

There is no such thing as a recipe for good teaching. If there were, teaching would be as simple as following that recipe. The aim of tertiary teaching is that the students learn something, at first, so that they are able to pass their exams, but also – and

[1] The term *didactical* is here – and throughout the article – not to be understood in an Anglo-Saxon sense, but according to a German/Nordic tradition as regarding issues related to the aims, goals, methods, content and evaluation of teaching and learning.

this is far more important – so that they will acquire knowledge that will be helpful in their future professional life. This is widely agreed upon. Hence, the disagreement does not concern the aim of tertiary teaching, but to a large extent the means to reach this goal. (See Roth's discussion, Chap. 6).

8.1.1 Effective University Teaching

The literature on good university teaching deals primarily with the question of effective teaching and is influenced by research using a phenomenographic approach (Marton, 1981). Therefore, we have some very general knowledge about what effective teaching is. Biggs (1999) uses the concept of alignment as a criterion for effective teaching, stating that there has to be a coherent orientation of teaching activities, learning activities, and exams towards the explicit curricular aims. Biggs (1999) characterises good teaching practice in four points:

- A well-structured knowledge base
- An appropriate motivational context (alignment)
- Learning activity, including interaction with others
- Making self-monitoring possible.

On the basis of reviews of research Ramsden (1992) has formulated six principles for effective teaching:

1. Explanations of content and the encouragement of student involvement.
2. Interest and respect for the students' learning.
3. Appropriate feedback on students' learning activities.
4. Clear aims and intellectual challenges.
5. Student autonomy, control and active engagement in the learning process.
6. Learning from the student.

8.1.2 Teachers' Concepts of Teaching and Learning

The description of the concept of university teaching or ways of thinking tends to be dichotomized in a *teaching centred* conception and a *learner centred* conception, where the latter is preferred (McManus, 2001). Others describe differences in teachers' concepts as different in degrees of complexity (Entwistle and Walker, 2002; Ho, 2000; Trigwell and Prosser, 1996). Entwistle et al. (2000) state that only teachers with sophisticated and complex concepts of teaching and learning have the expanded awareness of learning and teaching that makes a strategic alertness to classroom events possible, so that they can profit from their experience in classrooms.

One of the consequences of the paradigm shift from teaching to learning is the focus on teachers' conceptions of teaching and learning in staff development in an attempt to reform teaching and learning culture at universities towards a learner centred approach. Some studies have shown that science teachers especially tend to be more teaching centred in their conception of teaching and learning than university teachers in other areas (Prosser et al., 2005).

Kane et al. (2002) conducted a review of the research concerning teachers' beliefs or concepts of teaching, including a study of individual teachers, a qualitative analysis of groups (10–25) of teachers, and large-scale quantitative investigations. They concluded that there is great confusion about the concepts of teachers' beliefs and intentions. First, the problem is that most of the studies only tell half the story about teaching because they do not take into account the difference between the *theory espoused* in an interview or questionnaire and the *theory used* in teaching practice. Trigwell and Prosser (1996) also make a distinction between teachers' *concepts (belief or narrative)* of teaching and learning and the teachers' *approach* to teaching in practice, and they state that it is the quality of the teaching approach that can improve students' learning outcome. Second, the dichotomy between a *teaching centred* conception and a *learner centred* conception cannot capture the complexity of teaching: when and why student activity is relevant is not considered, and the academic discipline and the nature of the knowledge content is not reflected (Kane et al., 2002).

In order to prevent a gap between teachers' concepts of teaching and their approach to teaching, Trigwell and Shane (2004) suggest a praxis-oriented model of teaching scholarship as an activity. The scholarly knowledge is related to the sensibility towards the resonance between the teacher's knowledge and the student's learning process and the learning outcome for both student and teacher. They anticipate that the activities of teaching and learning scholarship are based on a partnership between teachers and students about learning instead of a relation based on the transfer of knowledge. Their model includes three interrelated components in a teaching and learning activity system:

1. Knowledge about teaching and learning in a specific topic or discipline.
2. Practice, including the planning, conduct and evaluation of teaching and learning processes.
3. The documentation of students' learning outcome.

The communication and reflection on the interrelation between teaching knowledge, practice and the documented learning outcome is a second order activity producing teaching and learning scholarship.

8.1.3 Learning to Teach

The idea underlying the LTHE programmes is to teach new academic staff how to teach. An underlying issue is whether it is possible to teach someone (the new academic staff) to teach, so that they learn to teach someone else (the students) so that they learn. Thereby, we distinguish between "learning to teach" and "teaching in teaching" (Kruse, 2001). On the one hand, it is stated that teaching in teaching is founded on an educational rationale: that it is possible to gain insight into common knowledge about teaching quality (Dale, 1989). On the other hand, it is stated that teaching is learned by practicing (Lortie, 1975; Schön, 1987). Martin and Lueckenhausen (2005) have shown that one third of university teachers change their teaching style during a semester. Half of these changes are not planned but a result of new experiences in teaching situations. Cope and Prosser (2005) have concluded that an important part of didactical knowledge is learning from students' learning experiences, being able to see learning situations through the students' eyes. In the reinterpretation of scholarship they find a relationship between practicing and developing scholarship: "If teaching is to be seen as a form of scholarship, then the practice of teaching must be seen as giving rise to new knowledge" (Schön, 1995, p. 31; see also Trigwell et al., 2000). It is evident that experienced teachers get better results than novice teachers, but after 3–5 years experience alone has no significant effect. So experience in teaching is necessary, but not sufficient for developing high quality teaching. The conclusion is that good teaching is promoted by a combination of conceptual reflection and experience through experiments (Entwistle, 2003; Martin and Ramsden, 1993).

It is documented that some staff training programmes can have a positive effect on the teaching and learning outcome. (Biggs, 1999; Gibbs and Coffey, 2004; Trigwell et al., 1999). Traditional staff development activities in tertiary education are largely concerned with teaching skills and methods, including, for example, how to improve lectures or how to conduct tutorials. Our investigation confirms that new academic staff members expect that programmes will provide them with descriptive methods or recipes. Experiences from many programmes have suggested otherwise, namely, that many participants defend their usual methods (Ho, 2000; Trigwell and Prosser, 1996). Reviews by Ho (2000) and Entwistle et al. (2000) show that traditional staff development programmes with focus on methods and recipes do not lead teachers to change their teaching practice. They have to change their concepts of teaching and learn to make fundamental changes in their way of teaching.

8.2 The University Teacher Project

Funded by the Centre for Educational Development in University Science, we conducted a project entitled *Professionalization of the university teacher* in 2001–2003 (Kruse et al., 2004). The background for the project was the set of LTHE programmes for new academic staff at science and technical faculties at Danish universities and how new academic staff considered themselves in their new role as teachers.

8.2.1 The Aim of the Project: Main Questions

The LTHE programmes for new academic staff reflect a political aspiration to continually raise the standards of teaching at Danish universities. Based on this aspiration, the overall aim of the project was to identify the potential and problems that new academic staff encounter in becoming professional teachers. To reach this goal we have conducted three studies:

- Theoretical investigative work on the background for this professionalization of teachers
- A didactical analysis of the LTHE programmes
- An empirical investigation among new academic staff in science and technology

Here we will mainly focus on the third study of the project. The empirical investigation is aimed at describing how the didactical framework set up by the institutions as being utilized is perceived by new academic staff and how they perceive the framework per se.

Also relevant to the discussion of the framework of LTHE programmes and the new academic staff's perception of what they might learn from attending such programmes, are questions about how the concept of "good teaching" is conceived by new academic staff. Accordingly, it is also an aim of the empirical investigation to identify views (the beliefs and concepts) of new academic staff toward good teaching and how they strive to produce good teaching. The identification of new academic staff's understanding of good teaching gives rise to a debate about the notion of "good teaching" at universities that the LTHE programmes are and should be communicating. Kane et al.'s (2002) review of the research on teachers' beliefs and concepts of teaching and learning makes a clear statement that this only tells half the story about teaching. There might be significant differences between the *espoused theory* (beliefs, concepts, narratives) and the *theory used* by teachers in their teaching practice. To investigate the theory in practice is very difficult to observe, if possible at all. But observation of teaching and teaching documents can give information about the correlation between teachers' concepts and their approach to teaching. In our study we have only asked the teachers about their views on good teaching, which limits the interpretations to saying something about their beliefs and concepts of good teaching and not about their teaching practice or approach to teaching.

8.2.2 Methods Used in the University Teacher Project

Throughout the project we have used a wide range of methods to reach an understanding of the problems and potentials encountered by new academic staff in their professionalization as teachers: theoretical considerations, didactical analyses and empirical investigations. Since this chapter focuses on the attitudes of new academic staff toward the concept of good teaching, we will at this point discuss the specific methods used in obtaining data with regard to this issue.

In order to identify beliefs and concepts on good teaching in university settings, we have conducted both quantitative and qualitative investigations: a questionnaire and an interview study.

The aim of the questionnaire was to consider our research questions in a quantitative manner. For this reason the questionnaire consists of four sections that correspond to the main objectives of the project:

- Background information
- On your teaching and role as a teacher
- Views on tertiary education and teaching in general
- On your experiences with Learning and Teaching in Higher Education programmes.

The participants in the quantitative study were new academic staff in science and technology at Danish Universities who had either participated in a LTHE programme the year before this study or were participating in a LTHE programme at the time of the study. Almost 130 new academic staff (the total number of persons who attended LTHE programmes either in the year 2001/2002 or 2002/2003) were asked to join the study and 64 questionnaires were returned. The returned questionnaires were processed using SPSS. The questionnaire study was conducted in the winter of 2002/2003.

As a supplement, an interview study was conducted with the intention of varying and clarifying the quantitative data. Two group interviews were conducted in the early spring of 2003 with three participants each from a total of five different universities. The participants had, in other words, experienced five different LTHE programmes. Each group interview was structured around three themes: "Presentation of the participants," "On your experience with a Learning and Teaching in Higher Education programme," and "Learning and Teaching in Higher Education programmes of the future (what should they consist of?)." The interviews were audio taped and relevant parts were transcribed.

8.2.3 Staff Development Programmes in Learning and Teaching in Higher Education

Pedagogical education of scientific staff is internationally recognized as the most important strategy for living up to the above-mentioned expectations for tertiary teaching (Berendt, 1998). In Denmark, the ministerial demand for LTHE programmes reflects this recognition. The Danish University Act of 1993 stated that all new academic staff should be offered supervision by an experienced colleague and should receive a written statement about their pedagogical qualifications. Over a 10-year period all Danish universities have developed educational programmes for their new academic staff.

As mentioned above, we have conducted a didactical analysis as part of the overall project. In the didactical analysis we analyzed the guidelines for teaching in teaching, which are set out by the universities. We have analyzed the formal curricula of the LTHE programmes in order to see what kind of values are being marked as *good* LTHE programmes (aims, objectives and content of the programmes) and as *effective* LTHE programmes (form and methods of the programmes).

Analyses of the formal curricula and syllabi at six universities reveal similar aims and objectives. The aim is to develop, strengthen or improve the new academic staff's ability to plan, run and evaluate teaching and to gain experience with, and knowledge about, principles and theories about learning and teaching methods and didactical knowledge. The general objectives/content are:

- Planning of teaching activities
- Teaching methods
- Peer observation (supervision)
- Evaluation
- Tutoring of students
- Experiments and investigations in own teaching practice
- Learning and learning processes
- Aims and objectives
- Exams and tests
- Problem-based project work/problem-based learning.

The methods described are more diverse, with the exception of the formal expectation of supervision. The similarities are different kinds of experience-based approaches, exchange of experiences and the stimulation of gaining new experience. Typically a functional view on educational theory is taken. Theory is a tool for reflection and communication about teaching experience. Theory is considered a valuable tool for professional reflection and not a scientific interpretation or explanation of educational phenomena. In other words, this didactical analysis describes the framework and the resources used by each Danish university on LTHE programmes for its new academic staff.

To sum up, the conclusion of the didactical analysis of the eight LTHE programmes is that the expectations (aims) related to the professionalization of new academic staff can be described as:

- The new academic staff attain concepts and pedagogical theories that allow them to participate in didactical and pedagogical discussions in a qualified manner
- The new academic staff are able to consciously choose scientific content and form with respect to the learning and development of the competencies of the students, and to motivate these choices explicitly within the framework set up by the aims of the particular education
- The new academic staff are able to continuously develop and improve their pedagogical practice as teachers, counsellors on projects and so on. (Kruse et al., 2004).

8.3 Danish Novice Science and Mathematics Teachers' Views on "Good" Teaching

We now focus on how the concept of good teaching is conceived by new academic staff in mathematics, science and technology. As part of the earlier mentioned questionnaire, two open questions try to uncover how both good and less positive aspects of their own teacher role are viewed by the respondents.

8.3.1 What are You Good at as a Teacher?

The first open question is: *What are you good at as a teacher/counsellor?* 59 new academic staff answered the questions, but with more than one sentence/statement (107 in all). We have gathered them in ten categories.

The numbers in brackets refer to the percentage of statements that we have placed in the different categories. However, we are aware that our categories are not clearly defined and that our judgement in placing the statements in one category or another has been subjective. The need for subjective judgement is based on the fact that we only possess the written statement from the respondent and are thus not able to ask the respondent to clarify or explain the specific context.

The top scorer in the categorization of the over 100 statements is the teacher's relation to the students in both teaching and supervision situations. The key statement is *understanding the students* – meeting them where they are, listening to them and spending time on them, giving positive response and showing appreciation, and creating a good atmosphere.

The second highest score concerns the teachers themselves as communicators. These statements can be gathered under the key concept *enthusiasm*. You cannot reach an audience of 120 students in a lecture unless you signal engagement, energy and presence. Enthusiasm and making the content matter alive and exciting is regarded as a direct path to motivating and engaging the students. Supposedly, the rationale behind this is that only if I myself as a teacher am motivated and engaged, can I expect the students to become engaged.

Content matter is ranked third. The teachers view themselves as being good at making the content matter understandable, designing it with respect to the level of the students and using different pedagogical tools. This content knowledge is followed closely by another kind of content knowledge: the ability to structure and prioritize the chosen content. In contrast, only two statements concern the necessity of having a strong knowledge of content matter. This can, however, be understood as part of the ability to prioritize and communicate pedagogically.

The ability to activate the students and enter into a dialogue with them is mentioned by some of the new academic staff as an important capacity for themselves as teachers.

The categories can be joined in three groups:

- Focus on the students: categories 1, 5, 6, 7
- Focus on the content matter and the communication of it: categories 3, 4, 8, 9 and 10
- Focus on the teacher: category 2 (19 statements).

This joining of categories can, of course, be discussed. As an example, category 1: *Understanding the students*, which was the absolute top scorer, was hard to place. Of course, the students are important in this category, but it is the teacher who understands the students' situation, is empathetic, and so on.

All in all we interpret the statements from the new academic staff as if they acknowledge the overall message that university teaching deals as much with students' learning (in the sense of active participation) as with the communication of subject matter. Hence, the most important aspect of good university teaching is the focus on the students.

8.3.2 What is Difficult for You as a Teacher?

In a similar way, the new academic staff answered the question: *What is difficult for you as a teacher/counsellor?* (N = 61, 83 statements). In this case, we have categorized the statements as follows (Table 8.1):

There are two categories that the teachers find hard to handle. They both deal with the content matter in the sense of structuring, prioritizing and making the content matter understandable for the students. The teachers experience problems in identifying the students' prerequisites and therefore have a hard time finding the right teaching level. In all, statements concerning content matter add up to about one third of all 83 statements.

The third biggest category concerns personal appearance, which covers a wide range of statements like: "oral communication (speaking loudly and clearly)," "being funny in appropriate situations," "keeping track of my writing on the board," "avoiding getting confused and unstructured." This uncertainty on a personal level has, for some of the respondents, to do with uncertainty about the content matter. That is, uncertainty and anxiousness before the actual teaching situation: are the students going to ask subject matter questions that you cannot answer, or are you going to succeed in being a patient, pedagogical and motivating teacher?

Many of the new academic staff are concerned that they spend too much time on teaching and preparation for teaching, as they are obliged to qualify themselves as researchers too: "to have the time to show enough interest in the students' projects." The concerns deal with structuring and prioritizing one's use of time: "to balance the amount of time spent on teaching in relation to the quality of the outcome (and my other activities)."

Table 8.1 What is difficult for you as a teacher/counselor? N = 61, 83 statements

A. Structuring and prioritizing the content matter. Delimitation, planning (17%)
B. Making the content matter understandable. Finding the level of the students (14%)
C. Personal appearance (14%)
D. Limiting the time use. Prioritizing the time available (11%)
E. Making and maintaining demands and limits. Making the students responsible (10%)
F. Motivating and engaging the students (7%)
G. Feeling insecure of the content matter (6%)
H. Entering into a dialogue with the students (6%)
I. Lectures (6%)
J. Activating the students in the actual teaching situation (5%)
K. Using audio-visual aids (5%)
L. Exemplifying, involving practice (4%)
M. Other (6%)

Another aspect that many of the respondents find difficult is maintaining demands and limits in order for the students to become responsible for their projects and their own learning: "to experience that the students are able to stand on their own two feet – I spend a lot of time supervising."

Finally, a minor group of respondents point to problems regarding their technical communication abilities, like organizing the blackboard and using their voices effectively, and a similarly small group state lack of imagination to conduct anything other than a "dull" lecture in the traditional sense as part of their experience of the negative side of the teacher role.

Similarly to the statements on positive aspects of the teacher role, the negative statements can be joined together in three groups:

- Focus on content, both regarding the subject and the pedagogy: categories A + B + G + L (34 statements)
- Focus on the students: categories E + F + J (23 statements)
- Focus on the teacher: category C (12 statements).

Again, the placing of categories in these three groups can be discussed, but the fact is that almost one third of all statements deal with problems regarding the teacher's relationship to the students.

The discussion above draws a perhaps unsurprising picture of the fact that the aspects that some teachers regard as their strong points are at the same time what others regard as the most difficult aspects of being a teacher. We still do not have a recipe for good teaching, but there is obviously a certain degree of consensus about its characteristics among new academic staff as listening to the students, motivating them, being enthusiastic concerning the subject and teaching and being good at organizing and explaining the content.

8.4 Discussion of the Complexity of Good Teaching

There is also consensus among researchers and experienced practitioners, at least on certain characteristics of the good teacher (see e.g., Biggs 1999; Ramsden, 1992), but also regarding teaching as a complex phenomenon.

A research group from New Zealand (Kane et al., 2004) has tried to gather the characteristics agreed upon into a model. The model builds on an empirical investigation that, among other things, contains interviews with 28 excellent science teachers and researchers in New Zealand (Fig. 8.1).

The wheel-like model consists of five interrelated dimensions with reflective practice as the hub:

1. Subject knowledge
2. (Pedagogical) skills
3. Interpersonal relationships
4. Research/teaching nexus
5. Personality

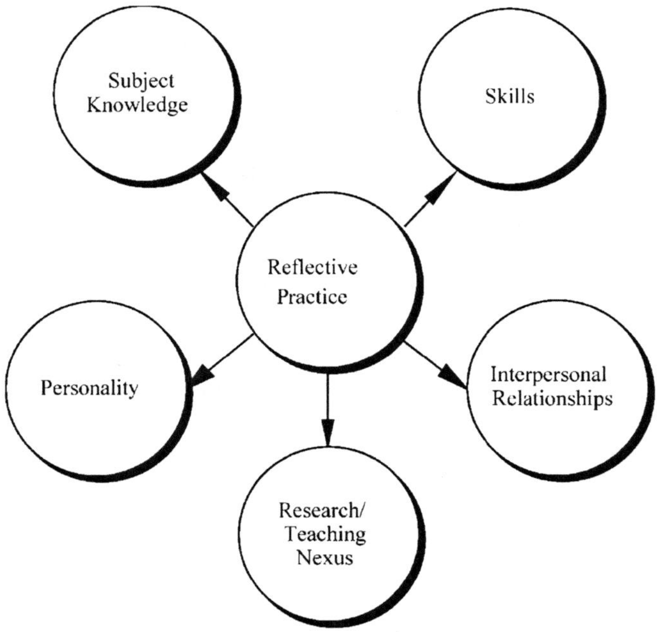

Fig. 8.1 Model – dimensions of tertiary teaching (Kane et al., 2004)

While these dimensions may not be unexpected, the ways in which the dimensions are interrelated are critical.

There are many differences between the New Zealand study and our survey. First, our respondents cannot be termed excellent teachers at first sight. Second, we only have data from a questionnaire and a few group interviews independent of the respondents' teaching context, whereas the teachers from New Zealand have been interviewed, observed and re-interviewed in connection with a certain teaching situation. Nevertheless, we find the comparison between the two investigations meaningful because Kane et al. draw attention to some important dimensions that we can recognize in our material when our new academic staff point them out. The model of the five dimensions creates in our opinion an overview of which aspects of teaching and of the teachers themselves it is relevant to focus on in educational programmes.

8.4.1 Subject Knowledge

Subject knowledge is a prerequisite for good teaching. It is important to continuously keep up-to-date, be able to implement new ideas and knowledge, and never to be content with the current level of subject knowledge. This dimension is similar to category 9, Table 8.2: *Having a strong knowledge of subject matter*. Only two of our new academic staff mention this, but we actually believe that it is implied – a belief that is partly confirmed by answers to another question in our survey: "Point out three of the

Table 8.2 What are you good at as teacher/counselor? N = 59, 107 statements

1. Understanding the students, listening, being patient, establishing contact, being open, being positive, creating a good atmosphere (25%)
2. Being an engaged teacher. Enthusiastic, dynamic, entertaining, alive, curious (18%)
3. Making the subject understandable, communicating the content matter. Explaining, reducing, being pedagogical (15%)
4. Structuring and prioritizing the subject. Showing what is most important and relevant, creating overview and coherence, continuity ... (12%)
5. Motivating and engaging the students (11%)
6. Activating the students. Asking questions, giving assignments, making the students think (8%)
7. Entering into a dialogue with the students (5%)
8. Being aware of teaching form and audio-visual aids (5%)
9. Having strong knowledge of content matter (2%)
10. Other (11%)

above-mentioned statements on the aims of science education that are the most important for you in your teaching." To this question 50% (32 out of 64) of the respondents answer that *central methods, concepts and theories of the subject* have the highest priority. This statement is by far what most of the respondents can agree on.

8.4.2 Pedagogical Skills

This dimension covers a range of skills, e.g., good communication skills, being heard and understood, making a clear presentation, good diction, and putting forward good examples, as well as being able to motivate and engage the students. Kane et al. bring together the range of skills in four words: clarity, organization, motivation and preparation. This dimension aligns with categories 3: Making the subject understandable, communicating the matter; 4: Structuring and prioritising the subject; 5: Motivating and engaging the students; and 8: Being aware of the teaching form. Categories 6: Activating the students, and 7: Entering into a dialogue with the students could also be added. In this dimension, most of what is mentioned above can be learned and, in time, mastered. The answers to another question in our survey: "Which of the content mentioned below would you consider the most important in an educational programme?" show that the new academic staff agrees that most of the pedagogical skills can be learned and hence should be taught in an educational programme. *Introduction to teaching methods* is considered important (30) or very important (28) by 58 out of 63 respondents as part of the content of en educational programme.

8.4.3 Interpersonal Relationships

In this dimension the importance of establishing a good relationship with the students is underlined, whether one is counselling one student or lecturing 150 students. Interpersonal relationships require that the teacher shows the students respect, by showing that she cares, is empathetic, and is able to understand the students' problems. Category 1: Understanding the students, creating contact, which is mentioned by a large part of the respondents, covers this dimension. However, many of the new members of the academic staff also mention interpersonal relationships as something they find very difficult (see Table 8.1, category E: Making and maintaining demands and limits. Making the students responsible).

8.4.4 Research/Teaching Nexus

A number of myths exist on the relationship between research and teaching. These myths point in at least two different directions: one type argues that if you are an excellent researcher, then you will automatically become an excellent teacher; whereas another type of myth argues the opposite, namely, that if you are a brilliant researcher you are unlikely to succeed as a teacher. A third type argues that there is no relationship between research and teaching whatsoever (Kreber, 2000). The participants in the New Zealand study (Kane et al., 2004) have all been chosen because of their excellence in teaching. They mention in the interviews that one of the reasons the students consider them excellent teachers may be that they attempt to communicate research skills to the students and that they often use results from their own research in their teaching. They actually strive to emphasize that research and communication of research are two sides of the same coin.

In our categorization of answers to questions on new academic staff's views of good and bad aspects of their own teacher role we have no category that matches this fourth dimension. However, another question in our survey: "Does your teaching and/or counselling inspire you in your research work?" tries to elucidate the respondents' views on the relationship between research and teaching. 16% of the new academic staff responded that their research is often inspired by their teaching and 43% that this sometimes happens.

8.4.5 Personality

It is a well-known fact that some teachers get through to their students while others do not. What is the difference? There are undoubtedly many answers to this question, but personality is definitely one of them. A key characteristic of the New Zealand teachers is enthusiasm. Other aspects of the personalities of excellent teachers include a sense of humour, not being afraid to give something of themselves, being approachable and human, and so on.

This dimension is widely acknowledged by our respondents as can be seen in category 2: Being an engaged teacher, enthusiastic and dynamic. The question is whether an educational programme like a Learning and Teaching in Higher Education programme or other types of courses for teachers and pedagogical developmental projects can do anything about that. As mentioned earlier, it is widely accepted that with a reasonable amount of effort teachers can learn different teaching techniques, including involving the students more in the teaching; but it is a lot harder to change teachers' personalities. It may also be questioned whether this would be meaningful. Part of one's personality is daring to show who one is. There is no doubt that teacher personality has a huge effect on the students' perception of the teaching situation. This is due partly to the opportunities for identification the teacher offers her students in the sense of her personal approach to the content – the subject – and the form of the teaching situation, and partly to the teacher's interaction with the students.

The question is what kind of qualities do we connect with a good teacher's personality and whether it is possible to act on these qualities through pedagogical counselling. The part of the teacher's personality that can be worked with consciously and targeted through educational programmes is perhaps better designated as the teacher's *role*. The teacher's role summarizes, to a great extent, an approach to teaching that can undergo experiments – not least because it deals more with the teaching form and students' expectations to the teaching than with the teacher personally.

8.4.6 Reflection as the Connection Point

In a direct comparison between the model and our investigation with special focus on what new academic staff members are good at and what they find difficult as teachers, we are not surprised to discover that most of our statements fit into the pedagogical skills dimension, indicating that the good teacher has good pedagogical skills (61 statements). Most of our respondents are new teachers with limited experience.

In contrast, our new academic staff know very well that good teaching does not come from skills alone, which can be concluded from the fact that the second and third largest categories in our investigation are the relation to the students (27 statements) and significance of personality (19 statements).

Kane et al.'s (2004) main point is that the excellent teacher does not only demonstrate all five dimensions (abilities, skills) but is also able to bring them into action by engaging in an ongoing reflection on their teaching both while they teach and after they have taught. Here, too, our material matches theirs: *Reflection on my own teaching* has by far the highest priority as possible content in an educational programme, with 38 new academic staff marking this content as very important and 20 marking it as important out of 62 answers. It is an expectation of new academic staff that the educational programme will give them the time and opportunity to reflect on their teaching. Our investigation, however, tells us nothing about how they regard reflection: as a dimension parallel to the other dimensions, or as an integration of the five dimensions of the model.

Neither we nor Kane et al. conclude that the teacher should possess similar amounts of all five dimensions, but our point is that attention to the existence of these five dimensions can serve as indicators for both supervision and for the teacher's self-reflection in order to make the teaching better. Once again, it is confirmed that teaching is a complex phenomenon and that a recipe for good teaching does not exist. Much of the literature about the scholarship of teaching (Entwistle et al., 2000; Healey, 2000; Trigwell et al., 2000;) states that the ability to reflect upon the teaching experience is the key to a flexible and critical practice. We will focus on that in the next section: How do LTHE-programmes support reflection?

8.5 Discussion of the Implications for the Development of LTHE Programmes

Regarding educational programmes, it is our opinion that the means should include considerations of what factors facilitate the development of good and effective teaching (and by implication, relevant learning) and what competences the teacher should have or ought to develop in order to become a good teacher. An important competence is to be able to reflect on one's own teaching and to react to that reflection, possibly by changing your teaching. We think that this reflection process is facilitated, most effectively, by incorporating peer counselling and supervision by experienced university teachers in educational programmes.

In the interpretation of our results, we conclude that most new academic staff still refer to the lecture as an important teaching form when they reflect on what good teaching is; one possible reason being that they have experience with lectures and because their didactical room for manoeuvre is often closely connected with lectures for a large group of students. The values related to good teaching represented by the new academic staff are in this way closely connected to the universities' tendency to structurally and organizationally reproduce the lecture form. In order for the new academic staff's "politically correct" views on teaching to be realized in practice, there may emerge a need for them to be acquainted with other forms of teaching to a higher degree. The weakness in the actual educational programmes, where reflections on your own practice through peer counselling and supervision are important features, is perhaps that it tends to reproduce the existing, structurally decided teaching forms. According to this argumentation it is important to stress that the existing forms have many qualities, but our point is, that the new academic staff should be able to vary and experiment with these teaching forms.

A highly generalized interpretation of our findings could be that the programmes make an initial reconceptualization possible, leading to more complex concepts about learning and teaching. The communication in the programmes leads to the stabilization of some general norms about good and effective teaching that is consistent with the research literature. Our conclusion is that these positive norms and concepts correspond to the traditional forms of teaching that still dominate universities. The question is whether these norms and concepts are enough to change and develop practice. The logical answer is that individuals alone cannot change an institutionalized and socially constructed practice, but that faculties need teachers with complex and coherent norms and concepts of teaching that are qualified by our common knowledge about good and effective teaching.

To focus solely on teacher's beliefs, concepts and intentions is not enough. To develop a more transparent and coherent knowledge about good and effective teaching which promotes a teaching and learning culture that stimulates a positive learning outcome we have to consider the complexity of a teaching and learning system. First of all, changing concepts or beliefs doesn't necessarily lead to the change of teachers' approaches to teaching, primarily because teachers' experience, perception and interpretation of their situation and the teaching and learning context interfere with both concepts and approach (Trigwell and Prosser, 1996). Second, we have to consider the students' concepts of learning and teaching and their approaches to studying and learning. Research has shown that we cannot blame the student, because their approach to learning is primarily an effect of the teaching and learning system they are part of. Third, the intentions of developing or changing teaching practice (approach) tend to use ideological reasons (norms and values of good teaching) as the guiding ideas for good or bad approaches. Instead of ideological talk and fighting, development must confront the relation between teaching approach and students learning outcome, to produce knowledge about effective teaching. Educational scholarship is promoted by investigating and reasoning about the normative what and why questions in higher education. But it is also a question of how to build experimental knowledge of how the teaching and learning approach can effectively lead to positive learning outcomes. The teachers' *interpretation* of the educational cultural context, their *concepts* of teaching and learning or their *approaches* to teaching are not individual constructions (Trowler and Cooper, 2002). Interpretation, concept and approach are part of a social practice and discourse depending on the academic culture among colleagues and the institutional culture.

So how do we promote educational professionalism and scholarship? To sum it up, Kruse (2006) has suggested a model. Professionalization is built on a self-referential and circular process in which teacher's implicit knowledge about teaching leads to teaching plans, new experience and interpretation leading to new implicit knowledge. But this implicit learning process is not professionalism, just practice building a praxis theory to conduct teaching. This praxis theory is necessary but not sufficient to create knowledge of good and effective learning and teaching systems. Professionalism starts with the explicit reflection and reasoning of teaching practice, making the implicit knowledge more explicit on aspects such as:

- Concepts of learning and teaching, including both values, beliefs and knowledge
- Didactical rationale: what, how, why etc
- Perception and interpretation of the structural frames around the teaching situation
- Expectations towards being student and teacher in the learning and teaching system.

Yet this is still not enough. To be a professional educator one constantly has to confront teaching practice and explicit knowledge with the learning outcome and with the knowledge that educational research can support. So the development of educational professionalism is promoted by:

- Making the teachers' implicit rationale and concepts more explicit
- Investigating the relations between concepts (rationale), approach (practice) and students learning outcome, and judging if the relation is reasonable
- Being informed of the knowledge produced by educational research
- Experimenting with teaching and learning from the students
- Having a community of colleagues communicating about the practice and quality of learning and teaching
- Educational leaders promoting professionalism.

8.6 Conclusion

To conclude, high quality teaching is based on the teacher's conceptual (theoretical) and methodical preparedness to reflect on and understand possible problems in teaching and learning processes in order to develop personal practice. Today, there is a general agreement that development of practice must be based on practice and on reflective processes that support each teacher in integrating theory, methods and practice. Above all, the pedagogical theories ought to be considered as conceptual tools that support and qualify the teacher to reflect on her teaching.

In order to promote teaching scholarship the hypothesis is that teaching in teaching is successful if it manages to support staff members in making systematic reflections about their own teaching practice, and to help them to compare their experience both with their own concepts of teaching and learning and with the common knowledge about teaching. This scholarship is both about knowledge and values, and also about power and status in academic establishments and institutions.

Therefore, a final conclusion could be that the LTHE programmes will not reform university teaching. Such a reform would demand much more than the education of the youngest staff members. But we would like to question whether the aim of the programmes is to reform universities. Every reform is a matter of policy, administration and, of course, of educating the educators – individually, in groups, and above all, in the complex contexts in which they work.

References

Andresen, L.W. (2000). A Suitable, Trans-Disciplinary Conception of Scholarship. *Higher Education Research & Development*, 19(2), 137–153.

Berendt, B. (1998). How to Support and Bring About the Shift from Teaching to Learning Through Academic Staff Development Programs: Examples and Perspectives. *Higher Education in Europe*, 23(3), 317–329.

Biggs, J. (1999). *Teaching for Quality Learning at University*. Buckingham: SRHE and Open University Press.

Boyer, E.L. (1990). *Scholarship Reconsidered: Priorities of the Professoriate*. Princeton: The Carnegie Foundation for the Advancement of Teaching.

Cope, C. & Prosser, M. (2005). Identifying Didactical Knowledge: An Empirical Study of the Educational Critical Aspect of Learning About Information Systems. *Higher Education in Europe*, 49, 345–372.

Dale, E.L. (1989). *Pedagogisk Professionalitet*. Oslo: Gyldendal.

Entwistle, N. (2003). *Concepts and Conceptual Frameworks Underpinning the ETL Project. Occasional Report 3*. Edinburgh: Higher and Community Education, School of Education, University of Edinburgh. Retrieved 28 May 2008, http://www.etl.tla.ed.ac.uk/docs/ETLreport3.pdf

Entwistle, N., Skinner, D., Entwistle, D. & Orr, A. (2000). Conceptions and Beliefs About "Good Teaching": An Integration of Contrasting Research Areas. *Higher Education Research & Development*, 19(1), 5–26.

Entwistle, N. & Walker, P. (2002). Strategic Alertness and Expanded Awareness Within Sophisticated Conceptions of Teaching. In N. Hativa & O. Goodyear (Eds.), *Teacher Thinking, Beliefs and Knowledge in Higher Education* (pp. 15–39). Dordrecht: Kluwer.

Feldman, K.A. (1987). Research Productivity and Scholarly Accomplishment of College Teachers as Related to Their Instructional Effectiveness. *Research in Higher Education*, 26, 227–291.

Felten, P. & Pingree, A. (2003). Centres for Teaching Improvement in Colleges and Universities. In J.W. Guthrie (Ed.), *Encyclopedia of Education*, Vol. 1 (pp. 254–257). New York: Macmillan.

Gibbs, G. & Coffey, M. (2004). The Impact of Training of University Teachers on Their Teaching Skills, Their Approach to Teaching and the Approach to Learning of Their Students. *Active Learning in Higher Education*, 5(1), 87–100.

Hattie, J. & Marsh, H.W. (1996). The Relationship Between Research and Teaching. A Meta-Analysis. *Review of Educational Research*, 66(4), 507–542.

Hattie, J. & Marsh, H.W. (2002). The Relation Between Research Productivity and Teaching Effectiveness. *Journal of Higher Education*, 73(5), 603–641.

Healey, M. (2000). Developing the Scholarship of Teaching in Higher Education: A Discipline-Based Approach. *Higher Education Research & Development*, 19(2), 169–189.

Ho, A.S.P. (2000). A Conceptual Change Approach to Staff Development: A Model for Programme Design. *The International Journal for Academic Development*, 5(1), 30–41.

Kane, R., Sandretto, S. & Heath, C. (2002). Telling Half the Story: A Critical Review of Research on the Teaching Beliefs and Practices of University Academics. *Review of Educational Research*, 72(2), 177–228.

Kane, R., Sandretto, S. & Heath, C. (2004). An Investigation into Excellent Tertiary Teaching: Emphasising Reflective Practice. *Higher Education*, 47, 283–310.

Kreber, C. (2000). How University Teaching Award Winners Conceptualise Academic Work: Some Further Thoughts on the Meaning of Scholarship. *Teaching in Higher Education*, 5(1), 61–78.

Kruse, S. (2001). *Pædagogisk professionalisme i universitetsundervisningen*. Aalborg: Dansk Center for Naturvidenskabsdidaktik.

Kruse, S. (2006). Udvikling af universitetslærerens pædagogiske kompetencer – en didaktisk skitse. *Dansk Universitetspædagogisk Tidsskrift*, 2, 36–44.

Kruse, S., Nielsen, K. & Troelsen, R. (2004). Pædagogisk uddannelse af universitetsadjunkter: Analyse af adjunktprogrammerne på de tekniske og naturvidenskabelige fakulteter ved højere læreanstalter i Danmark. *DCN Papers nr. 20*. Aalborg: Dansk Center for Naturvidenskabsdidaktik.

Lauersen, P.F. (2003). Universitetspædagogikkens barndom og DUN's. In DUN (Ed.), *Universitetspædagogik år 2003 – inkl. Kortlægning af universitetspædagogiske centre i Danmark*. København: Dansk Universitetspædagogisk Netværk.

Lortie, D.C. (1975). *Schoolteacher. A Sociological Study*. London: The University of Chicago Press.

Martin, E. & Lueckenhausen, G. (2005). How University Teaching Changes Teachers: Affective as well as Cognitive Challenges. *Higher Education*, 49, 389–412.

Martin, E. & Ramsden, P. (1993) An Expanding Awareness: How lecturers Change Their Understanding of Their Understanding of Teaching. *Research and Development in Higher Education*, 15, 148–55.

Marton, F. (1981). Phenomenography – Describing Conceptions of the World Around Us. *International Science*, 10, 177–200.

McManus, D.A. (2001). The Two Paradigms of Education and the Peer Review of Teaching. *Journal of Geoscience Education*, 49(5), 423–434.

Prosser, M., Martin, E., Trigwell, K., Ramsden, P. & Lueckenhausen, G. (2005). Academics' Experiences of Understanding of Their Subject Matter and the Relationship of This to Their Experiences of Teaching and Learning. *Instructional Science*, 33, 135–157.

Ramsden, P. (1992). *Learning to Teach in Higher Education*. London: Routledge.

Schön, D.A. (1987). *Educating the Reflective Practitioner*. San Francisco: Jossey Bass Publishers.

Schön, D.A. (1995). The new scholarship requires a new epistemology. Change Nov./dec. 27-34

Shulmann, L. (1993). Teaching as Community Property. *Changes*, 25, 6–7.

Trowler, P. & Cooper, A. (2002). Teaching and Learning Regimes: Implicit Theories and Recurrent Practice in the Enhancement of Teaching and Learning Through Educational Development Programmes. *Higher Education Research & Development*, 21(3), 221–240.

Trigwell, K., Martin, E., Benjamin, J. & Prosser, M. (2000). Scholarship of Teaching: A Model. *Higher Education Research & Development*, 19(2), 155–168.

Trigwell, K. & Prosser, M. (1996). Changing Approach to Teaching: A Relational Perspective. *Studies in Higher Education*, 21(3), 275–284.

Trigwell, K., Prosser, M. & Waterhouse, F. (1999). Relations Between Teacher's Approaches to Teaching and Students' Approaches to Learning. *Higher Education*, 37, 57–70.

Trigwell, K. & Shane, S. (2004). Student Learning and the Scholarship of University Teaching. *Studies in Higher Education*, 29(4), 523–535.

Part III
Changes in Structures and Organisations

Chapter 9
Conceptions of Universities as Organizations and Change in Science and Mathematics Education

John A. Bowden

This chapter draws on my experience as a change agent in universities over three to four decades and is partly autobiographical. I appreciate the willingness of the Editors to allow me to write in such a reflective and discursive way. The chapter outlines a problem, posits some theoretical explanations but offers no concrete solution. The kind of comprehensive change in students' experience of university learning that I believe in seems hardly closer now than when I adopted it in its primitive form as my mission more than thirty years ago. Perhaps it appeared to be getting closer at times but more recent years have seen the dream fade. This chapter attempts to analyze why and to ponder whether such an outcome is inevitable. That self-centred portrayal of the problem will be balanced by further analysis using distributed leadership theory[1]. I want to emphasize that the change I am talking about is not concerned simply with the degree to which discussion of teaching and learning among stakeholders in university education has increased *per se* (it has increased considerably) but rather whether progressive changes have been made to the learning environment with consequential, beneficial effects on learning outcomes. There has been a lot of discussion in recent decades but the rhetoric has not always been matched by the outcomes. Also, the nature of the activities undertaken by graduates has become much broader, in some situations much less sophisticated, as universities become institutions of mass education, and in other situations more complex and demanding. In a sense the goal posts for university undergraduate education have shifted and it is no longer clear where they are standing. That has made the task of creating the ideal learning environment more difficult to envisage and design.

If you continue with your reading of this chapter, doing so may catalyze your thinking and encourage you to reflect on whether the problem described is real and whether you have a way of solving it. My goal is that you, the reader, should use what I write to see if you and others can do in the future what, in the last few decades, my colleagues and I could not.

[1] I am grateful to Dr. Gloria Dall'Alba (University of Queensland, Australia) for alerting me to the existence of the literature on distributed leadership.

J.A. Bowden
Professor Emeritus, RMIT University and Adjunct Professor, Swinburne University of Technology, Australia

O. Skovsmose et al. (eds.), *University Science and Mathematics Education in Transition*,

9.1 Theoretical Base

The analysis in this chapter combines a number of perspectives that include those of the educational change agent, the student learner, the academic teacher, the university as an organization and the interests of external stakeholders. The theoretical base for this chapter lies in the following:

- The research by Marton's group (for example, Marton and Säljö, 1984) on student approaches to learning and the relation between approaches and learning outcomes has had significant influence. The notions of deep and surface approaches to learning were fundamental to educational thinking from the 1970s and beyond. The deep-surface dichotomy also has relevance to academics' responses to their teaching responsibilities. They can be undertaken in a minimalist, surface way or they can be approached more meaningfully, depending on the circumstances (Bowden, 1988).
- Ramsden's work (1992) on the relation between approaches to learning and the nature of the learning environment has provided part of the framework for educational development in recent decades. It also provides a useful perspective when examining academic staff activities in relation to the organizational environment in which they work. The influences on how students learn that Ramsden has discussed, such as heavy workload, reward for more or less sophisticated outcomes and limits placed on independent choice, apply in a parallel way to academics in their workplace.
- My own, more recent work on capabilities-driven curriculum design (Bowden, 2003, 2004) has played a part in influencing educational practice in a number of universities.

All of these have provided a theoretical underpinning for much of the educational development of the past few decades but they can also be adapted to analyze the decisions and practices taken up by academic teachers. There are other factors, such as:

- The influence of the institutional environment on academic behaviour, related to the aims-demands theory elaborated in The University of Learning (Bowden and Marton, 1998, p. 228). Aims set at higher levels in an organizational hierarchy can be interpreted as demands at lower levels and the theory provides an explanation for the common failure of top-down change processes.
- The importance of a networked approach to management of universities so that, while individuals have their own way of seeing each situation, they become aware of others' ways of seeing, with a consequential increase in collective consciousness (Bowden and Marton, 1998, pp. 189–209).
- The ways that the attempts of external stakeholders' to influence teaching and learning often clash with university teachers' perceptions of academic freedom (Bowden and Marton, 1998, pp. 220–224).
- The notion of distributed leadership (Bennett et al., 2003; Henry, 2006; Spillane et al., 2006). Distributed leadership theory provides a different framework from which to analyze all the influences referred to above and will now be described briefly.

9.1.1 Distributed Leadership

Distributed leadership is represented variously as an aspiration (Bennett et al., 2003) and as a framework for analysis (Spillane et al., 2006; Henry, 2006). I am principally going to draw on the latter but will use the former to illustrate its relevance. Bennett et al. suggest that "distributed leadership highlights leadership as an emergent property of a group or network of interacting individuals" rather than as "a phenomenon which arises from the individual". To them, the notion suggests "openness of the boundaries of leadership" and a distribution of a variety of expertise "across the many, not the few". If these are brought together, a "concertive dynamic" is created that represents more than the sum of the parts. This is reminiscent of the collective consciousness notion developed in The University of Learning (Bowden and Marton, 1998, pp. 189–209). We argued that within a group of people there is likely to be variation in the ways that different members of the group see any particular phenomenon and that collective consciousness refers to the degree to which each individual has awareness of the different ways of seeing by other members of the group. We used that notion to argue against models of university management such as the hierarchical model and the market model as not being conducive to the development of collective consciousness (Bowden and Marton, 1998, pp. 273–274). Instead we argued for a network model that seeks to enable those with expertise to be responsible for action and to avoid situations in which formal authority does not correspond with capability to act effectively. The importance, when analyzing for distributed leadership in an organization, of looking at the roles of both formal and informal leaders, is emphasized by Spillane et al. (2006).

9.1.2 Framework for Analysis

Bennett et al. (2003) suggest several variable features:

- Control/autonomy is a major variable involving constraints from higher levels in the hierarchy related to goals set by formal leaders. Those formal leaders are often in turn accountable to external stakeholders for institutional performance. Bowden and Marton's (1998, p. 228) aims-demands theory highlights the same issue. This contrasts with an emphasis on a greater degree of autonomy for those who contribute to leadership. While distributed leadership theory was developed from study of schools, it is quite relevant to universities. They have in common the fact that employees necessarily act independently when engaging in the business of the organization, *viz.* teaching students and doing research and yet there are institutional goals and processes that by their nature curb independent action. The balance between control and autonomy is a central question for universities.
- Organizational structure and agency. The emphasis on formal structure can be limiting as can the other extreme, the focus on informal agency. Perhaps universities half a century ago could be depicted as a loose alliance of a large number of effectively

independent agencies. Today some of them are tight corporations with a fierce profit focus, detailed business plans and an expectation of corporate allegiance. Most of them have a mix of the two and a common tension in universities in recent times has concerned academics' fear that the shift from the traditional to the corporate university has gone too far. A focus on formal leadership and reduced independence of action for members of the university are at the heart of such claims.

- Social and cultural context. Apart from the external culture in which the organization lives, the internal organizational culture has a significant effect on the uptake of distributed leadership. Past history of passive acceptance or of innovative uptake has an effect on the contemporary response to opportunities for distributed leadership. In universities, historically staff have had a broader independence of thought and action through the loosely held concept of academic freedom. While there is variation in the ways that different stakeholders view that concept, it is part of the culture of universities in one way or another. Hence the notion of distributed leadership seems appropriate when looking at universities. The traditional concept of the university mentioned in the previous dot-point above could be viewed not so much as collective consciousness but rather as uncoordinated, widely distributed leadership. The extreme but hopefully rare corporate model might well be represented as *collective unconsciousness* – the different ways of seeing among members of the organization are expected to be subjugated to the prevailing corporate perspective. In contrast, the notion of collective consciousness, as well as the distributed leadership model, aims at a culture of coincidence of expertise and responsibility with shared understanding and acceptance of the value that it brings to the university (Bowden and Marton, 1998, pp. 259–261).
- Source of change and involvement of formal and informal leaders. The impetus for development of distributed leadership often occurs through structural reorganization, initiated by either internal or external agencies, and the emergence of strong formal leaders who facilitate distributed leadership despite their formal power. It can also develop from a bottom-up response to some formal policy requirement coming from either internal or external sources. New informal leaders champion such a bottom-up development and this places demands on senior formal leaders to respond. In modern times, the top-down imposition of independence (admittedly this is an extreme statement of the Bennett et al. comment but it has the same flavour) or even the encouragement of independence by the university president,vice-chancellor or rector has often been missing in universities (from now on I will refer to the president,vice-chancellor or rector as the CEO – Chief Executive Officer – even though this "corporate" term sometimes has a pejorative connotation in universities). Few CEOs are recruited because of their commitment to distributed leadership. Rather, CEOs who are able to articulate some strong new vision of their own that they expect staff to implement are the ones who in recent years have been headhunted by competing universities. I am a little pessimistic that the idealistic vision of formal leadership that Bennett et al. depict is even possible in the world that universities have to inhabit in the twenty first Century.
- Dynamics of team working. Collaboration, openness, trust, sharing of expertise, common goals and processes, mutual support, networking, open communication and primacy of the group goals are seen as attributes of teams that best support

distributed leadership. Most university academics have experienced this kind of dynamic style of working and find it exhilarating and rewarding. At the best times, this is how academic life goes on. However, most academics would comment that over recent decades the opportunity for such an idyllic existence has reduced. The competitive systems that have been set in place, both between universities (for status and for funds) and between academics (for recognition and for career enhancement), militate against these values of trust and interdependence. How this can be reversed is one of the challenges.

- Institutional and spontaneous forms of distributed leadership. Distributed leadership is best maintained through reliance on expertise rather than position, and ad hoc functional groups rather than establishment of ongoing committees. Such fluid leadership depends on trust and mutual support, which must become part of the organizational social and cultural context. It also blurs the line between leaders and followers. This is what all of us have experienced in the moments that correspond to the ideal elaborated in the previous dot-point. How we can expand those moments into our whole working life is something we need to investigate.

9.1.3 Broad Implications of the Theoretical Base

There are a number of conclusions that we can draw from the theoretical base that has just been described. They concern student learning in general, learning for the unknown future in particular, environmental influences on academics' response to change processes, negative influences of top-down or external demands, and the special creativity needed for good teaching.

How students approach their learning tasks affects what is learned. If students are studying with the intention of memorizing as much as possible so as to regurgitate it in an examination, then they are unlikely to develop much understanding or draw much meaning from the material they study. Indeed, the latter is likely to occur only if they are going about their study with the intention of developing the meaning and trying to understand it. We also know that students adapt their approaches to learning in response to the learning environment they experience, as well as their own past learning experiences. If they face heavy workloads, an examination system that rewards rote learning and a teaching and learning environment that is alien to learning for meaning, then they will be more likely to take a surface approach. In turn, the learning environment experienced by students is affected strongly by university practices, particularly in relation to curriculum design and the activities of the teaching staff with whom students interact. What the university and its staff do about those elements of the environment does matter. It has an effect.

Students respond strongly to the nature of the examinations and other forms of assessment, which for many comprise the principal environmental factor shaping their approaches to learning. Encouragement by teachers for students to achieve particular learning outcomes will be more or less effective depending on the level of consistency between the intended outcomes, the teaching and learning experiences and the assessment practices. Similar issues are raised in Chapter 4 (p. 85).

Student learning at university needs to prepare them to deal with situations in their future life that cannot be predicted in any detail. The capability that students need to develop in order to cope with the unknown future includes the capacity to judge what aspects of any new situation they have to deal with are relevant, the ability to decide what knowledge that they have or can access is necessary and the capability to deal with all the relevant aspects simultaneously (Bowden, 2003).

There is a danger that the hierarchical structure of a university will lead to aims at one level being interpreted as demands at a lower level and converted into alternative aims that may be inconsistent. There are examples discussed in this chapter in relation to senior management aims for curricula to be designed with capabilities to deal with an unknown future as the central feature. Some academics may interpret those aims as demands to structure curriculum documents in a particular way. Their response is to edit their existing curriculum documents so that the new keywords like capability and unknown future are included, without any substantial change to the meaning or the practice. This response can only add to confusion within the university, among both staff and students. External stakeholders can also make demands that inadvertently have a negative impact on the learning environment. The way that some governments have imposed industrial models of quality assurance on university teaching and learning functions is one example discussed in this chapter.

Teaching cannot be optimized merely by imposition of a set of rules and regulations. It is a creative process that requires the development of appropriate ways of seeing. That, in turn, requires time, resources, opportunity and a supportive environment. Given that characteristic of the teaching process, as well as the sensitivity of the aims-demands relationship and the relative autonomy of academic staff, distributed leadership theory is quite relevant to organizational change within a university in relation to teaching and learning.

9.2 My Background

For two reasons I thought it would be useful to readers for me to summarize my education and work experience. My particular background and experience have shaped the thinking I reveal in this chapter. In addition, this account serves as an example of the way in which many people become academic teachers and some become educational developers. Readers may well recognize aspects of their own stories in my personal account.

I began as a school student who wanted to be a scientist. I had taken an occupational orientation test at about age 14 which concluded that I was most likely to succeed as a social scientist rather than as a scientist. That had no impact on me at the time although it subsequently proved to be prescient. I went on to university and graduated with honours in chemistry, following a combined degree in chemistry and mathematics. I had envisioned myself in a lab coat working for some company, but my image of the future never took a more specific shape than that. My success as an undergraduate and in my early research activities led to an offer to enrol in a Ph.D. in chemistry and an opportunity to teach undergraduates.

My teaching career began when, as a new research student, I was thrust into looking after laboratory groups in undergraduate chemistry at the University of Melbourne. My role was to make sure that the various science, medical, dental and engineering students safely completed the designated laboratory work in the allotted time. During that process I taught them both practical laboratory skills and theoretical explanations. After each laboratory session, they wrote a report and I gave them a grade.

My qualification for being appointed to such a role was that I had completed an honours degree in chemistry and that I was now enrolled in a postgraduate research degree. My capabilities to undertake the role were not part of the decision-making process that appointed me and whether I was good at such teaching or not was incidental, being totally unknown to those appointing me. In a similar vein I spent many hours interacting with tutorial groups of up to 15 students. The fact was that the chemistry school staffing structure would have been inadequate without the postgraduate students taking on the teaching load in the undergraduate laboratories and tutorial rooms.

After one year, I was appointed to a position as a senior demonstrator, which initially changed neither what I taught nor how I was trained to teach but merely increased the number of sessions I had to teach. A year or two later, I was given the opportunity to be one of four people giving lectures in parallel to four groups of first year chemistry students (about 150 students in each group). We followed the same curriculum in all four groups and the training I received for this activity comprised a weekly meeting with the other three lecturers to ensure that in the following week we would all cover the same ground.

I found that I enjoyed teaching very much, but I became painfully aware that, while I intuitively chose certain strategies because I thought they would help students to learn better, I had no fundamental knowledge of educational theory on which to base such strategies. I was doing research at the laboratory bench for my Ph.D. that was based on more than seven years of study of fundamental chemistry. Yet I was involving myself in the learning of hundreds of young students without any study of the processes of learning and teaching. Hence, while working as a full-time academic and doing my Ph.D. part-time, I also enrolled part-time in a Diploma in Education course that focused on university teaching and learning.

Over the next few years, I took courses in education to Masters level at the same time as completing my Ph.D. in chemistry. The courses in education and my daily teaching experience led me to think about the ways in which what we do as teachers impacts on what students do as learners and, ultimately, what they learn. This eventually became my life's work as I soon became a lecturer in an educational development unit and learned afresh to become a researcher, this time in education. I had spent more than ten years as a student and academic in the Chemistry School at the University of Melbourne and yet, as predicted by a psychology test when I was 14 years old, I had by age 28 taken an academic position as a social scientist.

There are two aspects of that story that are relevant to the argument of this chapter. The first relates to the unplanned ways in which academics prepare for a career that involves significant responsibility for the learning environment of hundreds of students per year and eventually, over a full career, many thousands of students. While the number of teacher education courses for university academics and the number of universities seeking evidence of teaching experience in academic appointments have increased, the average academic appointment is still largely based on research performance. In the same way, while promotion processes in most universities pay lip service to the importance of the teaching role, actual practices fall far short of the rhetoric and often lead academics to adopt practices that are detrimental to learning. This resonates with arguments by Kruse et al, in this volume.

The second aspect relates to my capability theory (Bowden, 2003, 2004). In retrospect I can see that my own education in science, my research experience in chemistry and my teaching experience as a chemistry academic have contributed to the way I have acted as a social scientist over the past three decades or so. However, that my career would take that trajectory was not predictable while I was learning about undergraduate science or chemistry research, not even to me let alone to my teachers. So in a real sense, as an undergraduate, I was learning for the unknown future (Bowden and Marton, 1998, pp. 24–27) and my capabilities-driven curriculum model (Bowden, 2003, 2004) reflects my thinking that this is largely true for all undergraduates. This will be referred to again later in the chapter in relation to the need to design curricula around the development of capabilities to handle the unknown future. No one knows what any particular undergraduate will be doing professionally in 5–10 years time.

9.3 The Problem of Sustainable Educational Development

Universities are complex systems in which change in one part of the system can influence activities in many other parts. This both inhibits planned change and catalyzes unintended change. The problem as I see it is that there is very little chance for change agents to have a significant, comprehensive and long-term impact on the way that universities facilitate undergraduate learning. As I reflect on my career since the 1960s, I have comfortable feelings of satisfaction with the work I have done. I have been a researcher in both chemistry and education, engaged in educational development activities, reflected on both of those and written many articles and books. Many of the things I have done or have written about have received favourable comment from peers and, at least in the short term, they have had an effect on the way the universities in which I have worked have framed their educational activities, to greater and lesser extents in various places. Certainly my work has had significant impact on the way a number of individual academics developed their interaction with students in the learning environment. I have always felt that it was worthwhile to be doing the things I was doing. Ironically I still feel the same way, as I continue to work with doctoral students during my retirement.

However, I think the fate that has befallen me is one that affects many like me and it is that, some years after my departure from the various campuses, the ongoing activities of the universities concerned show decreasing, if any, effects from my earlier presence. In some cases, subsequent change processes have been implemented that erode the effects of the previous change without necessarily having been designed to do so. I don't make that statement in any bitter or regretful way or with any churlishness, but with a sense of ironic realism.

For instance I recently described (in Bowden, 2003, 2004) the development of a capability-driven curriculum design process that had been introduced at RMIT University in Australia before I retired. It was a response to the point made earlier – that undergraduate students are necessarily learning now for a future that can't be predicted in any detail. They need to develop the capability to deal effectively with previously unseen situations. The capabilities-driven curriculum model I designed was my response to that need – a way of organizing university curricula and learning experiences for students in ways that lead to the development of those capabilities. It was decreed from the vice-chancellery that all new course design and all five-yearly course renewals would from then on be capabilities-driven. The early processes of curriculum change were based heavily on the documentation I had produced about capabilities and I was directly involved in the early design processes. It was a new concept that required different thinking and produced quite different course structures and approaches to teaching.

While earlier literature (e.g., Biggs, 2003; Entwistle, 1998; Marton et al., 1993) still provided valuable, supportive background, it did not address some of the fundamental implications of the new capability-driven approach. Now, just a few years later, the guiding principles for educational programmes run within the portfolio of the Pro Vice-Chancellor, Science, Engineering and Technology (SET) at RMIT University state explicitly (SET, 2007) that programme design reflects "a capability based curriculum". That appears to indicate that the capability approach has remained intact. However, in the SET document, there follows a range of statements referring to elements of the educational literature and a list of 17 references. Not one of those references is about the capability-driven approach that the whole university, including the SET portfolio, claims to have adopted.

So soon after this fundamental shift in curriculum design theory and practice has taken place, any science or engineering academic seeking guidelines for curriculum design will not be directed towards any literature that actually addresses the issue of capability. In March 2007, RMIT advertised and ran a workshop titled "Designing and Teaching for Graduate Capabilities". It is likely that anyone who attended that workshop would get a better insight into the curriculum and learning issues around capability development. However, science and engineering academics who read the advertisement would have had no reason to believe that it went beyond the SET guidelines and that they should attend. Anecdotally I am aware that many such academics are merely re-packaging their existing curriculum by including the term "capability" without making any fundamental change to their own practice or the learning experiences of their students. For such academics, this is a rational response. In educational jargon, they are taking a surface approach in a context that is confusing, just as students do in their learning. In what they have been advised to read, there is nothing that enlightens them about this new approach and so they revert to what they know. This example will be taken up again later but it is illustrative of the power of the individual response to what appears as a top-down demand.

9.4 Variation in Underpinning Educational Philosophies

9.4.1 Efficient Information Transfer

There is a range of educational philosophies on which curriculum development may be based. When I began my career as an academic, the theory-of-practice (Argyris and Schön, 1978) in place appeared to be about efficient information transfer. The emphasis in the Diploma in Education course that I completed was on structure, clarity and comprehensiveness of lecture delivery and on encouraging discussion in tutorials. The former involved, among other things, working with elocution teachers and the latter paid a great deal of attention to recording the proportion of time spent talking by the tutor compared with the students. Much less attention was paid to the nature of the learning activities of students in relation to the content of learning. When students were considered, it was in relation to psychological studies of attention spans and capacity to handle several ideas at once. Hence, Ausubel's well-known advance organizers and attention to retroactive inhibition were the kinds of issues discussed (Ausubel, 1968).

Academics experiencing this kind of educational development activity were effectively being led to focus on themselves and their own performance or behaviour, rather than on students. If only they could get their own act together (pun intended), students would learn. The learning process itself was not problematized in any depth. Nevertheless, those of us who had a predilection towards the learning of our students saw it as a great step forward that now the university was paying attention to the teaching function and was funding efforts to make improvements. However, this was in a context in which many senior academics saw such efforts as an inappropriate shifting of resources from research into teaching, beyond the minimum required to fulfil contractual obligations, and as a distortion of university functions. My direct experience of a head of department advising a new lecturer against participating in the course I just described underlines that point. The head told the lecturer that he had been appointed because of his research ability and that he should get on with more rather than less research.

"If you are the researcher, I thought you were, then you will have no trouble teaching in your field; if you have trouble teaching and need this course, then maybe you aren't as good a researcher as I thought" (paraphrased from recollections).

9.4.2 Facilitating Conceptual Change

That was around 1970 or so. Later, in the 1980s, the emphasis shifted towards student learning rather than teaching (Marton and Säljö, 1984; Bowden, 1989; Ramsden 1992). Of course, teaching and learning are always linked, but in a complex way and not exclusively. The shift in emphasis opened up the discussion to allow for consideration of the learning of students away from the classroom, separate from the performance of teachers, of seeing learning as a complex process of responding to a series of related experiences. The important influence of assessment or examination practices on learning began to be considered more carefully, as well as the different ways that learning might take place in a variety of contexts and the effects of those on what is learned. Hence surface and deep approaches to learning were considered in relation to rote learning and learning for meaning. Perhaps the notion of learning for conceptual change (Johansson et al., 1985) summarizes the central aspect of this era's approach to educational development. Teachers were being urged to fashion their interactions with students with conceptual change in mind, not just rote learning of so-called facts. They were being advised to assess students for learning outcomes associated with changed understanding of key concepts.

9.4.3 Capability to Deal with the Unknown Future

More recently, from the 1990s to the present, a more holistic approach to educational development is being promoted by a number of people, including myself. In my book with Ference Marton, we advocated that curriculum design at university level should take account of the difficulty in knowing what graduates would be doing in their professional, social and personal lives even just a few years after graduation, let alone later in their lives (Bowden and Marton, 1998, pp. 114–126). We argued for "learning for the unknown future" and I subsequently developed the notion of "capabilities-driven curriculum design" (Bowden, 2003, 2004) that was ultimately implemented at RMIT University. My own experiences were used as an illustration for this idea earlier in this chapter.

The emphasis is on working out, in each undergraduate degree field, the capabilities needed to handle previously unseen situations. The emphasis is not on learning one hundred or one thousand solutions to one hundred or one thousand known situations. The emphasis is on experiencing a range of previously unseen situations (from the student's perspective) in ways that lead to a capability of handling previously unseen situations in the future. If students are learning to develop capabilities to deal with the unknown future, it is not necessary that the situation they have to confront is in a list of one hundred or one thousand predicted situations; it can be any one of a million unpredictable situations.

9.5 Educational Development and Teacher Behaviour

The focus of educational development activities has varied over the last three or four decades (see Fig. 1 in Bowden, 2003, p. 15). My initial role in the 1970s was to work with individual teachers at universities to help them to find ways of improving their students' learning. In addition, I collaborated with the management of the universities in which I worked to create conditions that would allow those teachers to take the steps they needed to improve their teaching, to remove the barriers to such an effort, if you like. At this time, the interest of the government and the community in the quality of university activities meant that many of those barriers could well be replaced by other barriers, different and often well-intentioned but frequently impenetrable, such as an industrial model for educational quality assurance. Then I became head of an educational development group at another university and approached my role differently. I devoted the next decade to working at the institutional level, aiming to shape quality processes so that they were not just administratively efficient but also pedagogically sound and more likely to improve student learning.

One crucial issue is that teaching, as a process for enhancing learning, does not produce intended results merely by making sure each of a prescribed set of elements is in place. It is a creative process that requires the teacher to interact with the learners and to make judgements based on those interactions about what to do next (Bowden, 1989). Those judgements are often required every minute in face-to-face interactions or they may require careful planning such as in curriculum design or preparing examination papers.

So, if anyone or any organization wants to improve teaching and learning, it is not simply a matter of developing a set of rules of behaviour. It is the creative response of the teacher to the situation that is crucial. You can have a whole lot of elements in place but it is how they are implemented that counts. That will always involve individual teachers interacting with individual students or groups of students, either face-to-face or remotely through mediated interactions and, in a real sense, those teachers are free to act in any way they deem appropriate in the circumstances. That they are free to respond *in situ* is a necessary element of good teaching. However, it is not sufficient because it is not just important that they are free to respond. Of crucial importance is how they respond and educational developers over the decades have devoted themselves to helping teachers think through the underlying principles of learning so they can make the most effective responses, whatever the context. Just as good teachers are looking to their students to learn for meaning and to develop capabilities for deal with varying situations, so too are educational developers looking to academics to understand learning and to develop their capability to respond in any teaching and learning situation. The literature on student learning is applicable to the development of teaching capability.

I think that teaching university students and parenting have some similar characteristics. Parents are always older and more experienced than the children they are looking after. Teachers are usually more experienced in the subject matter they are teaching, although there are exceptions with some adult learners. One issue is that the children or the students are engaging in activities that are expected over time to help them develop capabilities they didn't have or that they had in less sophisticated and less powerful forms at an earlier time. A second issue is that both parents and teachers have an acknowledged responsibility to do things that assist those development processes. A third issue is that both parents and teachers face a tension – how to teach about what they know in ways that assist development of appropriate capabilities and, at the same time, how to foster independence. In some respects the former almost inevitably contributes to the development of the latter but it is in getting the balance right that the art of parenting or teaching is to be found.

9.5.1 Parenting Analogy

Now comes the analogy. I want to focus on the need children have in an urban environment to develop the ability to cross a busy street safely. All of us who are parents recall that learning process as a crucial one. Indeed as children develop that capability, their horizons widen and their independence is enhanced. How does a parent go about helping a child to develop that capability of crossing a busy street? I suggest three models for discussion.

Model A is, I hope, only rarely found. That is for the parent to take the attitude that it is a harsh world that everyone needs to survive in and, if children really want to cross the street, they'll find a way. So such a parent would be accepting the inevitable outcome that there is a high probability of the child being hit by a passing car. This seems far-fetched (although water safety experts in recent decades have still felt the need to warn against the sink-or-swim method for teaching children to swim – see Hegland, 1993). However, it could be regarded as a reality in situations where parents don't bother to keep track of their children's comings and goings and, effectively by neglect, allow them to fall into the circumstances described. The focus in Model A is on children becoming totally independent through lack of attention by the parent to nurturing, developmental processes or safe practice.

Model B is found very commonly. That is for the parent to make sure that every time the child needs to cross a busy street, the parent is present, takes the child by the hand and steers the child (even "drags" the child on some occasions) across the street at the appropriate moment. The problem with this model is that children may never develop their own capability to deal with such situations and certainly there can be no development of independence in such circumstances. However, the child certainly remains safe each time.

Model C is certainly my preferred model. Here the parent may steer the child across the street a few, even many, times as in Model B. However, there would be a related conversation about what judgements were being made and why the timing to cross was chosen. This is something Mary and I are doing when we walk with Max, our two-year-old grandson, across a busy street so he can play on the playground equipment in the park. Max walks between us holding our hands and we stop at the curb and talk through our action – saying that we have to check before we cross the street that there are no cars moving towards us that could hit us. We ask him to look to see if there are cars coming. When we see a car coming, we say we can't cross yet because the car might hit us. Max is some years away from such independent action himself but we believe we are contributing to the development of that capability. Model C would later on incorporate further steps in which the child was encouraged to make the decision to cross using his or her own judgement but with the parent nearby and ready to grab hold of the child if there is imminent danger. This model is focussed on three things: the development in the child of the capability to make appropriate judgements through direct experience; the consequential feeling of independence which is also a necessary part of the development process; and the taking of responsibility by the parent for the welfare of the child.

The most challenging aspect of parenthood is to get the balance among these three objectives at an optimal level. Model A is unacceptable because it imposes independence on the child without paying any attention to welfare or the process of development. Model B is not helpful in the long term because it focuses on welfare and ignores independence and, because of that, limits development of relevant capabilities in the child. Model C is preferable because it focuses on all three together – capability development, independence and welfare – and tries to get a balance amongst them. This is part of the development of capability as a parent. It is not easy and all of us are learning all the time, even as grandparents.

9.5.2 Application of the Parenting Models to Teaching

Now it won't surprise any reader when I claim that in each of the eras described earlier in this chapter, academics were being urged to do quite different things. In the 1970s, they were being urged to pay attention to their own performance, with curriculum content being defined primarily by traditional, discipline-based, knowledge structures. Learning would take place through efficient transmission of discipline knowledge and would be absorbed in a linear fashion as the efficient teaching took place. This has parallels to parenting Model A as learning per se is neglected. All of us who experienced those times remember academics justifying their lack of interaction with students by claiming their role was to define the syllabus in lectures and the students' role was to learn it – "it's not my job to teach you, it's your job to learn" paraphrases a common (Model A) sentiment of the time.

In the 1980s, teachers were being urged to turn that thinking on its head and to focus more on what was happening as students learned, both within and outside the classroom. The intention now was on promoting experiences that encouraged learning for meaning and the development of conceptual understanding. The curriculum content was now not just the topic structure of the discipline but also an elaboration of the key concepts within that content. Nevertheless a stepwise acquisition of such key concepts was assumed. This approach was often criticized by the Model A teachers as spoon-feeding. Certainly some teachers were led into adopting a Model B equivalent approach – "if I don't say it in the lectures it won't be on the exam" or "learn these ten items this week and I'll test you on Monday" or "the examination will contain only questions that have been on the weekly tests".

I remember a dispute at a Swedish University in the late 1980s involving an examination paper that contained a question that went beyond the lectures. Students mounted a successful protest and the evidence put forward was that the question called for knowledge not contained in the subject notes issued to students by the lecturer. This had a profound, and I would say, educationally detrimental effect on the institution for some time afterwards.

The capabilities-driven curriculum model of more recent times is consistent with Model C and is more demanding on academic teachers. No longer are they being urged either to "teach the facts" as they know them or simply to help students develop the conceptual understandings that the lecturers have. Now they are being urged to help students become capable of handling future situations that not only the students are unaware of but about which the lecturers also know little. Part of the capabilities-driven curriculum design model is the notion that integration of learning across the whole degree structure is essential so that students learn to draw on the various aspects of their learning in ways that can help them deal with previously unseen situations. It is the capability to figure out what aspects are relevant and how to use those aspects together to address the situation that needs to be learned. In this model, separate subjects taught and assessed independently are inadequate. A more holistic curriculum approach is required with learning experiences and assessment that cross subject boundaries without restriction. One such curriculum design is described in Bowden (2003).

Now in a controlled situation, academic teachers exposed to this kind of educational development over a number of decades might well progress through each model and become expert with and committed finally to a capabilities-driven curriculum approach. Unfortunately the world is not so linear. In any given university, all three models, as well as others, may be present at any one time. What is an academic teacher to think? The different models lead to different practices. What if some in the university are following one model and others a different one? How does the academic seeking cooperation with colleagues in implementing an integrated capabilities-driven curriculum with assessment across subjects deal with the likelihood that many of them may be focusing on the more subject-oriented conceptual development model or even on the more teacher-focused efficient transmission model?

How do you get comprehensive change? Well at RMIT, the then Deputy Vice-Chancellor (in 2000) was committed to the need for a capabilities-driven curriculum model and organized a number of one and two-day symposia to enable academic teachers to become familiar with the concept, to find out about it in detail and to try to apply it to the courses they taught. In addition, the relevant university committees endorsed a requirement that in future, all curriculum design for new courses and for the regular five-year renewals would use the capability-driven model. In Bowden (2003), the first curriculum design project at RMIT University is described. On paper the outcome looks fine. The capabilities intended as learning outcomes are defined in detail and appropriate cross-subject learning experiences and assessment included. However, the implementation was difficult. It took some time for the curriculum design team, which had members from all the relevant faculty areas, to come to terms with what was required. By the end of the design process, however, most if not all members understood the process and were comfortable with the design outcome. Then came a management decision in the Faculty that undermined the whole activity (as was both predictable and was predicted). No one who was on the curriculum design team was allowed to take any management role in the teaching of the curriculum – not as course coordinator, subject coordinator or year-level coordinator. I presume the rationale was that this would be a way to get a broader involvement of staff in the process. However, the consequence was that none of those now responsible for teaching the capabilities-driven curriculum had experienced the difficult but transforming discussions that members of the design team had. They knew about the model but didn't have the same "feel" for it that team members had. My observation was that teachers quickly drifted into the subject focus they had been used to and the integrating framework was fragmented.

If the underlying philosophy of the new change is at odds with the philosophy of the previous approach, then the new practices developed may be inconsistent with those previously in place. For instance, my own philosophy that curriculum design and the learning environment should be shaped to enable students to develop the capability to deal with a continuum of previously unseen situations in the future requires quite different educational practices from those required for the earlier focus on developing conceptual understanding of discipline knowledge. When a new regime takes responsibility for educational practice in the university and their own educational philosophy is the one that had already been replaced, they are likely to try to restore it and then the educational environment turns on itself and all coherence is lost. It is even worse when the changes implemented have no recognisable educational philosophy underpinning them. We have an idealized image of each generation standing on the shoulders of the previous one and reaching for the stars. The reality is different from that idealized model. Educational practices are just as likely to be determined by other drivers such as fashion, economic rationalism, politics and the ambition of both institutions and the academics within them. Often with such drivers the educational theory-in-practice, if it is discernible at all, may be limited, regressive (to an earlier less sophisticated theory) or confused (if there are competing theories simultaneously in operation).

It is almost like someone building a castle in the wet sand when the tide is out. No matter how intricate the design, no matter how favourable the comments from passers-by, no matter how many others have been encouraged to build their own spectacular castles alongside, no matter how more striking this part of the beach has become, no matter how much pleasure the shared construction of the sandcastles provided to both the builders and the observers, the simple truth is that, come the high tide, the sandcastles won't be there any more. Now that analogy may be thought of as rather glib. Surely attempts to improve teaching and learning in a university, if they are of sufficient quality, should have long-lasting effects. The inevitability of the next high tide doesn't have its equivalent in the life of a university, does it? Well, maybe it does. The example I just very briefly outlined earlier is one illustration and it has significant implications for academic teachers, educational practice and student learning. In the case of that first capabilities-driven curriculum at RMIT University, a striking sandcastle had been built but soon the rough water created by the Faculty staffing rules washed it all away and later on the SET high tide came in and found only bare sand, with no sandcastles to be seen anywhere.

9.6 A Further Look at the Aims-Demands Model

In this section I am going to extend the commentary from The University of Learning (Bowden and Marton, 1998, p. 228) regarding the tendency for people to interpret aims expressed at one level in the organization as demands at another level. For example, government interest in achieving certain outcomes in the university system might be interpreted by university CEOs as demands for certain action. This is then re-formulated as a set of new aims for the university, which are interpreted by the staff as demands for certain performances. Eventually, students' learning experiences may be changed in ways that seem very distant from the original, broad goals of the politicians. One prime example for me is quality assurance. A commendable goal by a government became a tool (in Australia at least) for control by public service administrators and in turn an opportunity for vice-chancellors to more closely monitor staff performance. Unfortunately, a context of this kind sometimes leads to the appointment of senior and middle managers who thrive on such opportunities. In the case of the quality assurance systems, academic staff eventually learned well how to fill out the required forms without allowing any of it to interfere with their teaching. However, the raft of evaluation sheets constantly required to be filled out by students often led to cynicism and alienation among students that had the opposite effect to that originally intended. In a similar way one can well imagine that some European academics will attempt to respond to the Bologna process by not changing teaching at all but by introducing a formally designated "pause" part way through – to satisfy the formal requirements without interfering with the academic processes.

I see the increasing interest of the community and governments in more detailed involvement in what goes on in universities as a potential gain but with the danger that ignorant interference can produce negative outcomes, for students say. Academics have a great opportunity to embrace community interest and use it beneficially. However their individual autonomy allows for and often leads to apparent compliance without real (or at least without positive) change. The issues are about (1) the way aims at one level are interpreted at a lower level in the system, (2) the extent to which those at the more senior level have the expertise to be specific about required activities at a lower, more narrowly professional level, and (3) how those at the lower level use their degree of autonomy to support the spirit of the intended change, to subvert it or, more dangerously, to respond to the perceived demands in ways that are detrimental to function, e.g., student learning.

It seems to me that the solution relates to the development of a culture (or its enhancement if it already exists) that, if embraced by everyone in the university system, would link responsibility with expertise. The decisions about what should happen would be made by those who best understand the processes but, within such a culture, they would also take account of the goals of those higher up in the system and the consequences of their own actions for those lower down in the organization. The string in regards to teaching might look something like: politicians – bureaucrats – CEOs – senior managers and boards – faculties/schools – departments – academic teachers – students. It would not always be linear but the kind of thing I have in mind would have CEOs talking to the bureaucrats about, say, quality assurance and acknowledging that they understand the need for government to be assured of the quality of university learning. However, if such a culture existed, they would be able to convince the government that the imposition of, say, an industrial model of quality assurance is not the best way. If such a process were successful and, just as importantly, known about by academic teachers, then those teachers may engage more enthusiastically in an internally designed quality assurance process aimed both at improving student learning and providing evidence that it has been successful. The issue is to make it possible for everyone to be able to work towards shared aims rather than to be merely responding to what appear to them to be imposed demands. I don't think I have a generalizable process as to how to do that. Rather I would be asserting the need for individuals, groups and organizations to work towards such a culture. But they would have to find the best way within each of their own contexts.

I need to say here that I am pessimistic however. This chapter has concentrated so far on the internal university relationships. If universities were left alone, I think it would be a difficult task but it would be possible to develop the kind of culture I just outlined. However, universities are part of the community and, in particular, are exposed to political influences. These political influences have direct effects on university functions and their presence also influences the internal culture. Both of these factors militate against the development of the kind of culture I believe is essential for the kind of distributed leadership that is essential, for coincidence of expertise and responsibility.

I want to emphasize that universities are both fragile and resilient organizations. They are resilient in the way I have already described. Many changes imposed from outside might change the formal processes but the activities that are in the independent domain of the academic may not change in substance at all. This can be both a positive and negative attribute. It means that the worst attempts at political interference might not succeed. It also means that it is difficult for change agents to influence positive change. On the other hand, universities are fragile because of the following:

- The growing interest of government and the community in the details of university functions
- The increasing expectation for a university president, vice-chancellor or rector to be like a CEO of a business, rather than the chairperson of a community of academic peers
- The increasingly vertical management structures within universities
- The normal human trait of those providing money demanding a voice in how it is spent

- The expectation by such external stakeholders that measurements will be made to ensure that performance is at the level required
- The varying levels of expertise among different stakeholders in regards to issues of learning, teaching, assessing performance, managing people and responding to stress

Political interest in some specific aspect of on-campus life can be driven by a philosophy that is alien to the philosophy espoused and being acted upon by the university community at the time. The attempted imposition by government of an industrial model of a quality assurance system on universities in many countries is an example. This produced either no change in the most resilient universities or greatly distorted teaching and learning relationships in the most fragile. Few were totally unscarred by the experience.

A topical example comes from Australia. It is not about science or mathematics education but it could easily have been and still provides a relevant message. For quite a few years, then Australian Prime Minister John Howard had frequently criticized the teaching of history (mainly at school level). Shanahan and Healy (2000) quoted him as saying that there was "a little too much emphasis on the issues rather than on exactly what happened". Well an educationist can live with the fact that the Prime Minister believes that reading history can tell us "exactly what happened", provided that he doesn't try to change what goes on in the classroom. However action did follow. As Campbell (2007) reports:

> In January (2006) he [Howard] called for 'root and branch' renewal of the teaching of Australian history. He accused teachers of abandoning a 'methodical narrative style' and condemned them for promoting a 'fragmented stew of themes and issues'. This was followed up in August by a History Summit, hosted by Federal Education Minister Julie Bishop.

The historians invited were of those responsible for the curricula being criticized were included. Not surprisingly the conference concluded that changes needed to be made in the direction the Prime Minister had indicated. More recently Howard had attacked again by denigrating the number of subjects being called by "ridiculous titles" such as "Time, Continuity and Change". They should be called History according to Howard because history is what should be taught. Of course, I find the first title wonderfully descriptive of what a study of history might be about although I might have alarmed the Prime Minister even further if my preference "Time, Context, Continuity and Change" had been used instead. As Peter Mac (2006) stated:

> In fact, thematic teaching of history is crucial for imparting an understanding of subjects such as the dispossession of the Aboriginal people, or Australia's involvement in US-led wars. As Vietnam veteran and Labor MP Graham Edwards remarked caustically after a national history summit in Canberra last week, 'Teaching history [...] might keep us from being involved in more follies and more wars'.
>
> In contrast, a Howard government history course would comprise a narrow narrative of government-defined nation-shaping events. As an example, it would emphasize the dates of Cook's exploration of the Australian coast, but would reveal very little about the motive forces for British colonialism.

The outcome depends upon how the curriculum is taught and examined but I would have more optimism for a curriculum that focused attention on the influence of situation and context on action at any given time than on a curriculum that emphasized the listing of so-called facts. The former is more likely than the latter to lead to understandings that affect contemporary thinking – a key objective of any study of history.

Now I have treated this example in detail because it demonstrates most of the points I have made already. In the first place, an individual's response to any situation is related to their way of seeing it (Bowden and Marton, 1998, pp. 6–8). Howard appears to believe that there is some verifiable evidence that allows a teacher to say, to a student, "X happened in year Y for reasons Z". My own position and that of most educators would be that you need to look at a whole range of issues from a variety of sources before you can estimate the credibility of any one piece of evidence. It is developing that capability that studying history is about, to realise that the writings of any one historian represents their way of seeing the situation, the evidence. A stark example is the way that past historians have written about contemporary events as if women played no role. Yet an examination of past events with a theme related to women's roles can demonstrate that women indeed played an important but largely unreported role and that their absence from the "narrative" that Howard wants to cling to is a consequence of the narrow thinking of the (mainly male) historians at the time. Given Howard's beliefs, it is not surprising that he scorns thematic treatment of history. His view of what history is about is different from mine and so it is to be expected that he and others like me would not have the same view on how it should be taught. There is nothing remarkable or problematic about that. However, whether the view of the Prime Minister on such an issue should influence the conduct of educational institutions is another matter. I think it should not and yet universities are pressured from time to time to make changes that government members or their agencies believe should be made. They assert their right to bring about the change despite their lack of professional expertise in the relevant field.

The second issue is the power of funding. So far, the Australian Government has not taken the history debate into the funding arena. However, there have been many other examples where they have. The requirement for quality assurance and curriculum development in teacher training are two areas where funding has been contingent on compliance, often using scores on national scales. Such measures and their funding consequences are common in the United Kingdom. The Swedish Government's funding of universities was made contingent during the 1990s on student pass rates. Once the intentions of government become embedded in funding arrangements, university CEOs are encouraged to implement the scheme devised by the government. Then all of the aims-demands issues arise in the university and any attempts to produce or preserve the kind of idealized culture I described earlier are destroyed. In the general model I described earlier, such government intervention in the day-to-day workings of the university comprises an example in which authority and expertise do not coincide. The gravity of that is elaborated in Bowden and Marton (1998, pp. 259–261). It is the opposite of the kind of arrangement that Bennett et al. (2003) describe as being ideal for distributed leadership.

Universities are dynamic and complex organizations that are part of the community. They are both fragile and resilient. As a consequence they can be harmed, on occasion, by either internal or external forces. In different circumstances they can be resilient and absorb the onslaught without harm to their functions. From another perspective they can sometimes embrace positive change and on other occasions resist it. An ideal university would have a culture that was open, trusting, reflective and optimistic. It would ensure that decisions were the responsibility of those with the expertise to make them. They would in effect be examples of a pure form of distributed leadership. That model would apply not just to the university internally but also to its external relations with the community and government. It is a valuable target to aim at but I doubt that it is an attainable goal. I hope the readers of this chapter go on to prove me wrong.

9.7 Afterword

This section has been written after the chapter was completed, reviewed and revised. I have written it because there is a personal aspect throughout the chapter that at first glance doesn't make sense. In several parts I indicate my pessimism that any comprehensive and coherent change to the university learning environment can be accomplished in ways that foster for graduates the kind of learning outcomes that I value. I suggest that educational developers like me have not, over the past three to four decades, witnessed any successful change process at the system level. I express the hope that academics in the future will be able to achieve what we have not and yet I express my doubts that it will happen. There is a consistency to all of that. Yet at the same time I have said that I have been satisfied with the work I have done since the mid 1970s and that I enjoy the work I am still doing in retirement. How can that be so? How can I be satisfied with what I have depicted as failure?

All of the statements in the previous paragraph are true in that they represent my true beliefs. Yet it seems that they can't all be true as stated – hence the questions at the end. The questions are not just for readers. They are questions I am putting to myself. How could have I been satisfied with my work during those many years? Well, I was thinking about that as I went on my daily seven kilometre walk through the Yarra Valley (Melbourne's principal river) this fine autumn morning (Easter Monday) – a delightful opportunity each day for reflection, despite our being in the grip of a drought entering its eighth year. I began by thinking about the Kuhn-Popper debate concerning scientific development (Kuhn, 1970; Popper, 1963) and trying, in Popperian or Kuhnian terms, to make sense of the apparent inconsistencies cited above. My first efforts failed but then I began to adapt the Kuhn and Popper theories to take account of the fact that university educational development is a different entity from natural science, and the following represents my train of thought. I could have gone back and revised the whole chapter and pretended that, as I was writing it, I knew what the outcome would be. However, that would have masked the actual development in my own thinking that has taken place through the writing and I prefer to leave my thinking trails uncovered and visible. Readers may find them useful for their own reflection.

Popper's position was that any scientific theory is tentative and that it survives only for as long as it resists ongoing attempts to refute it. As more evidence becomes available, it is likely that a new theory is needed to explain both the old and new evidence together. He depicted science as being in a constant state of change. Nothing is totally accepted: scientists are constantly trying to refute their theories and build new ones.

Kuhn suggested that Popper's depiction of scientific change was not historically accurate. He pointed to the ways in which various branches of science cling to established theory long after some of its inadequacies have been exposed, and the theory only changes when the inconsistencies become so numerous and problematic that everyone agrees that the theory has to be replaced. He depicted such an event as a scientific revolution. In between such revolutions, he saw scientists as being engaged not so much in the task of questioning and trying to refute the current theory but rather in puzzle solving, with the theoretical base remaining stable. Innovation in between revolutionary periods is related to better and more effective puzzle-solving approaches rather than to theory change. A way of rationalizing the two perspectives is to see researchers in science generally taking a Popperian perspective and practitioners normally behaving as Kuhn describes. Of course this represents scientific advance as an essentially social process – a perspective that I believe is accurate.

When I tried to apply those ideas to pedagogical change in universities I noted that there were several points of difference. The first was that pedagogical change involves human activity both in terms of who changes and of what is changed; the focus of the Kuhn-Popper debate was how people's understanding of the physical world changes. People are still involved (the social process) but what they are involved about is different (natural science phenomena versus how people learn). Scientific theory is very specific about the contextual elements and so there is no doubt about its intent, whether or not it precisely reflects the range of human experiences of a phenomenon. On the other hand, explanation of human activity is significantly context dependent and so the notion of a universally accepted theory related to the human activity of learning, let alone a theory that is stable across time and context, is somewhat fanciful.

Secondly, scientific theory is developed within a social framework that seeks and accepts a sanctioned theoretical form at any particular time. Within studies of human activity like learning, if any such social framework exists, it is merely one among many competing frameworks, each with their own different theoretical perspectives.

Thirdly a Kuhnian revolution does not usually occur across all of science unless there are exceptional situations such as with Einstein's relativity theory. This is related to the structure of the scientific community. In fact it is a series of micro-communities, each focused around some specific aspect of the natural world. When revolutions occur, they are generally confined to a specific micro community and a small number of "neighbouring" micro communities.

In education, most efforts to advance teaching and learning are aimed at a system level. Any successful revolution would need to be relatively universal to be recognized. So the Kuhnian notion of occasional scientific revolutions with stable periods of puzzle-solving in between doesn't work for university educational development as we have practiced it so far. When I tried to think of the shifts from teacher-centredness to conceptual change to capability development as Kuhnian paradigm shifts at the system level, I found it impossible to incorporate the partial nature of those changes and the co-existence of all the paradigms, even today.

Yet there are some insights to be gained. A general case can be made that educational researchers such as myself and my colleagues have taken a Popperian stance. In my own case, my early embrace of conceptual change learning and its application to professional development (Bowden, 1989) and my subsequent investigations into issues of competence and competencies (Bowden and Masters, 1993; Bowden and Marton, 1998) and my recent focus on capability-driven curriculum design (Bowden et al., 2000; Bowden, 2003, 2004) have all been undertaken within a context of developmental support, each time, for the very perspective that I was trying to refute and supplant.

On the other hand, the Kuhnian model seems to work for individual teachers. Good teachers listen to their students and make judgements about how to respond best to help students learn as intended. However, after they have responded, they continue to listen and are ready to revise their thinking and respond again differently as they take in new evidence, perhaps from educational researchers, among others. Good teachers are doing that constantly and it is important because teaching and learning are so context-dependent. In a different time, place and situation, new responses are required. I would suggest that good teachers are engaged in Kuhnian puzzle solving within a stable theoretical framework. That stable framework is their personal theory of teaching and learning. This personal theory can in time be supplanted by a more powerful theory, a revolution at a personal level rather than a universal one. That is a difference from scientific revolutions where the new theory is codified by the relevant element of the scientific community. This doesn't happen usually universally in fields involving human interactions. Another parallel relates to the observation I made earlier in this chapter about academics' tendency to resist change. A Kuhnian scientific revolution occurs only after the natural resistance to change within the relevant micro community is overcome. Similarly even good university teachers resist having to overturn their framework; like all Kuhnian puzzle-solvers they keep making more and more adjustments to their current framework until change becomes inevitable.

I can see that my satisfaction with my educational development work of past decades has derived from the fact that many teachers with whom I have interacted and some parts of universities in which I have worked have experienced revolutionary change at the personal and local levels. I can also see that perhaps educational developers like myself have over the past few decades, whether consciously or not and whether naively or not, been hoping for a Kuhnian revolution at the system level. In the education field, as we deal with human interactions and a community that is not built on a culture of a single codified theory, such a revolution will not occur easily. Hence some of the pessimism I have expressed in this chapter.

Perhaps the lesson for me to learn here is that the future may be about re-thinking the goals – not to lower expectations but to become more realistic and perhaps more specific as a consequence. Camilla Rump, in an emailed comment on my first draft of this section, suggested that:

> …a more viable base for changes in the future would be to try to build stronger international, national or at least cross-university knowledge cultures; that is, if there is indeed a true culture of knowledge about teaching a certain subject with some sort of quality assurance system going on – like peer-reviews in the sciences – then this culture would be more stable […] but also more subject to 'revolutionary insights'.
>
> (Rump, 2007)

His reminds me that, in The University of Learning, we proposed what we called "a bold undertaking" that included allocation of one per cent of the university budget to studies in knowledge formation in each knowledge area and to "invite all departments to open up their doctoral studies for specialization on knowledge formation in the discipline or domain" (Bowden and Marton, 1998, p. 288). In this way, we would be opening up the possibility that theories of learning could be developed and codified in mathematics and science fields, making use of the same or similar micro communities (or sets of neighbouring ones) that support Kuhnian scientific revolutions. Over the past three to four decades, many personal and local changes have occurred but have not become more comprehensive because of the lack of an educational community structure that values, sanctions and implements such revolutionary changes.

Perhaps that is what my challenge to readers is actually about. The task in the future is to develop structures within mathematics and science education that enable educational developments to be more comprehensively embraced, put into practice and then, in turn, be supplanted by even more appropriate theory and practice. However, the other edge on the sword that is being handed over is that the social context in which universities sit is also undergoing rapid change and the interactions between responses to growing pedagogical need and to changing political interest in university life have to be balanced. I wish readers well in their endeavour.

References

Argyris, C. & Schön, D. (1978). *Organisational Learning: A Theory-Of-Action Perspective.* Reading, Massachusetts: Addison-Wesley.

Ausubel, D. (1968). *Educational Psychology: A Cognitive View.* New York: Holt, Rinehart & Winston.

Bennett, N., Wise, C., Woods, P. & Harvey, J.A. (2003). *Distributed Leadership.* Oxford, UK: National College for School Leadership.

Biggs, J. (2003). *Teaching For Quality Learning At University.* Buckingham: SRHE and Open University Press.

Bowden, J.A. (1988). Achieving Change In Teaching Practices. In P. Ramsden (Ed.), *Improving learning: New perspectives* (pp. 255–267). London: Kogan-Page.

Bowden, J.A. (1989). Curriculum Development For Conceptual Change Learning: A Phenomenographic Pedagogy. Paper Presented At The 6th Annual (International) Conference Of The Hong Kong Educational Research Association, November (Also Available As RMIT EQARD Occasional Paper No 90.3).

Bowden, J.A. (2003). Why Do We Need Reforms, Which, And How Do We Implement Them? In J Carter, K Eriksen, S Horst & R Troelsen (Eds.), *If Reform Of Science Education Is The Answer – What Were The Questions?* (pp. 9–29). Centre for Science Education: University of Copenhagen.

Bowden, J.A. (2004). Capabilities-Driven Curriculum Design. In C. Baillie & I. Moore (Eds.), *Effective Teaching And Learning In Engineering* (pp. 36–47). London: Kogan Page.

Bowden, J.A. & Masters, G.N. (1993). *Implications For Higher Education Of A Competency-Based Approach To Education And Training*. AGPS: Canberra.

Bowden, J.A. & Marton, F. (1998). *The University Of Learning: Beyond Quality And Competence*. London: Kogan Page. (Paperback edition published by RoutledgeFalmer: London, 2003).

Bowden, J.A., Hart, G., King, B., Trigwell, K. & Watts, O. (2000). *Generic Capabilities Of ATN University Graduates*. Retrieved April 10, 2007, From http://www.clt.uts.edu.au/ATN.grad.cap.project.index.html.

Campbell, D. (2007). Teachers Enjoy Holiday … From Pollies. The Age, January 8: Melbourne.

Entwistle, N. (1998). *Conceptions Of Learning, Understanding And Teaching In Higher Education Scre Fellowship Presentation*. Retrieved February 16, 2007, From http://www.scre.ac.uk/fellow/fellow98/entwistle.html.

Hegland, J. (1993). "Sink or Swim?" – Swimming Lessons Can Be Scary For A Child. *Mothering, 68*. Retrieved February 2007, from http://www.thefreelibrary.com/%22Sink + or + swim%3f%22 + (swimming + lessons + can + be + scary + for + a + child)-a014055083.

Henry, S. (2006). *A New View: Distributed Leadership. Usable Knowledge*. Harvard Graduate School of Education. Retrieved February 2007, from Http://www.uknow.gse.harvard.edu/leadership/leadership002a.html.

Johansson, B., Marton, F. & Svensson, L. (1985). An Approach To Describing Learning As A Change Between Qualitatively Different Conceptions. In A.L. Pines & T.H. West (Eds.), *Cognitive Structure And Conceptual Change* (pp. 233–257). New York: Academic Press.

Kuhn, T.S. (1970) *The Structure Of Scientific Revolutions*. (2nd Ed.). Chicago: The University of Chicago Press.

Mac, P. (2006). *Howard's History Is Certainly Bunk*. The Guardian, 30 August: UK

Marton, F. & Säljö, R. (1984). Approaches To Learning. In F. Marton et al. (Eds.), *The Experience Of Learning* (pp. 36–55). Edinburgh: Scottish Academic Press.

Marton, F., Dall'Alba, G. & Beaty, E. (1993). Conceptions Of Learning. *International Journal of Educational Research*, 19, 277–300.

Popper, C. (1963). *Conjectures And Refutations: The Growth Of Scientific Knowledge*. Routledge: London.

Ramsden, P. (1992). *Learning To Teach In Higher Education*. London: Routledge.

Rump, C. (2007). Personal communication.

Science, Engineering & Technology Portfolio (2007). *SET Portfolio Guiding Principles For Educational Programmes*. Retrieved February 2007, From http://mams.rmit.edu.au/ioohvmm4o44wz.doc.

Shanahan, D. & Healy, G. (2000). *PM's Timely History Lesson*. The Australian, 28 April: Canberra.

Spillane, J.P., Camburn, E. & Lewis, G. (2006). *Taking A Distributed Perspective In Studying School Leadership And Management: Epistemological And Methodological Trade-Offs*. Draft paper prepared for the Annual Meeting of the American Educational Research Association, San Francisco, April 7–11, 2006.

Chapter 10
Reforming University Studies

How Well Does New Legislation Modify Entrenched Behaviour?

Sebastian Horst and Kjeld Bagger Laursen

This chapter is a case study of a change of "legislation," the formal conditions under which degree programs operate at "classical" research-driven universities. Before setting up our analytic and conceptual tools a brief history of our case:

The pressure was on at the Science Faculty at the University of Copenhagen in the middle 1990s: the Faculty had trouble making economic ends meet, and had been forced into budgetary cuts that reduced the scientific staff, in some departments by a quarter; the science degree programs were experiencing a drop in recruitment numbers; a large evaluation of these programs, covering the mathematical, physical and chemical fields, and all Danish universities who produced candidates with a Master of Science degree, had revealed rather high drop out rates, in some cases more than 50% (Danmark's Evaluerings Institut, 1999). There was an impression, among politicians, perhaps even among voters in general, and certainly voiced rather forcefully by influential lobbying organizations, such as the Confederation of Danish Industries (Dansk Industri), that many educational institutions were not fulfilling their share of the responsibility for educating the Danish youth, and that these universities were too costly, as well as too far removed from modern educational demands.

It was not easy to maintain high spirits within the academic staff under these circumstances. Nobody can stand prolonged and serious criticism, and nobody enjoys having to be on the defensive indefinitely. So the mood became rather mixed. On the one hand, there was pronounced irritation at the apparent lack of respect for the value of maintaining classical working conditions for university scientists. On the other hand, there was a growing understanding of the need to change some things, as well as a rather basic urge to respond proactively, and not just reactively, to the challenges voiced via the criticism.

In the spring of 2001, the Dean formulated a strategy for how to generate a response. This strategy was of course based on the rules of governance of degree programs at the Science Faculty: all decisions on scientific and scholarly content must be made by a study board, not by the Dean, so it was clear that centrally coordinated changes could only be recommendations, if they attempted to deal

S. Horst and K.B. Laursen
University of Copenhagen, Denmark

O. Skovsmose et al. (eds.), *University Science and Mathematics Education in Transition*,

with educational content. The study boards, which are responsible for degree program content, are mostly "single subject" boards, and tend to be associated with individual academic departments. Thus there is a board for the chemistry degrees and a board for the physics degrees etc., not withstanding the fact that many students would (and will) pursue two-subject degree programs. On the other hand, the structure of these degree programs could be decided in a rather more centralized and coordinated fashion. The upshot of these considerations was the formation of a coordinating committee, chaired by the Dean, and consisting mostly of the chairmen of the nine study boards, supplemented by student representatives, representatives from the Board of the Science Faculty (the Academic Senate of the Science Faculty), as well as the two centres that had to do with teaching, the Centre for Science Education, and the Natural Sciences IT Competence Centre. This coordinating committee was given the task of defining the framework for all future bachelor and Masters degree programs at the Science Faculty.

The guiding principles for this task were that the degrees in questions should be made more attractive through increased transparency, cohesiveness, relevance and flexibility. After an initial round of rather freewheeling debate in this group, the Dean summed up these goals in this way:

- There should be fewer separate bachelor degree programs at the point of admission
- There should be greater freedom of choice and flexibility in the course of studies
- Replacing the semester system, the academic year should be divided into four short sessions ("block structure"), to enhance mobility and flexibility
- It should be possible for the student to change his/her main subject in advancing from the bachelor to the Masters degree
- The programs should allow for increased differentiation of perspective, e.g., depending on individual career plans
- New methods of teaching and assessment, including increased use of IT, should be introduced
- Interaction with the surrounding society should be encouraged, for example by means of work-study or international exchanges. (Fundamental document, October 2002)
- Significantly, the resulting plans for the new degree programs should be expressed in terms of intended educational outcomes, or competencies, so that the individual candidate's academic profile would be clearly visible to prospective employers (including admissions officers of graduate degree programs).

Taken as a whole this agenda would appear to be an example of a blueprint for *fundamental change*. However, as we are about to define terms, this is not necessarily the case. To describe and evaluate the ensuing process we specify the analytic and conceptual framework, which is a modest modification of L. Cuban's typology (Cuban, 1999, pp. 62–75). By means of this framework we will be able to classify the items in the above list, as well as the entire process. We shall then pick out some of the particularly significant and interesting items above and examine these in greater detail. These items primarily have to do with the block and scheduling features, and with the competency descriptions.

Before we continue we should make a remark on our own roles in this case study. Kjeld Bagger Laursen has followed the reform of the degree programs all the way through; first as a faculty member and head of the study board at the Institute of Mathematics, and later also as Head of Centre for Science Education. As Head of Centre he has been a member of the coordinating committee of the reform: he has not had any formal responsibility for managing the reform. Sebastian Horst has worked as a Consultant at the Centre for Science Education since 2003. We have both monitored the reform and the changes via our participation in ad hoc groups reporting to the Faculty about the implementation of the reform. We have also carried out a handful of projects on specific courses dealing with their specific problems and challenges.[1]

10.1 The Framework: A Typology of Scope of Deliberate Change

As indicated by the headline, we consider deliberately initiated processes of change. In such processes we may distinguish three stages: proposed changes, changes that have been decided on, and executed changes. These distinctions are partly made to enable a nuanced description of the political processes, and partly simply to introduce the time dimension into our discussion. More about this later.

The typology is four-dimensional. The dimensions are described by the terms depth, breadth, level and time. Figure 10.1 shows the two dimensions depth and breadth, and a point in this diagram would indicate a classification of a given change with respect to these dimensions. Examples will follow shortly.

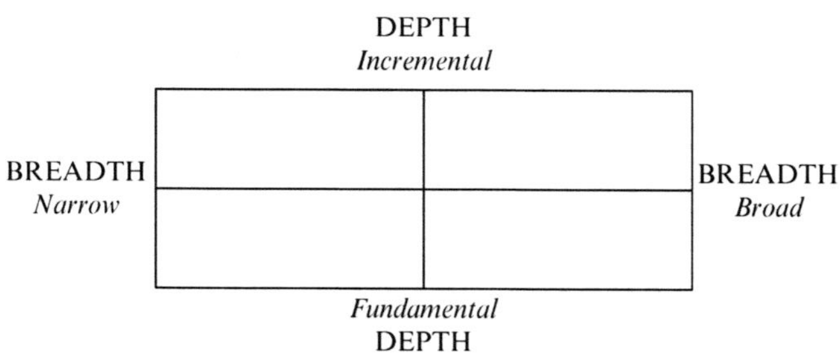

Fig. 10.1 The two dimensions BREADTH and DEPTH provide tools for analyzing changes (from Cuban, 1999, p. 67)

[1] All project reports available in Danish at http://www.ind.ku.dk/side21727.htm.

Depth of change indicates the degree to which the designers of intended change seek to make minor, modest or major transformations of key features of the university (or, in this case, of a Faculty). The extremes of the depth of change will be described by the terms incremental, and fundamental, respectively. Key features could be the concept of "academic freedom" as it expresses itself in the freedom of faculty to choose research topics. Or it could be the fundamental rules of governance of the university, as set out in legislation or charter.[2] An incremental change could be a small change in the teaching workload, or a change in the placing of two courses in the study program. Characteristic of incremental change is the implicit belief in the soundness of the existing structure or basic values – a belief leading to only initiating corrections of the existing structures and processes. Note that because we also have a time dimension in our typology, there are no implications between the use of "incremental" versus "fundamental" and the time span of a change process, although often, of course, incremental changes in depth will be executed over a short period of time. The concept "reform" would normally indicate a fundamental change, irrespective of the time perspective.

To give an example of a change process in which the depth dimension is useful: if a university wants to make it clear to all involved that research and teaching are to be viewed as two equal parts of its business, then this attitude must show up in the way these criteria for identification and appreciation are defined, and how they deal with those two kinds of academic activities. In practical terms, this means that the university must start by defining what it views as good teaching, and how it intends to measure it. The university must, in short, set criteria for good teaching that operate at the same level of seriousness as the assessment systems for judging research effort. This will almost certainly involve concepts such as peer assessment, criteria for career promotion, and pay raises, relying on teaching indicators, as well as ways of developing and using meaningful ways of judging teaching qualifications (not just teaching experience) when hiring academic staff (Gibbs, 2003). This would all amount to fundamental changes, although some of the constituent parts would be broad and others narrow.

These terms – narrow and broad – refer to the range of the dimension *Breadth of change*. This dimension is an indicator of how many structures and processes the change involves. The shift of a mandatory course from the second year of a particular

---

[2] In the current context of discussing the conditions for teaching at research-driven universities probably the most significant key feature is the sheer competition between research and teaching. That there is competition is demonstrated in many ways. Not only is there some uncertainty concerning how to maintain high scientific standards within the maintenance of teaching programs nowadays, there is also simple competition for the teachers' time and attention. We see it in the way future researchers are educated, in the way they are recruited, and the terms under which they are employed; in the criteria for promotions; and in the development of assessment schemes for performance: how is good research identified and appreciated – and how is good teaching identified and appreciated?

degree program to the third is an obvious example of a narrow change. But narrow changes can easily have large consequences. When City University of New York introduced an open admissions policy in 1970, it was a narrow change, but its effects on the composition of its student body and thus the conditions for teaching were enormous (Traub, 1994). A broad change, on the other hand, would modify several elements or the entire system. But a broad change does not have to be fundamental: it is easy to imagine changing a lot, e.g., requiring all courses offered by the entire university to have assigned a course weight (e.g., ECTS points) without altering the contents of the courses. This would be a broad and non-fundamental change.

Level of change simply refers to the organizational level at which the change takes place. Our case study obviously refers to the faculty level overall, but many of its individual points, such as "fewer separate bachelor degree programs at the point of admission" are executed by a kind of middle level, that is, a particular study board at a particular department. The "bottom" level is located in the lecture hall, classroom and laboratory where the professors decide what to teach, and how to teach it.

The fourth dimension is *Time*. This dimension will not play as large a role in this study as it deserves. We can draw some conclusions about the initial effects of the changes, but they are somewhat tentative, because of the short time span of our observations. Thus, one of the main points of the Cuban study on which we base our study, namely that research-driven universities tend to exhibit a certain change resistance when it comes to changes that interfere with their fundamental features, will not be possible to confirm or refute conclusively at this stage. We have a few comments at the end of this chapter (as does Bowden, Chap. 9). We hope to return to this point in a couple of years.

A particularly interesting aspect of the level dimension (which may well be specific to universities) is its potential bearing on the depth dimension. In any organization in which change is being implemented, the interaction between the "top-down" aspects (the decisions made by management) and the "shop floor" aspects (how these changes, decided on by management, are actually carried out) is decisive for the success of the change process. In universities there is a special twist to all of this: the faculty at a university works in a tradition of academic freedom. This tradition expresses itself in two related ways, both of which may well result in de facto resistance to change. First of all, as a researcher, a university employee has (some) freedom in choice of research topic, and certainly has considerable freedom in choice of approach to the solution of the chosen research problem. Secondly, as a teacher (and as an advisor), a faculty member works in a strong tradition of independence. In a way, all of this originates in the very basic notions of how universities best go about the business of pursuing truth.[3] Rump and Winsløw (Chap. 11), also present some observations on this.

[3] A good case-within-a-case that displays how 'ordinary' academic staff were involved in this local process, is provided by the new degree program in chemistry. Here it was decided to amalgamate two programs, one in chemistry and one in environmental chemistry, into one. Efforts were made to engage a wide segment of the academic staff in the planning: the study board decided on a way

Before discussing the problems that a top-down approach may run into in a university setting, let us take a look at some of the reasons why the exercise of leadership (i.e., "top-down") is indispensable in our case. The degree programs at the Science Faculty are all the result of traditions that are remarkably specific to the individual degrees. The programs in the biological sciences, the earth sciences, and in physical education are accustomed to being run as single-subject programs, while the programs in physics and chemistry, as well as those in mathematics and computer science, are more based on coexistence, probably because physics and chemistry, cannot be dealt with as advanced and research-driven subjects without mathematics – and as a result these subjects are geared to being run as combination studies.

This could of course continue, if no changes were needed. But since the germination of this underlying feeling that "something had to be done," and that this "something" should produce programs that would give every future student a better chance to construct his/her education as they saw fit (with some restrictions of course), including the option of mixing several subjects, it was evident that there would be a strong need for coordination.

Moreover, with the study board structure being as we mentioned before, no attempts to bring these rather different programs under some common umbrella is likely to succeed without real effort: with a body of study board chairs, who are each others' peers and cannot dictate anything to one another, only work by persuasion, no decisions are easily reached. Indeed, around the year 2000 there was an attempt to make a new set of study programs, but it failed. So by the time the process under scrutiny here was being set up it was clear that the Dean would have to exercise his influence and simply insist that an agreement be reached.

Even with this realization emerging, it is not clear that this process could have succeeded, were it not for the changing climate for leadership at Danish universities. This change was mostly externally induced. The Danish parliament and government were losing patience with the perceived inability of university management to effectively handle the altered economic and demographic conditions, and a major revision of the laws governing the operation of these institutions was underway. The trend was in the direction of more forceful leadership. And, even though

of dividing the chemical sciences into seven separate groups: analytical chemistry + applied spectroscopy, biological chemistry, physical chemistry, environmental chemistry, spectroscopy and theoretical chemistry, structural chemistry, nano-science and macromolecular chemistry, and synthetic chemistry. For each such scientific discipline a working group was formed, with a mandate to formulate how this particular topic should preferably (ideally, but at the same time realistically) figure in the study plans. The groups were deliberately made with overlapping membership, and the total membership spanned the entire department. Each group contained members who worked inside the discipline and outside the discipline. Thus active involvement by everybody was ensured, although influence from the local management was in clear evidence. It seems safe to predict that this change process at the local level is likely to be quite stable under the risk of 'passive resistance' by the members of the department, and its impact may well not dwindle to something minor over time (see also Bowden, Chap. 9).

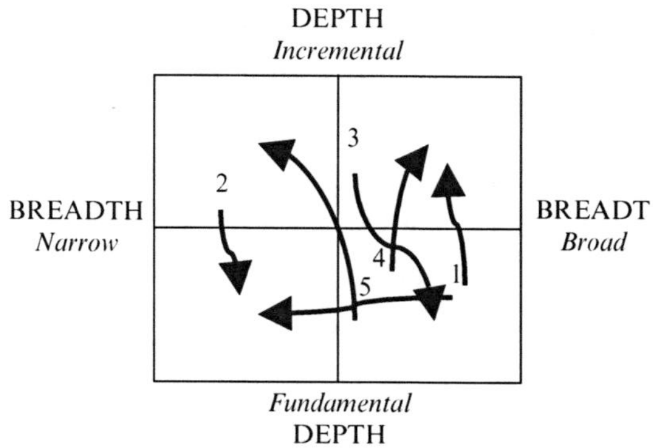

Fig. 10.2 The time development of the scope of change placed in the two dimensions BREADTH and DEPTH. The numbers refer to the specific intended change, and the arrows indicate the change in classification of the implementation up till now. Two arrows indicate that the change is being implemented in two different ways. (1) Redesign of courses. (2) Increased intensity. (3) The scheduling structure. (4) Outcomes defined in competency terms. (5) Assessment integrated in courses

academic staff valued (and probably still value) the principles of self-governance and democracy that have been in the rule books of Danish universities for a very long time, there was a growing realization of the need to sing in unison.

So the necessary willingness on the part of the local study leadership to accept decisions made in the coordinating committee was present. However, this readiness to act is still within the "top-down" part of the process. Shortly we shall get to the "bottom-up" part and the interaction between the two. In this interaction the time dimension is clearly present: policy and reform decisions are made by the "top" and carried out, over time, by the levels below the top. But, as the following sections of this chapter will show, it is the implementation by the "bottom" layer that determines the outcome of the process. The graph in Fig. 10.2 represents this relationship.

10.2 Characterizing the Over-All Features of Our Case

As listed above, the intentions of the Faculty of Science leadership in 2002 were quite wide ranging: It had produced a blueprint for broad change: there should be fewer separate bachelor degree programs at the point of admission, to give each incoming student a broader and simpler initial view of the possibilities and a better and more flexible basis on which to select further specialization. The two-semester academic year with January and June used for examination was to be replaced by four short sessions each running for nine weeks, and separated by one week periods without classes ("block structure"). The last two weeks of every block are intended for examinations. New methods of teaching and assessment, including increased use of IT, should be introduced and. Work-study or international exchanges should be made more readily available to all students. And last, but surely not least, the contents of all degree programs should be specified in competency terms.

It is still too early to conclude much about how the new features work practice and internationalization worked out, therefore in our examination of the significant facets of the character of this blueprint we shall concentrate on:

- The block structure: redesign of courses and increased intensity
- The scheduling structure
- Outcomes defined in competency terms
- Assessment as an integral part of courses.

It borders on the trivial to observe that a process of change has to be allowed to run its course before any persuasive assessment of its results can be made. So the fourth dimension of our typology, time, obviously has to be allowed to run. However, this trivial observation is pertinent here, because our case is still young enough and only tentative conclusions are available. But even preliminary conclusions are useful, because they are indications of the extent to which our fundamental thesis is supported by evidence. So we present what we can see now, and pledge to return to the entire issue in due course.

However, before describing in some detail the above mentioned four features of the process let us explain why the entire blueprint should be characterized as a plan for broad but not fundamental change. Broad it certainly is, since it encompasses all degree programs, both at the bachelor and at the Masters level (years 1–3, as well as years 4 and 5). But even though it attempted to bring about fundamental change, e.g., by making bachelor degrees single-subject degrees, and even though the students' freedom of choice was increased rather dramatically (the initial plans called for the single major subject taking up to only two years of the three year degree program), this was immediately watered down to accommodate joint programs, and to maintain certain prerequisites when preparing for admission to a post-bachelor degree. Moreover, ministerial rules later put some restrictions on student-electives. And equally significantly, many basic conditions were not changed at all: the efforts to "retool" teachers and give them real opportunity and time to address the new teaching challenges ("new methods of teaching and assessment") were sporadic and more or less random, only supported by rather modest efforts of study boards and their chairmen and by somewhat sub-critical activity by the faculty's educational centres, and without very visible and effective extra funding from the Dean. This approach is not one to create fundamental change, so we choose to describe the total plan as one for broad change of modest depth, even though, as we shall see, there were some aspects of the process that have the characteristics of fundamental change. We hasten to add that our conclusion – that the overall process until now is one of incremental change – in no way precludes that the stipulated faculty goal of making its science degrees more attractive through increased transparency, cohesiveness, relevance and flexibility can be reached.

The observations that we now present, are culled from a study of a good part of the first year of the reform (2004/2005), when the faculty and the Centre for

Science Education had a monitoring group working on determining how the initiative had worked out. The period under observation covers most of the first year of operation of the new study plans (Monitoring group, 2005).

10.3 The Block Structure

One of the first major decisions made by the coordinating committee was to switch to the block structure, i.e., replace the two semester academic year with four short sessions each containing 15 ECTS of workload for the students and running for nine weeks, and separated by one week periods without classes. The report from an ad hoc working group is dated in July 2002. It followed a spirited debate in the committee and in the working group and, even though the decision was collectively taken, it certainly constitutes part of the "top-down" end of the entire process. It was also a decision that was made on rather skimpy evidence. The arguments in favour of it had to do with the possibility of making student exchanges with Swedish universities easier. But the real argument for the change was probably even simpler: to make a rather dramatic change that would force the redesign of all courses, and thus enable improvements! Certainly no thorough and professional investigation of experiences elsewhere was undertaken. Many of the arguments against changing to the block structure – such as the risks of rushing learning processes that ought to be allowed considerable time to "settle" (the complex concepts of quantum mechanics kept cropping up as an example of material that would be best taught and learned over longer periods than nine weeks) – were never really countered, and certainly not to an extent that became common knowledge among the regular academic staff. Thus a "confrontation" between top-down and bottom-up was looming: how was the redesign of new courses, made to fit the new block structure, going to come off?

Not surprisingly many of these "new" courses were restructured on the basis of existing activities. Typically, courses with a size of 10 ECTS were to be refitted into a half block, i.e., shrunk to 7.5 ECTS. This did not work well in all cases; some of the resulting "new" courses would still contain too much material, perhaps because many teachers traditionally think of course contents in terms of "what material should I be able to cover in my lectures," (also see Bowden, Chap. 9). It could also be because the following course did not open up for a shift of material into this next course, or perhaps simply because communication between teachers was inadequate. In other cases the very definition of ECTS as a measure of student workload was not taken seriously, in other ways: one course achieved its reduction in size by reducing the number of contact hours, it simply cut out the exercise class hours! On the other hand, many of the completely newly designed courses did not experience this problem. A variant of this issue arose in the cases where a full block, 15 ECTS, was made up with ingredients from several "old" courses. This construction made scheduling rather easy, because the students were full time participants, but sometimes the constituent bits did not get coordinated particularly well.

One aspect of the block structure is especially interesting; it is a narrow, but potentially fundamental change. With a minimum course size of 7.5 ECTS, a deliberate attempt was made to ensure a certain measure of "calm": no more than two lines of thought (i.e., two separate courses) for the student to pursue in any given block. This was intended as compensation for the increased intensity that should be expected from the shortening of the teaching period. The initial evidence seems to show that this feature has worked out well: everywhere we have asked students about the advantages of the block structure this is named as an important one.

10.3.1 The Increased Intensity

The shorter blocks (nine weeks compared to twenty weeks in a semester) have made the increased intensity an important theme in the way students and teachers perceive the reform. One of the intentions was to squeeze more effort out of the students by making it clearly necessary to study seriously from the very beginning of the block – and this seems to have worked! Everybody, teachers and students alike, have to buckle down and get going from the first day of the course, when only nine (and in some cases only seven) weeks are available. The students do not necessarily like this effect – but there are many reports of considerably increased efforts. Similarly, the teachers experience the increased intensity as an increase in the demands on their time and attention. This increase is probably a significant contribution to the increase in quality of teaching delivered, even if the teachers involved do not necessarily like it immediately. Most people do recognize, however, that there is compensation to be found for each individual teacher in the possibility of having blocks, every year, in which no scheduled teaching is required. But this particular part of the overall change, again in itself a narrow change, may well be another example of a nearly fundamental change – if the block structure is allowed to last! Certainly the students feel some ambiguity about the intensity. Some explain very happily about how they enjoy getting the work done from the beginning – they acknowledge the need to be pushed. Others focus on the potential problem of being stressed because of the higher speed at which new material is being presented. This is without doubt an issue that displays the heterogeneity of students.

The intensity of every course has also increased the importance of teacher collaboration. There are surprisingly many courses at the Faculty of Science that employ teams of lecturers (as well as teaching and/or lab assistants). This is either due to tradition or because of the variety of topics dealt with (or both) within the course. The number of students taking the course may also justify it. But even so, the new structure with its rather low number of concurrently running activities (from the view point of the individual student) has increased expectations of better teacher coordination, both from students and from the "system." The greater intensity of the teaching also means that just one miscommunication between the teacher in charge of the course and the other teachers could result in the loss of a significant part of the entire course. So the initial incremental change in intensity could perhaps, with time, result in a more fundamental change: the change of the teaching culture from an individual egocentric one to a more collaborative and common one. This may well be more of a modest hope than a realistic prediction, if this change is not promoted by other incentives supporting teacher collaboration. Bowden, (Chap. 9), has even sharper things to say on this.

Another aspect of the change to block structure might be called the "transitional issue." By transitional issue we mean the coordination between contents and outcomes in a given course and that of later courses, for which the former is a prerequisite. The total number of courses that a typical student will take in the complete degree program is probably not very different from what it was earlier, so the number of transitions should not be significantly changed. But at the beginning of the studies the course size has typically decreased slightly, so that many students will take eight courses in the new structure in the first year, while earlier it might have been a matter of passing only four, or perhaps six courses: and there are four transitions every year instead of two, because of the four blocks instead of two semesters. The biggest transition problem probably arose from the lack of homogeneity in the structure (and expectations) of the courses, where some were completely new designs, and others were run as very close copies of earlier courses from the previous study programs. All in all the initial phase of the whole process has displayed this aspect as a somewhat deeper change than anticipated. This also, in all likelihood, has to do with the fact that the teachers generally did not have the energy to familiarize themselves with the changes made in other courses than their own. So we expect this problem to decrease with time.

On the issue of block structure we conclude tentatively that the courses taken as a whole have not changed as much as was expected. There is a lot of inertia in the way the studies are traditionally divided into courses. This has only changed in a few subject studies, most significantly in chemistry. But all courses had to renew their internal structure to cope with the increased intensity. So we classify the change to block structure as a broad and moderate change.

10.4 The Scheduling Structure

The reform has imposed a scheduling structure which gives the individual course complete control (in principle!) if the course is a 15 ECTS unit, and one full day plus one half day (plus study time away from the university) if the course takes up 7.5 ECTS units (one half of a block). The scheduling structure was chosen partly to facilitate the classroom scheduling itself and to create more opportunities for the students to choose courses across subjects, and partly to allow for teaching approaches that went beyond the classical lecture. Although some actually tried it, nobody can lecture for eight consecutive hours in a row, at least not with great student outcome!

The resulting reviews are mixed. It has become clear that the successful use of a full day's student presence, without formalized teaching at all times, depends a great deal on rather careful planning. It also depends on a certain amount of restraint on the part of the teachers: the temptation to schedule too many lectures is great. Many university teachers still think that the best way to ensure proper learning is to lecture on the material, (cf., Bowden, Chap. 9). However, in cases where this trap was avoided, it has been reported that students were significantly better prepared and much more active during class sessions. But again, as an aspect of a mandate for change, this is an interesting example of a formal and narrow administrative change, and a narrow, but potentially fundamental change, when viewed as an aspect of the basic conditions for teaching and learning. It is our view that this change is heading in the direction of being more and more fundamental because it makes it impossible to use lecturing as the only teaching method.

10.5 Outcomes Defined in Competency Terms

When we now describe the process resulting from the decision to make the degree programs outcomes driven, i.e., to make them defined in competency terms, we also get a good example of the discrepancies between the top-down and the bottom-up processes. One of the overall goals of the blueprint was to sharpen the educational profile of the degree programs. The educational intentions should be made explicit. When these intentions included aims like "enhanced mobility and flexibility," and "increased differentiation of perspective, e.g., depending on individual career plans," as well as "better interaction with society," the traditional ways of defining a university degree as the union of certain courses and projects – most of which are defined by means of lists of topics – would necessarily have to be discarded in favour of more outcomes based specifications. The coordinating committee decided instead to make use of competency specifications. It was realized from the outset that such descriptions would have to be created on several levels, with the general ones, covering entire programs, to be formulated at the level of the coordinating committee; while specific professional competencies, in particular those at the individual course level, would have to be dealt with by subject specialists. This approach to educational design is well known and has been rather thoroughly tested, for instance at Australian universities (Bowden, 2003), but it must be said that experiences from other universities were not studied very much here, and consequently not relied very much on, either. The results have been mixed. But where the competency specifications have been formulated in concrete and operational terms, the people involved have found them useful. From chemistry comes the statement that these specifications were useful for the students because "this way they become acquainted with what they are actually expected to learn."

Not surprisingly, this "homemade" approach has created its own problems, and notable among these is a certain gap which has yet to be adequately bridged: a working group reporting to the coordinating committee in November of 2002 proposed, and had accepted, the following general competencies (we have chosen to translate the Danish terms from Science Faculty, 2002 without regard for the many descriptive terms already in use in English; the working group was not relying very much on established terminology or research. We suggest that the reader's attention is on the defining statements):

- *Professional competencies*, referring to knowledge of theories and methods within the particular subject area
- *Application competencies*, referring to the mastery of putting professional competencies to use in other professional fields (e.g., the use of mathematics in physics)
- *Societal competencies*, referring to all kinds of interactions between the individual possessing (or not!) this competency and his/her surroundings, and, finally
- *Personal competencies*, referring to personal communicative skills, ability to cooperate, etc.

The individual study boards were requested to comment on these concepts, and particularly to assess their usefulness when it came to translating them into specific educational efforts. A rather typical response said: "…So, can we use these four competencies to describe and to distinguish among our bachelor and candidates degrees? We feel that the answer ought to be "yes" at least in the sense that we haven't been able to come up with anything better." This board then proceeded to make use of another, already existing set of subject specific competencies! Or, in some cases, to make competency goals that not only failed to establish a relationship to the general competencies, but did not even use this existing set of subject specific competencies: "At the conclusion of this [mathematics] course the student should be able to follow and reproduce proofs, to solve problems with and without the use of [a particular mathematics computer system], and he/she should master the subjects described below [referring to a list of topics]". The use that students and teachers on this course wound up making of this description need not concern us here, but the distance in this instance between the "top" and the "bottom" of this reform process is rather obvious.

The introduction of competency-specified degree programs must be classified as a broad and potentially fundamental change. Fundamental because its goal is to make every teacher choose all content with due respect for – not tradition or opinion – but the specific, desired competencies of that particular educational element. But at the same time, because its execution depends on what many faculty members view as a radically altered mode of thinking about educational goals, its success is likely to require a much more thorough and professional approach. Not only is there the risk of conflicts of top-down versus bottom-up type, but this risk is also present because there is undeniably a major task of raising the level of consciousness among faculty of the significance of making degree plans (and their implementation!) outcomes based. Perhaps the introduction of good quality assurance programs will succeed in such a monumental task. Otherwise the current effort is in danger of being another example of broad but modest change, and one supporting the claim that research-driven universities have a way of turning even great educational schemes into incremental changes.

10.6 Assessment as an Integral Part of Courses

As described earlier, it has been one of the goals of the reform to change the rather traditional use of assessment to a more diverse and modern one. It has been common at the Faculty of Science to assess at the end of a course, most often by a standard four hours written exam. The exam could be open-book or it could be closed book. All exams took place in January and June, and there were very rarely any classes in the exam periods. With the block structure this is being changed in an interesting way: if a course wants to make use of the traditional kind of examination – written or oral – at the end of the nine weeks, the course has seven weeks for teaching. But if the course is able to integrate assessment into the normal teaching – for example by using project work or papers written during the course – then the course is allowed to make use of all nine weeks. This means that the teachers have an incentive to change the examination forms. Why? Simply because they – or some of them – want to get more time for teaching!

The blueprint called for new methods of teaching and assessment, and in the case of assessment there were expectations of the adoption of many new forms to supplant the classical written exam at the end of the term. It has been clear that this classical form has not disappeared, but the monitoring group found strong signs of the employment of a rather large variety of assessment methods. There was also evidence of increased activity level among students in the courses where continuous assessment (for instance weekly hand-ins of homework problem sets or essays) was used. This was not very surprising. Not quite so clear was whether the outcome was better, for instance as measured in performance later on in the studies. In some cases there was a discrepancy between the way in which a course was run (including the kind of issues and problems that were covered during term) and the way the final exam tested the students. This is not a new problem, but may well have been accentuated by a mismatch between efforts to reform the course itself and lack of efforts to design appropriate assessment. It is through observations such as these that we see that the entire change was not sufficiently thoroughly planned and prepared: efforts to inform about assessment options were probably not sufficiently extensive and pervasive. The result has been that, in some cases, new assessment methods have been abandoned again and the older approaches have reappeared. Thus we see indications that a change, intended as a rather broad and fundamental one, over time may well degenerate somewhat and in the end be classified as a narrow change – although it is still fundamental in those cases where assessment is carried out in alternative ways to the classical written exam, e.g., project work with an oral exam, or as continuous assessment.

Of course, it has been very interesting to see how course assessments really turned out. We now have data from two years of first year students and we can compare this data with the last year before the reform was started, in the academic year 2002/2003.

As shown in Fig. 10.3 the result at faculty level (last column) from 2002/2003 to 2004/2005 is rather good – and even better in 2005/2006, although the improvements are small. But when you look at 2002/2003 and 2004/2005 for the specific subject study, e.g., mathematics, computer science, chemistry and biology, there are great differences. Over all chemistry has the greatest improvement in completion rate while biology actually has a decrease from the 2002/2003-level. And when you compare 2004/2005 with 2005/2006 on subject level, the picture is getting even more diverse: what goes up one year, goes down the next, and vice versa. Why is this? We think that this is an important indication of the fact that it is impossible to draw any simple conclusions on the impact of the reform from completion rates. One cannot say that the quality of teaching has gotten worse just because the completion rate is going down and therefore you cannot say that a better completion rate is due to better teaching.

Our best explanation of the results in Fig. 10.3 is that "completion rate" is a very complex quantity. It is influenced by many factors, not all of which are in the hands of the university, e.g., student intake. Actually we have the impression that when intake goes up, the completion rate goes down – and vice versa. We know this has been the case for chemistry, actuarial science and geoscience in the years covered by Fig. 10.3. Of course there are many factors that influence completion rate which the university can control: changes in assessment, teachers, administrative support, etc. All in all, the complicated situation surrounding completion rates is highly worthy of study, but we should warn against making completion rates the only measure of success.

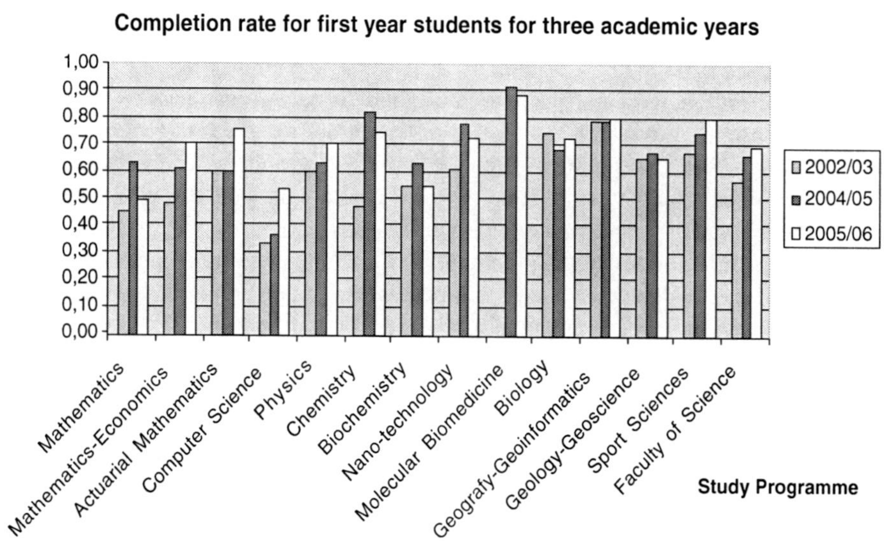

Fig. 10.3 Completion rate for first year students for the academic years 2002/2003, 2004/2005, and 2005/2006. This rate is calculated as the "Completion of Full Year Load" (STÅ) for first year courses divided by the student intake

10.7 Conclusions

As we have already indicated our conclusions are based on observations and reflections on a very complicated process, a process which is very much still in progress. So the next statements are rather tentative. Regardless of this, a good way of summing up the adopted and implemented changes described in this chapter is to place them in the field of depth and breadth as shown in Fig. 10.1. This is done in Fig. 10.2, where the arrows are to be taken only as *indications* of the change in scope of the changes.

We have already pointed out that our (preliminary) conclusion concerning the modesty of the depth of change is based on the inadequacy of the resources provided for the preparation and implementation of these changes. This affects the chances of success of the plans and intentions of the "top level," because the conservative tendencies at the "bottom level" have been given rather better chances of remaining present. Another contributing factor comes from the decision making structure at the Faculty level, combined with that at the local level (the many subject-specific study boards): here we refer to the fact that the scientific judgments concerning what the individual degree programs should contain rest with the study board, and that this power includes the power to determine the framework for teaching, if not the specific teaching methods, as well as the power to decide how to assess. These consequences include various results of a certain "we-can-do-this-and ourselves" approach that has been present in the process. Again we can observe the balance of power between the top level and those below, as it has expressed itself, also over time. It should be emphasized that this attitude has some significant and positive effects as well: it enhances ownership, and thus probably has tended to limit alienation from the process among the regular teaching staff.

Nevertheless, incremental changes are likely to be the result when change is initiated in a setting in which there is little if any tradition for formalized management of educational development. Both the study boards and the faculty administration are good at managing status quo with only small changes. But when it comes to leading big radical processes of change, they have no experience to draw from. There seems little doubt that the potential for moderate, perhaps even fundamental, change processes will be enhanced if more attention is paid to how to lead the processes in a way that engages and commits all staff in the process.

But enabling such a change of climate will not come easily: it will necessarily have to entail an acceptance, on the part of the faculty, of the gradual changes in the role of teaching at universities. The continental European university tradition has viewed students as independent adults, whose activities as students were only to be measured by their performance at exams. This tradition has been under political pressure, and quite a bit of the Anglo Saxon tradition, where rather more responsibility for the progress of students rests with the educating institution, is now becoming part of the philosophical luggage of institutional educational theory. Universities assume a share of the responsibility for students' success when they admit a young person to a degree program; thus the pressure to make use of a broader spectrum of relevant educational tools has increased – and the change process that we have examined here attempts to make this happen. It is clearly not an easy process. As we have pointed out, this reform must cope with:

- The generally low level of insight among academic staff into pedagogical options
- The grey zone between maintaining academic standards and modernizing teaching and assessment techniques
- The difficulties of establishing quality assurance criteria for teaching which are on the same level of "seriousness" as those that prevail within research
- A hitherto somewhat wobbly decision making process within the university.

How well this process will succeed remains to be seen. But we hope that we have provided at least a convincing, albeit perhaps partial answer to the question posed in the headline. Behavioural change is not easily achieved, and it depends very much on how thoroughly the changes are being prepared and carried out. At the base of this lies of course the substantial independence of university teachers regarding what goes on in the classroom, as also pointed out by Bowden, (Chap. 9). Without the commitment of the teacher to the change process it will not take place in the way intended. Bowden reaches a stronger conclusion than do we: no change in teaching and learning practices will be effective unless the teachers' ways of seeing their roles are changed. The question is, how to bring about such change.

References

Bowden, J. (2003). Why Do We Need Reforms, Which, and How Do We Implement Them? In J. Carter, K. Eriksen, S. Horst & R. Troelsen (Eds.), If Reform of University Science Education is the Answer – What were the Questions? Proceedings from the Third DCN Conference, May 2003, CNDs skriftserie No. 7, Copenhagen. Retrieved May 2008. Available at: http://www.ind.ku.dk/side22515.htm.

Cuban, L. (1999). How Scholars Trumped Teachers. Change Without Reform in University Curriculum, Teaching, and Research, 1890–1990. New York: Teachers College Press.

Danmarks Evaluering Institut (1999). De videregående matematik-, fysik- og kemiuddannelser. Copenhagen: EVA.

Gibbs, G. (2003). Incentives for Teachers – How Should We Reward Participation in Reforms? In J. Carter, K. Eriksen, S. Horst & R. Troelsen (Eds.), If Reform of University Science Education is the Answer – What Were the Questions? Proceedings from the Third DCN Conference, May 2003, CNDs skriftserie No. 7, Copenhagen. Retrieved May 2008. Available at: http://www.ind.ku.dk/side22515.htm.

Monitoring Group (2005). Studiereformstatus 2005. Copenhagen: Faculty of Science, University of Copenhagen. Retrieved May 2008. Available at: http://www.ind.ku.dk/side43282.htm.

Science Faculty (2002). On the Competence Criteria as Framework for the Study Reform at the Science Faculty, University of Copenhagen. Internal Document in Danish. Copenhagen: Science Faculty. Retrieved May 2008. Available at: http://www.nat.ku.dk/reform/.

Traub, J. (1994). City on a Hill: Testing the American Dream at City College. Reading, MA, Addison-Wesley.

Chapter 11
The Role and Means for Tertiary Didactics in a Faculty of Science

Camilla Rump and Carl Winsløw

Until a few years ago, university teaching of science was considered the specialists' private craft in Danish research universities. Excellent research remains the main parameter in young scientists' careers. But for the same reasons as elsewhere – including broader student populations and higher demands on efficiency of educational programmes – universities are increasingly preoccupied by the quality of their teaching, as are funding authorities. The notion of quality may be rather vague, and the means for improvement similarly unclear. This chapter is an attempt to analyze the potential roles and contributions of the didactics of science and mathematics towards the articulation and response to these demands for "quality teaching". In the continental European tradition, we talk about the *didactics* of a subject (such as geometry or physics) when we refer to the study of teaching and learning of that specific subject, as explained by, e.g., Chevallard (1999a). Notice that the subject area referred to may be very specific (e.g., "Newtonian mechanics") or very general (e.g., "(natural) science"), although it is often an institutionally established discipline (e.g., physics).

The pressure on universities for improved "throughput" manifests itself in numerous ways but primarily through a broad scope of "top-down" initiatives (see Horst and Laursen, Chap. 10). Political or administrative demands may or may not originate in genuine pedagogical considerations, and it may or may not result in initiatives that are helpful for improving teaching quality. However, if we leave it to the individual university teacher to interpret these demands, they may often be seen as unnecessary additional administrative burdens rather than helpful tools for improving teaching quality. Since the result of any initiative intended to affect teaching (or learning) depends, ultimately, on the actual educational activities initiated by teachers, un-mediated top-down initiatives are likely to have either no effect or even a negative effect, if teachers spend time on something they perceive as irrelevant extra administration and less time on students and teaching.

On the other hand, the past 30 years of experience show that if teaching is not focused but is left to the individual autonomous teacher, teaching is unlikely to be

C. Rump and C. Winsløw
University of Copenhagen, Denmark

O. Skovsmose et al. (eds.), *University Science and Mathematics Education in Transition*,

adapted to changes in student populations or to new didactical knowledge, for instance about more effective teaching methods for a given subject. Even if resources in the form of didactical research knowledge are present and readily available, incentives may be lacking to use these resources. Thus an exclusive bottom-up approach to improving university teaching is no more likely to be effective than an exclusive top- down approach. In some institutional contexts, the idea of distributed leadership described by Bowden (Chap. 9) may be a way to think about alternatives.

We will argue that if top-down initiatives are mediated by agents who are able to convey didactical knowledge and facilitate a common learning process involving the teachers on how to teach the particular subject knowledge in the particular institutional setting, quality of teaching may improve considerably in the sense of improved student learning. Furthermore, a didactical analysis, which will always be specific to subject matter, e.g., in the sense of core curriculum, core competencies and sequencing, are more likely to transfer to other institutional contexts than general pedagogical ideas about teaching methods, detached from contents. This is because university teachers' main interests and competencies are typically closely related to their subject matter knowledge, rather than to pedagogy at large.

Didactics could serve three different but related roles in a faculty of science within a university, of which at least the first two are similar to that of any other scientific field:

1. An *area of scientific research*, developing new knowledge – in this case about the teaching and learning of mathematics, physics, chemistry, biology and so on
2. A *subject to be taught* to students – in this case particularly to students aiming at a career as a teacher
3. A body of knowledge to be used for *developing the teaching of the faculty of science at large*, both through development projects of various types and through dissemination to faculty members in seminars, courses and so on.

These roles are potentially related. If research (1) is understood as pertaining to university level teaching, the potential interaction of (1) and (3) is concrete and direct. Our focus in this chapter will be mainly on this interaction. Didactical research pertaining to lower levels of education at which university students aim to teach later on will be more clearly related to (2).

If the didactics of a scientific discipline *X* is institutionally present in a faculty of science, it becomes important to relate it to the research and teaching of *X*. The simple picture in Figure 11.1 (adapted from Winsløw, 2006) will serve to guide our discussion.

Here, the horizontal arrows mean that the teaching of a discipline – especially at university – takes as its object the knowledge developed and maintained by researchers, although for a closer analysis this knowledge is necessarily transformed or *transposed* in order to become teachable (Chevallard, 1991). The third arrow indicates that the object of the research done by didacticians of *X* is, roughly,

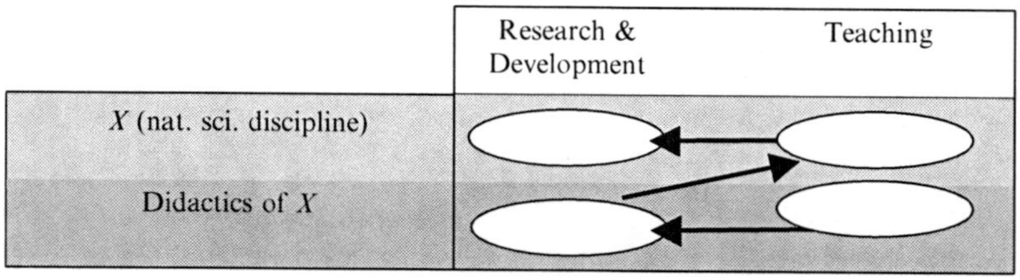

Fig. 11.1 Research and teaching in sciences and their didactics: What is researched? What is taught?

the teaching of *X*. In Sect. 3, we shall refine this description considerably. In particular, we elaborate the above model in order to explain what is in the empty bubbles.

We should however note that to us, research on the didactics of *X* is closely related to development of the teaching of X, although they are also distinct. A good model for this relation and distinction is that of *didactical engineering* (Artigue, 1994): while systematic development work may take on a local character, aiming to improve the teaching in a particular context (e.g., a course on mechanics), it can benefit from results and methods coming from research. But, conversely, research – with its demands for validity and meaningfulness beyond the particular context – can also be carried out in a development context, to the extent that the developmental design is constructed and evaluated with such generality in mind. This often implies a process of *cycles* of development, with careful attention paid to the *didactic variables*, i.e., teaching and content elements that are being worked on, as well as to *the constraints linked to the particular context*. The analogy with engineering lies in the fact that even results of didactical research can rarely be regarded as reproducible in the sense that designs can be simply transferred from one context to another – just like every construction of a bridge must take into account the relevant local conditions. Furthermore, engineering research also often arises from specific problems encountered in engineering projects. This is why careful analysis and declaration of local constraints is so important. It does not mean that we can know nothing about teaching (or bridge building) in general, but it means that we cannot hope for designs that can be directly transferred to any other context.

In this chapter, we will do three things in order to clarify and exemplify the above remarks:

- Describe briefly our engagement in the development of university level teaching, through two concrete cases from our work
- Introduce elements of the so-called "Anthropological Approach to Didactics" (which originates in the didactics of mathematics) and use it together with the cases to present a model of the role and means of didactics in improving university science education
- Give a perspective on the role of didactics in the institutional framework of a faculty of science.

11.1 Two Development Projects at the University of Copenhagen

In this section, we will briefly introduce two examples of development projects in which we have been recently engaged. Both projects arose from the so-called study reform at the Faculty of Science at the University of Copenhagen, where the structure is changed as follows (see also Horst and Laursen, Chap. 10):

- From two semesters a year to four so-called "blocks" of nine weeks including assessment
- During each block, two 7.5 ECTS points courses run in parallel, and
- A common week-schedule is introduced with 4-h slots; each course is assigned 1½ days a week, that is, a day with one 4-h slot and a full day with both 4-h slots. This means that students attend courses for a maximum of 24 h per week, with at least one scheduled day off.

11.1.1 A Development Project in First Year Physics

The re-design of the entrance course to the Bachelor of Physics programme of the University of Copenhagen, Physics 1, originated in a common ambition of a didactician and the main teacher of the course to improve students' understanding in general, and to increase the number of students who pass with an acceptable outcome in particular.

The Physics 1 course covers most of classical Newtonian mechanics. The need to re-design this course, as a consequence of the study reform, was seen as a most welcome chance to rethink the whole course.

11.1.1.1 Design Phase

The design phase started in Spring 2004 with regular meetings and informal discussions. The experience from the 4-h written examinations in the predecessor course (Physics 11) was that a group of students seemed to use a surface approach in problem solving, resulting in a lack of ability to solve problems posed just slightly differently from the "standard" text-book problems. The teachers wanted this group of students to gain a deeper understanding.

For the didactician, the design process could be seen as part of an ongoing exploration of what it takes to learn physics in a university setting. Being a didactical engineering project, the design group brought both general knowledge on physics learning and specific knowledge of the institutional context at the University of Copenhagen (KU) to bear. In short, the general knowledge pertains to the problems experienced in physics learning, obtained through quite comprehensive empirical studies and epistemological analyses of what it means to be a physicist. Kuhn (1970) and Bowden and Marton (1998) argue that the competence of a physicist is the capability to conceive of or "see" the world in a certain way: The physicist must first and foremost be able to *reduce* everyday complex problem settings to a physical model expressed as mathematical equations. The more specific knowledge of the context at KU came from studies by Ulriksen (2003) and Hasse (2002) who describe the dilemma in traditional first year physics between the teachers' emphasis on curiosity and creativity as important capabilities of a physicist and the typical overloaded curriculum with focus on standard problem solving and "cook-book" labs, from different perspectives. The design process and the underlying analysis are further elaborated in a report by Rump and Horst (2005).

Empirical studies show that Newtonian mechanics is difficult to learn, because the laws of Newtonian mechanics, in many cases, contradict our everyday knowledge of causal relations in nature. Many students' experiences of the laws are counter-intuitive. For instance, Newtonian theory takes as a starting point that if objects are not affected by a force they move at constant velocity. Contrary to this, every-day experience tells us that if objects are not affected by a force, they lie still. This is also the starting point of Aristotelian mechanics, which is why our everyday knowledge is sometimes referred to as Aristotelian mechanics. This means that most people will have to learn to conceive of everyday problems differently in order to solve physics problems. The studies show that this process is hard to bring about (see e.g., Kim and Pak 2002).

Analyses like the above led to the idea of "illustration exercises". The basic idea was to take the demonstrations out of the lecture hall and into the lab. The value of demonstrations in lectures for student understanding has been increasingly questioned by empirical studies (e.g., Crouch et al., 2004). Instead we thought that students would appreciate the hands-on experience of doing small experiments, which could illustrate simple physical causal relations. Furthermore the intention was to promote creativity and curiosity among the students.

Out of the discussions emerged a set of learning goals for the new Physics 1 course:

- Students should be able to *reduce* everyday complex mechanical problem settings to a physical model expressed as mathematical equations. The notion of "Everyday complex mechanical problems" is similar to the notion of "context rich problems" as used at the University of Minnesota (web page, "Context Rich Problems", see references). We use this term henceforth: students should be able to solve context rich problems.
- Students should be able to design and perform little experiments demonstrating central relationships and notions of the theory.

The schedule of the course and allocation of classrooms etc. was done by the study board, which set up a certain frame for the course. This resulted in a schedule made up of 2-h calculation exercises followed by a 2-h lecture twice a week and a 3-h lab with illustration exercises in between. In order to allow students some basic practice, it was decided that one of the 2-h calculation exercises should contain standard "end of chapter textbook problems". The other calculation exercise slot should contain context rich problems, which we suggested be solved in groups. The illustration exercises were formulated as open-ended design questions, like: "Design a clock, which can measure 2 min". The students would have to hand a box with springs, a ruler and some weights, a statement of Hooks law, but no description of how to do it. Furthermore, they were asked some thought provoking questions, like "How does the mass of the spring influence the precision of your clock?" or "Without a precise clock, how could Galileo demonstrate the period for the harmonic oscillator?"

11.1.1.2 The Implementation

The lectures were given by the main teacher and made some use of the so-called "peer-instruction" format, where the lecture is periodically interrupted by a conceptual multiple choice problem, like "How does rain affect the speed of an open train wagon?" Students then vote on the answers by raising a sheet with a number for their answer and if a large group of students vote for the wrong answer, they are asked to take a buzz meeting. Then a new vote is taken. Usually then, the large majority will vote for the right answer, which can then be discussed.

The calculation exercises were taught by a group of physics teachers mainly recruited from the permanent staff, i.e., physics researchers. This group of teachers was appointed only a few weeks before the beginning of the course and therefore only had a short period of time to get acquainted with the ideas behind the new design. The teachers particularly expressed some doubts on how to handle the illustration exercises. Consequently we held a workshop containing a role-play of an illustration exercise. Apart from some practical problems, these exercises went well. We also tried to argue that the traditional recitation classes which usually constitute the calculation exercises be at least partly replaced by problem solving in groups – especially with the context rich problems. In practice, the format varied a lot. One class out of five almost exclusively used recitation of the standard problems, but the other four mainly used one-hour group work, one-hour recitation.

The role of the didacticians in this phase was data collection and participation in the weekly one-hour lunch meetings. Two questionnaires were administered to the students at the beginning and at the end of the course, the so-called Force Concept Inventory (FCI) and The Maryland Physics Expectations Survey (MPEX). Results from these are reported elsewhere (Rump and Horst, 2005).

Our data also include a detailed written student evaluation and a focus group interview after 5 weeks and 5 individual interviews at the end. We also observed a number of classes and illustration exercises.

11.1.1.3 Reporting and Reflecting

All in all, the new format was successful. In particular the illustration exercises turned out as we had hoped. The students seemed to appreciate them and were working in an engaged manor and seriously or the full three hours, sometimes longer because they became inspired to make new designs and experiments. The teachers in the following lab course report a significant difference in behaviour of the students in the lab. Despite the fact that this is a traditional cook book lab, the students now perform the experiments in a much more experimental way (!) – much more along the lines of the original intentions with physics labs. Indeed, the experimental skills of the students seemed to have improved considerably compared with earlier years.

This time 82% of the students passed the course. The passing rate for the earlier version of the course, Physics 11, was typically 65%. Many circumstances, including the study reform mentioned above, make the direct comparison of these numbers problematic. In particular one cannot infer from these numbers alone that the students learned "more". In fact, measures from the FCI are similar to previous years. However, the students worked hard: on average around 20 h a week, including 10 h of homework. Moreover, it is certainly a positive reality for the university and for the students that fewer students failed in what is usually their very first university course.

The role of the didactician in this project has been mainly to study student learning and reflect on the results together with the teachers. In this case, the didactician involved did not have a background in physics, but in computer science engineering. She had, on the other hand, studied students' conceptual understanding of various engineering fields and students' learning of mechanics for some years. This was clear to the main teacher from the beginning, but it took a while for the group of teachers to learn how to use this in a constructive manner during the process. A certain hostile attitude was present to begin with: "Who is she to tell us how to teach physics, when she can't even do it herself?" Only later in the process did the teachers start to appreciate the access to knowledge of students' learning processes obtained from data collections and other studies. Although knowledge sharing may be the ideal of such collaborative processes, there is often not much experience at universities of collaboration among people with different backgrounds. Usually, an important characteristic of a research groups is that the background knowledge of the group members is extremely homogenous. In teaching this could profitably be different, although the tendency for increased competition among academics may be an obstacle to teamwork even in this area (cf. point (5) in Bowden's framework described in Chap. 9).

11.1.2 A First Year Course on Probability and Statistics

11.1.2.1 Context

SaSt1 is a basic course in probability theory and statistics, taken by mathematics and statistics students during the first three months of their university studies. For the mathematics students, it is one of two mandatory courses on these topics in their study programme (elements of probability and statistics are required in Denmark for qualifying to become a high school teacher in mathematics). For the students of statistics and related disciplines, including actuarial science, it is simply the first in a long series of courses on probability and statistics. In other words, the course is directed towards two distinct groups of students: some for whom statistics and probability is at the heart of their study programme, and some for whom it is a more or less a auxiliary requirement – in particular for the mathematics students who will typically not take more than the required courses in this area.

This course was running for the first time in 2004, the year we were engaged with helping the general design and following the implementation of the course. One of the reasons for creating the new course was the decision to give one course for the mixed audience just mentioned, as the two groups had previously taken separate courses on probability and statistics. The other main reason was the study reform at the University of Copenhagen described above. Besides these new challenges of a mixed public and a short, intense course period (eight weeks where the students work half time on this course), SaSt1 also faced the situation of being a part of participants' entrance to the university. For all these reasons, it was clear that simply repeating (parts of) the old courses was not an option.

11.1.2.2 Our Tasks

The two main problems for this project, defined in collaboration among two didacticians and the two main teachers, were the following:

- How to design the course so as to make use of the short but intense study period
- How to assess the students' work (the study plan requiring a "continuous", as opposed to "final" form of evaluation).

The teachers had heard about thematic projects (see Grønbæk et al., Chap. 4) and were interested in using a similar format to that in the present course, as a kind of simultaneous answer to the two questions. Besides contributing to these general aspects of the course design, the didacticians should also follow the development of the course and report on it to the teachers and the study board, with particular focus on how the designs work as solutions to the two problems.

In terms of content, the course should provide an introduction to basic notions of probability theory and statistics based on *discrete* distributions and (counting) data. In particular the binomial distribution was to be a leitmotif.

11.1.2.3 The Design Phase

The format for thematic projects (cf. above) presupposes a final oral exam, and was developed for a more advanced and theoretically oriented course. The teachers had two or perhaps three larger projects in mind for the course, but it was quickly agreed that it would be better with smaller, more manageable projects to be delivered and assessed each week in order for the students to have more continuous and less "high stakes" feed back on their work. The teaching had already been planned with 6 lectures (given by the main teachers) and 6 exercise classes (given by instructors); these 12 h of teaching completely fill the full and the half-day set aside for the course. It was decided that two of the exercise classes be set aside to work on the thematic projects (with guidance from the instructors) while the four remaining were used for doing and discussing more simple training exercises, partly as preparation for the weekly projects. The lectures would, as usual, mainly be concerned with introducing and exemplifying the material covered in the textbook (actually, lecture notes written by the lecturers in advance). The didacticians observed that it might be an idea to convert one or two hours of lecture to guided project work, but the teachers finally chose to maintain the original schedule.

11.1.2.4 The Implementation Phase

The actual formulation of the weekly projects – as well as all aspects of the teaching – was done by the teachers, and except for some discussions on project formulations, the main role of the didacticians in the implementation phase was to gather data on the development of the course (including focus group interviews with students) and to report back to the teachers. The didacticians had a few meetings with them during the course. The most interesting discussions were those concerning details of the formulation of weekly projects, based in part on observation of students' work with them. In particular, the use of a *Maple*-based software for simulating binomial distribution was an interesting but also difficult element in some of the projects. Also, although the hearts of the projects were naturally tasks for the students to solve, the formulations of them were in general rather long (two to three pages) with extensive instructions and explanations. In some cases the didacticians suggested a more concrete approach to some of the theoretical points, allowing a reduction of the instructions and guidelines for students. Typically, this would imply that they *begin the project with a characteristic example or data set* and increase the level of abstraction in later tasks (which could be partly optional), rather than introducing theoretical points first through "guided tasks", and then applying them to a case. However, despite the existence of such rough principles, it is not easy to formulate coherent projects with tasks that allow for a certain student autonomy and reflection, and at the same time remain realistic for students to solve meaningfully within a few days and in the first weeks of the students' university life.

11.1.2.5 Reporting and Reflecting

The observations, and other collections of data during the course, led to a number of recommendations, most of which were of a rather practical nature. It must also be said that in several important respects the course was a massive success:

- The students worked hard on pertinent matters and they generally felt that their profit from the course matched the effort, *even* if about half of the students felt the required effort was too large
- The completion rate (students who passed among those who enrolled) was considerably higher than in preceding courses (about 80% against 50–60%; in fact as many as 93% passed among those students who participated in one or more projects, where passing required at least six accepted projects among the eight which could be done). The comments made before on non-comparability of passing rates apply also here, because the whole programme had changed.

The major corrections suggested for future runs of the course are mostly concerned with developing the format and formulation of weekly projects:

- Projects should be more focused on specific, clear goals, with clearer criteria for acceptance and better feed back
- More self-differentiation and options in the project work, as well as more autonomy, could be achieved but would require that the guidance of groups be strengthened, both in quality and quantity
- A better coherence among different parts of the course, in particular between the projects and the lectures, would help to enhance the students' experience and profit from the course. Any changes should of course be made with due observance of the fact that the student workload is probably maximized, if not too high, in the observed run of the course.

The didacticians involved (namely the authors of this chapter) are not experts in the field of probability theory and statistics – although they surely know the subjects taught in the course. Their main function with respect to the essential – the work assigned to students – was to suggest alternatives, often in the form of questions, and to focus mainly on global aspects of the course within the given frames. Besides the suggestion to reduce the lecture time, these frames were not questioned in the project.

11.2 Analysis of the Cases Based on the Anthropological Approach in Didactics

The cases described above give examples of what can be found in the lower left bubble of Fig. 11.1. To do research and development projects on university mathematics and science teaching is an example of a particular *form of human practice* which involves articulated reflection on and justification of actions. In this section, we explore a model for such practices (or *praxeologies*), developed within the framework of didactics of mathematics. The point will be that *all* four bubbles of Fig. 11.1 are families of praxeologies and that the anthropological approach enables a closer analysis of their interaction, which will be demonstrated in the cases.

11.2.1 The Anthropological Approach to the Didactic of Science

This subsection gives some key points from work by Yves Chevallard (1999b) (see also Barbé et al., 2005), in order to make the chapter reasonably self-contained. However, readers who need a deeper understanding of these points – e.g., for use in their own work – would clearly also need to refer to these sources.

The starting point of the anthropological approach is very general: the object of study is institutionalized human actions, which are considered as responding to *types of tasks* while employing corresponding *techniques*. Notice that the term *task* here is to be understood in a very general sense, including almost anything that can be described by transitive verbs with a human subject. As types of tasks, we thus have e.g., *determine the forces acting on a given body*, *bake a loaf of bread*, or even *write a paper on didactics*. To execute or solve tasks of a certain type in an institutional setting, there will be one or more *techniques* (or "ways"). Clearly types of tasks can in principle be delimited freely, however the pair (type of tasks, technique) is not arbitrary in a given institution, as the examples mentioned suggest. Even if a given technique does not apply to *all* tasks of the given type, it can still be commonly thought of as one among others related to the type. The connected pair (type of task, technique) is referred to as a *practice block*.

While tasks and techniques may well exist with little or no use of language to describe them, it is a characteristic and occasionally crucial aspect of human practice that it may be explained and justified. A rational discourse about a practice block – to describe and explain it – is called a *technology* (etymologically, discourse bearing on techniques). Finally, the technology related to one or more practice blocks will in many contexts – not least scientific ones – need *justification and context* from a wider discursive level called a *theory*. As before, within a given institutional context, the pair (technology, theory) will be closely connected. It is called a *knowledge block*.

A *praxeology* consists of a practice block and a theoretical block, i.e., of four related elements: the *type of task,* which indeed is the identifying part, and a corresponding *technique*, *technology* and *theory*. For instance, *determining the forces acting on a given body* (task type) may, in some cases, be executed using the technique of *drawing a vector diagram of the forces*, which in turn can be explained as a way to visualize a (two-dimensional projection of) the forces acting on the body. Of course, to contextualize and justify the meaning of this, one must appeal to a wider *theory* where force is modelled by vectors in space, namely (the theory of) classical Newtonian mechanics.

As this and other examples show, a praxeology is often just a "minimal part" of larger, familiar forms of practice involving several types of tasks, several techniques for each task, possibly overarching technologies and one or more theories for them. Such a family of praxeologies is called an *organization*. It is still characterized by a family of task types that may, however, be quite huge, as the praxeologies associated with *classical Newtonian mechanics* (a physics organization). Notice that in this example – as is often the case with scientific organizations – organizations are often referred to, and thought of, as being primarily characterized by their knowledge blocks and in particular the theories involved; practice blocks seem to be "derivative" of knowledge blocks (e.g., when we construct student exercises in a physics course to "activate" certain elements of theoretical knowledge). However for any person faced with an actual task, the order is opposite and must be associated with one or more praxeologies familiar to that person – even if the larger organization is well determined, as when a student is about to solve an exercise in a course on Newtonian mechanics. Both the genesis and the learning of praxeologies, including knowledge blocks, are of course basically related to situations where we meet tasks for which we do not have a technique, let alone a technology, readily available.

The main point of introducing a more general framework of human practice to describe educational practice is that when we want to discuss the potential and actual roles of didactics within a faculty of science, we need to relate several forms of practice (as illustrated by Fig. 11.1) and hence to analyze them within an overarching epistemological framework. Furthermore, this framework is also suitable for studying *transpositions* of "knowledge" in the common sense (including all four praxeological levels), which is an important part of didactical research: from science to syllabi and textbooks (external transposition), and from these to teaching and learning in universities and other educational institutions (internal transposition).

11.2.2 Analysis of the Two Cases

In both cases previously described, we have three gross types of organizations involved:

- Organizations from the scientific discipline to be taught (Newtonian mechanics, discrete distributions, etc.), abbreviated here SO (scientific organization)
- Organizations of teaching the SO (involving tasks such as designing labs, formulating a week-project etc.); these are didactical organizations abbreviated here DO
- Organizations of developing, observing and evaluating the DO (involving the tasks described to some extent in Sect. 2, such as to design or develop techniques of the DO, document the teachers' execution of their tasks, etc.). We call these (didactical) engineering organizations (EO) with reference to the analogy mentioned at the end of first section of this chapter, bearing in mind that a particular EO may be more or less related to on-going research.

We now elaborate on each of these for the two cases as well as their interplay.

11.2.2.1 The Organizations of Case 1

The SO of "Newtonian mechanics" is reported to be complex and difficult to access. Among its most challenging task types are various forms of *reduction* and *modelling* where a given mechanical situation should be reduced to a mechanical problem and then encoded in diagrams, mathematical formulae, etc. Another grand type of task is to design experiments and instruments to measure physical entities by use of physical laws. In praxeology terms, the main challenge in teaching this

SO is that students already have a body of practice blocks for mechanical situations, but with no related knowledge (articulate) blocks: the so-called "everyday knowledge" of mechanics. These practice blocks do not "match" the knowledge blocks of the SO of Newtonian mechanics, but rather on Aristotelian mechanics, as argued above. These "everyday practice blocks" then become epistemological obstacles for many students in their attempt to learn Newtonian mechanics (cf. Bachelard, 1938/2002, Chap. 2). Students must learn a whole new set of scientific concepts (contained in the knowledge of the SO) to complement or correct conceptions generated from primary experience. This means they must learn to use the SO technology to interpret mechanical situations while gradually acquiring the SO discourse. Meanwhile, students must also learn the corresponding mathematical techniques and technologies (the precise epistemic status of mathematical discourse in physics can be argued). Thus the SO interacts with mathematical organizations (e.g., Winsløw, 2006).

Notice that the SO in question is, a small part of, the scientific praxeologies of the scientific community, whose main agents are professional physicists. However, before teaching it will be officially redefined in view of teaching, through text books, syllabi, course aims etc. This *external didactical transposition* of the SO turns it into an organization SO^E "to be taught". In the university context one will often assume a close relationship between the SO and the SO^E.

In mathematics and physics it is recognized that sometimes the relationship between the SO and the SO^E is not as strong: some of the SO^Es in the introductory years are "dead" in the sense that active research is no longer performed within the area of the SO^E. Introductory Newtonian mechanics is "dead" in that sense. There is research within mechanics, however it is statistical mechanics and this is only taught in later years of study. This raises a question of justification: why teach "dead" material at university? The most common and straightforward answer is: because the SO^E of Newtonian mechanics is a prerequisite for understanding the SO of statistical mechanics (and other areas of physics). However, this justification seems to be problematic to handle in teaching: both students and teachers find it hard to deal with "dead" material just for the sake of later use in the curriculum. Another, more fruitful and solid, justification has been argued: that the theory structure of Newtonian mechanics is *exemplary* for all other theory structures in physics (and in science in general for that matter). It is sometimes even seen as an archetypal scientific structure. Therefore, by learning about it, students begin to conceive of the world as a physicist.

At universities, there is a manifest tendency to specify SO^E by referring to knowledge blocks only (syllabi etc.), even if traces of previous teaching practice (such as old exam assignments), not to mention course competency aims, may give more precise information about the student competence which is expected by the institution. Recognition of the practice blocks of SO's is in general rare and this is arguably a problem when doing didactical transpositions.

The aim of the DO is now to transpose the SO^E to actual didactical situations, in order to enable students to enact selected parts of it as specified (more or less) by the external didactical transposition in the form of the syllabus etc. It is a main task of the teacher to enact this *internal didactical transposition* $SO^E \rightarrow SO^I$. Here, SO^I

designates the scientific organization that is actually proposed to students, whether presented in lectures or given through exercise assignments, lab work, exams etc. Certain aspects of these tasks are explained and explicitly emphasized in the learning goals set up for the development project. Thus the learning goals are part of the EO. By stating that students should be able to "model everyday complex mechanical problems" and to "design and perform small experiments", the goals indirectly state that these modelling and design tasks are among the grand types of SO^I tasks the students should work with.

In the concrete SO^I, we have the example of a (student) task, which involves both of the grand types of SO^I tasks, namely "design a watch which can measure 2 min". Such a *student task* (belonging to SO^I) is at the same time a technique to solve parts of the *teachers* task (belonging to the DO) and it could be further explained and justified by relating it to the pertinent parts of the SO^E (Hooke's law in particular) as well as an analysis of what types of SO^I knowledge the students must mobilize or develop in order to make use of information that is provided with the statement of the task. A part of the DO technique is the stipulated format for the students' work (here, group work), which could also be explained and justified by appealing to didactical knowledge (DO) about the benefits and challenges of student interaction with this type of task.

This description illustrates how the DO is intimately related with the SO^E and its transposition, both when more classical DO techniques (lecturing, recitation sessions) are employed, and when more innovative techniques (such as the illustration exercise labs) are used. The fact that the latter technique has been developed explicitly in discussions during the design phase, constitutes it as being part of a DO *technology*.

As already mentioned the basic task types of the EO are to support the design of the DO and to document and analyze its results. It is enacted by the didactician who employed a range of techniques, from formal and informal meetings with the teacher over class observations and focus group interviews to the FCI test and the MPEX questionnaire. Each of these techniques could be further explained and justified by reference to appropriate literature, namely relevant technology and theory from the didactics of physics or other parts of educational research, which in this case constitutes the knowledge block of the EO. In the description of how the EO was enacted, one perceives the role of institutional conditions that offer both resources and constraints for the use of its techniques. Moreover, the DO of the course Physics 1 cannot be regarded in isolation, since the subsequent course builds on its results and employs partly different DO techniques.

11.2.2.2 The organizations of Case 2

The starting point for this case is of a somewhat different nature than in the first case: a possible DO technique (thematic projects with reference to corresponding technology and theory for its application to mathematics) and the new general conditions for teaching and assessment imposed by the study reform as well as the locally fixed formats of teaching (six hours of lectures, six hours of exercise classes). The task of

DO was of course still the transposition of the given SO elements with respect to these constraints on the DO techniques to be brought into play.

An important delimitation of the SO^E which results from the transposition, is the restriction to discrete distributions, in particular the binomial distribution, through which the basic techniques and knowledge blocks of probability theory and statistics were to be presented. An important and given technical element of the DO was the use of the lecture notes previously developed by the teachers. The justification of this was reference to previous experience with similar courses, thus a form of informal technology. This also meant that the global delimitations of this part of the SO^I were already explicitly fixed and were not substantially focused on in the EO tasks, although the consideration of its relation to the inhomogeneous student population was in fact discussed during the process and in the final evaluation of the project.

The main tasks of the EO were then to support the design and document the results of a new DO-technical element and corresponding technology for continuous assessment which would at the same time enhance the students' possibilities to work more autonomously with selected SO^I knowledge blocks than is usual with standard SO^I tasks (as provided in the lecture notes and worked on in exercise classes). The knowledge block of the DO would rest on knowledge about thematic projects (Grønbæk and Winsløw, 2007), but had to be rearticulated in the current context. The format of *weekly projects*, as described in the case, is hence a global technology of the implemented DO, and at the same time implied a framework for the formulation of new SO^I tasks. An interesting part of these tasks calls upon IT-based techniques (SO^I), namely the use of certain *Maple* routines, which are different from standard SO techniques (involving dedicated statistics software).

The most crucial discussions in the design phase concerned the formulation of the weekly projects. Quite extensive instructions often resulted from the teachers' attempts to "derive" the tasks from SO^E knowledge blocks to be "covered" by them. Here, the DO technique suggested by elements of EO technology and theory (as imported from previous research on thematic projects) was to try to construct project assignments with an outset in (SO^I) tasks pertaining to a concrete situation (here, typically contextual data) in order to allow for a more autonomous student mobilization of techniques and technology. Another DO-technical problem was the provision for guidance of student work, which had to draw mainly on resources assigned for exercise sessions; it was to some extent informed by didactical technology and theory (Grønbæk and Winsløw, 2007) imported into the EO, both in the design and in the evaluation phase of the project.

11.2.3 *The Interaction Between Sciences and Their Didactics at University*

In this brief analysis of the cases from Sect. 2, we have encountered some important general traits in the interaction between the scientific and didactic organizations involved, and in the potential role of the EO in designing the DO based on the

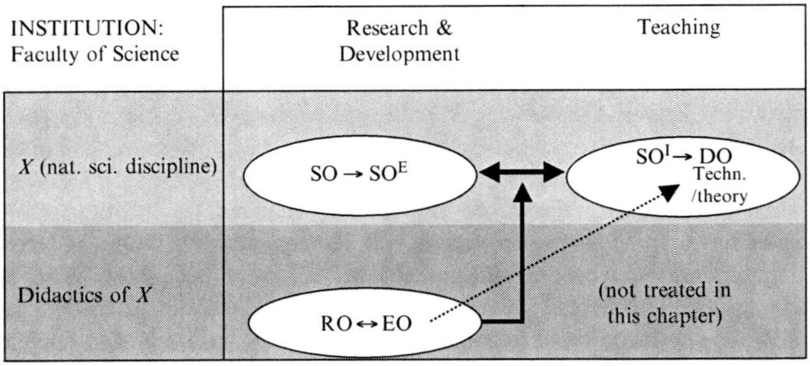

Fig. 11.2 Didactics studies the interaction of research and teaching at universities

internal didactical transposition $SO^E \rightarrow SO^I$. In fact, the DO is not only based on the transposed organization but *interacts with it*: in both cases, new explicit formats of teaching (DO-technologies) aim to allow for more autonomous student work with the SO^I, which is then itself locally modified in the process, typically with a wider range of SO^I task types. A main part of the interaction between the DO and the SO^I takes place, in both cases, between DO techniques and SO^I task types. Over time, the results of teaching may feed back to the official definitions of the knowledge to be taught (i.e., the external transposition) and this process could be enlightened by documentation and analysis of the products of teaching.

Indeed, the EO aims to support and document rational change; its object is the *interaction between scientific and didactic organizations of the discipline* rather than just the didactic organizations and the internal transposition. The EO is based on the research organization (RO) of the involved didactics (it is indeed a scientific organization, but not in the basic discipline such as physics although it is strongly related to it). The RO provides knowledge blocks for the EO, and *through the engineering intervention it also allows for a rational techno-theoretical discourse about the DO*, without which the latter would remain a "craft", as mentioned before. Conversely, it has already been mentioned that didactical research may draw on results from development work. These points are summarized in Fig. 11.2, which, as promised, provides an elaborate version of Fig. 11.1.

For lack of space, we refrain from considering, in this chapter, the praxeologies involved in the *teaching* of the didactics of the natural sciences, which is also an important part of our work.

11.3 Act Locally – Think Globally!

In the previous sections, we have described a possible role of didactical knowledge and research in a faculty of science. We have in particular described some cases of its development and use in the context of the internal didactical transposition. In these settings the scientific organization SO in terms of practice and knowledge in science has already been transposed, by others, into another scientific organization "to be taught", SO^E. At Danish universities, the external transposition usually happens at the institutional level of study boards. We have also indicated how this transposition may sometimes turn out to be in conflict with necessities or habits related to the internal didactical transposition, and consequently that it may be counterproductive even for furthering the intended goals for teaching. Why, one might ask, should the didacticians direct so much of their attention to the actual teaching and the didactical organization at the level of the internal transposition? Would it not be better to focus mainly on the higher levels of educational planning and policy, or even on the "bird's eye view" of science in society? We think not, and there are several reasons for this.

As we, and others (e.g., Chevallard, 2002) have argued, what actually takes place in the classroom depends on decisions at several institutional levels. For our context of university teaching, these levels may be illustrated as in Fig. 11.3. There is, however, no guarantee, that decisions made at one level lead to the desired result at lower levels. In fact, decisions taken about teaching design, that is the internal transposition at the lower level, sometimes seem to be quite immune to upper level demands. For instance, it has recently been decided at ministerial level that all Danish University programmes at bachelor and master level should have descriptions of the competence expected for the graduates. The intention was that this would eventually result in more emphasis on student competence as an outcome of learning, than on "passive knowledge". Of course study boards and teachers have made fine competence descriptions of programmes and courses, but they tend to be so general and vague that they are unlikely to have an effect on the educational practice that is supposed to achieve the goals. There are many ways to perform an internal didactical transposition so that it appears to be in accordance with the external one but is not at all.

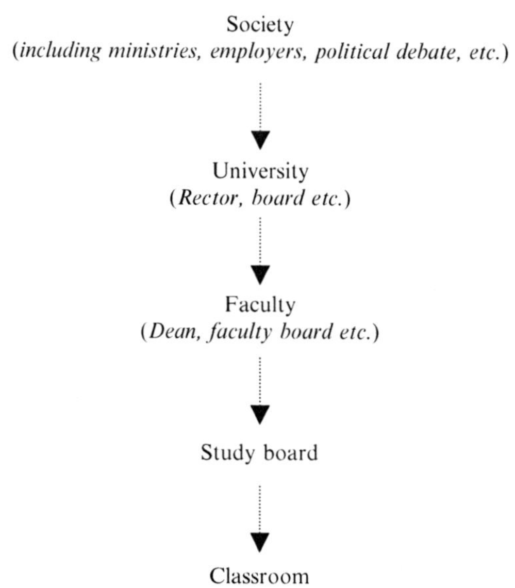

Fig. 11.3 The hierarchy of didactical determination (cf. Chevallard, 2002).

In development projects teaching does change. A contribution is made to the knowledge block of the didactical organization (DO) through the teachers' and didacticians' common learning process of the specific didactical practice in question – the teaching and learning in the developed course. This knowledge reaches beyond the specific course, for several reasons.

First, the course can be seen as an *exemplar* for a family of courses. As it has been argued above, the aim of didactical engineering is to obtain general knowledge through the study of the specific, an approach used in other sciences since Antiquity. Knowledge of possible teaching designs can be used by other teachers and in other development projects. Examples of concrete changes in local courses presented by dedicated and enthusiastic teachers are also very strong arguments in favour of further educational change in a local institutional context.

Secondly, concrete involvement in successful development projects at course level provides *credibility* in the institutional context to the didacticians involved in particular and to didactical knowledge in general. The case from physics described above shows how a certain suspicion toward "outsiders" interfering with the local teaching practice, can be overcome through an ongoing negotiation and learning process with a group of teachers.

Thirdly, systematic knowledge about institutional impediments to appropriate teaching design can provide constructive *feedback* to upper levels in the institution from the lowest level. If knowledge obtained through a development project shows that the framework given by the external didactical transposition SO^E is insufficient or misleading, the fact that this knowledge is obtained through concrete local course context (interaction between DO and SO^I), and is recognized and argued by the teachers themselves, gives a much higher chance that the study board will actually change their policy. Such changes could naturally affect other parts of the study programme as well.

Altogether these three aspects of the development projects – the exemplary status, the credibility obtained and the feed-back of knowledge from the lower to the upper institutional levels – seem to be crucial to develop didactical knowledge which is not confined to closed circuits of didacticians but which has *impact* on educational practice. Reform based only on a top-down strategy has a low chance of success. Reform based solely on bottom-up initiatives has a certain chance of success but may lose pace in the long run. Reform anchored at bottom level and with support from the top may eventually succeed. Local involvement and knowledge development must therefore be the foundation of any global strategy for university science didactics.

References

Artigue, M. (1994). Didactical Engineering as a Framework for the Conception of Teaching Products. In R. Biehler, R.W. Scholz, R. Strässer & B. Winkelmann (Eds.), *Didactics of Mathematics as a Scientific Discipline* (pp. 27–39). Dordrecht: Kluwer.

Bachelard, G. (1938; English translation 2002). *The Formation of the Scientific Mind*. Manchester: Clinamen Press.

Barbé, J., Bosch, M., Espinoza, L. & Gascón, J. (2005). Didactic Restrictions on Teachers' Practice: The Case of Limits of Functions at Spanish High Schools. *Educational Studies in Mathematics*, 59(1–3), 235–268.

Bowden, J.A. & Marton, F. (1998). *The University of Learning: Beyond Quality and Competence*. London: Kogan Page.

Chevallard, Y. (1991). *La transposition didactique: du savoir savant au savoir enseigné*. Grenoble: La pensée sauvage.

Chevallard, Y. (1999a). Didactique? You Must be Joking! A Critical Comment on Terminology. *Instructional Science*, 27, 5–7.

Chevallard, Y. (1999b). L'analyse des pratiques enseignantes en théorie anthropologique du didactique. *Recherches en Didactique des Mathématiques* 19(2), 221–265.

Chevallard, Y. (2002). Organiser l'étude 3. Écologie & régulation. In J.L. Dorier, et al. (Eds.), Actes de la 11e école de didactique des mathématiques. Grenoble: La Pensée Sauvage.

Context Rich Problems, web page for University of Minnesota Physics Education Research and Development. Retrieved May 2005. http://groups.physics.umn.edu/physed/Research/CRP/crintro.html.

Crouch, C.H., Fagen, A.P., Callan, E.P. & Mazur, E. (2004). Classroom Demonstrations: Learning Tools or Entertainment? *American Journal of Physics*, 72(6), 835–838.

Grønbæk, N. & Winsløw, C. (2007). Developing and Assessing Specific Competencies in a First Course on Real Analysis. In F. Hitt, G. Harel & A. Selden (Eds.), *Research in Collegiate Mathematics Education* VI (pp. 99–138). Providence, RI: American Mathematical Society.

Hasse, C. (2002). *Kultur i bevægelse – fra deltagerobservation til kulturanalyse – i det fysiske rum*. Frederiksberg: Samfundslitteratur.

Kim, E. & Pak, S.-J. (2002). Students do not Overcome Conceptual Difficulties After Solving 1000 Traditional Problems. *American Journal of Physics*, 70(7), 759–765.

Kuhn, T. (1970). *The Structure of Scientific Revolutions*. Chicago: University of Chicago Press.

Rump, C. & Horst, S. (2004). *Fysik 1, blok 1, E2004*, available through the authors or at http://www.ind.ku.dk/udvikling/projekter/

Ulriksen, L. (2003). Hvad skal de studerende lære i fysik? Et lærerperspektiv. In N.O. Andersen & K.B. Laursen (Eds.), *Studieforløbsundersøgelser i naturvidenskab – en antologi*. Copenhagen: Center for Naturfagenes Didaktik.

Winsløw, C. (2006). Research and Development of University Level Teaching: The Interaction of Didactical and Mathematical Organisations. In M. Bosch (Ed.), *Proceedings of the Fourth Congress of the European Society for Research in Mathematics Education* (pp. 1821–1830). Barcelona: Universitat Ramon Llull – ERME.

Chapter 12
Problem-Based and Project-Based Learning

Institutional and Global Change

Anette Kolmos

During the last 30 years, Higher Education (HE) has changed a lot. In the Northern part of Europe, until the 1960s, the traditional universities dominated with the ideal of an elitist university with freedom of academic research, which implied searching for the truth, and with an approach to education as presenting research. After the 1960s, with the students' rebellion, the university culture changed. At the traditional universities, students started to formulate demands, and became critical of the decline of the dominant research paradigm. At the managerial level, students had more influence on the boards, and they also applied pressure to develop new educational models such as the project- and problem-based learning (PBL) models that were implemented in Bremen University in Germany, Maastricht University in the Netherlands, Linköping University in Sweden, and Aalborg and Roskilde University in Denmark. During the 1970s and 1980s, the predominant picture was still that of traditional universities, but the university system expanded because of increased enrolment numbers (Carter et al., 2003).

Often in history, unexpected partnerships occur: one of these partnerships was between the students' movements and industry. Many of the students' criticisms and alternatives, such as peer learning and problem-based and project-based learning, went hand in hand with demands from industry. For example, Aalborg University in Denmark was established in 1974 as a PBL university, due to lobbying from industry as well as being a result of the students' movement. This partnership was founded due to the fact that the largest faculty at Aalborg University was the Faculty of Engineering. It has been well known that there is a need for the ability to manage projects and co-operate, particularly in the technological field.

During the 1990s, new competencies such as lifelong learning appeared on the agenda. A few HE institutions changed towards student-centred learning (case studies, problem-based and project-based learning, co-operative learning, etc.) on both small and large scales, in order to meet the challenges, but certainly also with a wish to decrease students' drop-out rates and to improve the quality of learning.

A. Kolmos
Aalborg University, Denmark

O. Skovsmose et al. (eds.), *University Science and Mathematics Education in Transition*,

Furthermore, employability was an issue, meaning that higher education should be more oriented towards the labour market in co-operation with industry.

The university of tomorrow represents a trend towards an entrepreneurial university (Clark, 1998) with a high degree of complexity concerning management, research and teaching. This development involves the risk that universities will be competing for students, funding and that there will be a tendency for trading knowledge (Bok, 2003). Barnett (1999) points out a scenario based on "supercomplexity," which has an impact on leadership, research and teaching, where collaboration with other groups of people outside of university is a feature. This type of university is based on a network society where people form networks across traditional knowledge boundaries, i.e., across disciplines and across universities and companies. Institutions will also need more institutional power in order to navigate in this supercomplex scenario. So, there is not only more autonomy and privatization of universities, but at the same time, increased competition between institutions to attract students and research funding. In more and more countries, the existence of external boards, with appointed leaders instead of elected leaders, has become a reality at all levels in the university system (Kogan et al., 2000; Sporn, 2003; Askling and Henkel, 2000).

The impact of this on the development of a university culture and the quality of teaching and learning is still to be investigated. However, there is a clear trend in Denmark that appointed leaders are not only given a more powerful position from which they can work, but they are able to use this power in order to accomplish change in the system. The question is, in which direction will future teaching and learning of engineering and science go?

12.1 Learning from Engineering Education

Since the 1990s, requests for new skills have been on the agenda for engineering education. The differences between engineering and science are no longer clear, so what happens in engineering might well be directly used in certain areas of science and inspire the development of hardcore science. Engineering education gets special attention as technology and science are the cornerstones of societal development, and the speed of development is increasing in order to secure a constant production of new goods. Because of this rapid development, technological and scientific knowledge might be outdated within a few years. This is a challenge for science and engineering education all over the world (see Christensen and Hansen, Chap. 13). Knowledge about yesterday's mobile phone might be superseded by tomorrow's communication technology. So, the challenge for engineering education is both to keep pace with innovation and to manage more and more knowledge in order to avoid overloaded curricula. Furthermore, globalization entwines with innovation and technological development. Technology and innovation have no national borders, leading to global knowledge sharing, which both science and engineering education have to address (Friedman, 2005).

The boundaries between engineering and science are unclear. Basic science is part of the biotechnological development as well as computer science, and is aimed at the end users. Gibbons et al. (1994) argue for a concept of "Mode 2 knowledge." They argue that the growth of complexity and uncertainty are linked phenomena, and that there is a decreasing belief in the cause–effect relationship between science and society. The relationship between science and society has changed radically: it is no longer a case of "science speaking to society," but rather of "society speaking to science." Science has become contextualized. Environmental technologies are especially context dependent. However, there are still knowledge areas that are not contextualized and are more basic, e.g., physics. Knowledge production is no longer taking place only in universities but at a range of institutions in society, both private and public; consequently the skills of the future graduates need to have a much broader scope. The future engineers and scientists will need to learn how to share knowledge within both a globalized academia and industry.

These new demands for knowledge and skills are reflected in concepts such as lifelong learning, the demand for core competence, transferable skills, etc. (Assiter, 1995). This trend can also be found in the development of accreditation criteria for engineering education. A representative of the American educational system is the Accreditation Board for Engineering and Technology (ABET). They have been practicing evaluation and accreditation for the last 70 years. The ABET criteria for accreditation were recently reformulated with an emphasis on learning outcomes and student-centred learning, which engineering education institutions have to fulfil if they want ABET certification. Of special interest is the strong emphasis on the scientific and technological aspects as well as on professional and process skills. Some of these ABET criteria are shown in Table 12.1.

Table 12.1 ABET criteria (2008) (from http://www.abet.org/)

- An ability to function on multi-disciplinary teams
- An ability to identify and solve applied science problems
- An understanding of professional and ethical responsibility
- An ability to communicate effectively
- The broad education necessary to understand the impact of solutions in a global and societal context
- A recognition of the need for, and an ability to engage in life-long learning
- A knowledge of contemporary issues
- An ability to use the techniques, skills, and modern scientific and technical tools necessary for professional practice

The corresponding European organization is EUR-ACE (Accreditation of European Engineering Programmes and Graduates). They have formulated programme outcomes on a bachelor and a master level. Table 12.2 shows the formulations at the bachelor level.

Table 12.2 EUR-ACE (2008): personal programme outcomes for the bachelor level (from http://www.feani.org/EUR_ACE/EUR_ACE_Main_Page.htm)

	First cycle graduate
1. Individual and team work	Function effectively as an individual, and as a member or leader in diverse engineering teams
2. Communication	Communicate effectively on *intermediate* engineering activities with the engineering community and with society at large, by being able to comprehend and write effective reports and design documentation, make effective presentations, and give and receive clear instructions
3. The engineer and society	Demonstrate understanding of the societal, health, safety, legal and cultural issues and the consequent responsibilities relevant to engineering practice
4. Ethics	Understand and commit to professional ethics and responsibilities and norms of engineering practice
5. Environment and sustainability	Understand the impact of engineering solutions in a societal context and demonstrate knowledge of and need for sustainable development
6. Project management and finance	Demonstrate an awareness and understanding of management and business practices, such as risk and change management, and understand their limitations

Shuman et al. (2005) interpret the development in the ABET criteria as a strong development of the professional skills, which is a term covering both awareness and process skills. This is in line with the argumentation by Kolmos (2006) saying that education needs to contain interdisciplinary aspects and the development of process skills, such as being able to analyze and interact during the creation of technology which involves cooperation, project management, communication, etc.

This emphasis on the trend for accreditation criteria in engineering education proves that new skills in the area of engineering education have already become institutionalized, and that engineering education is under pressure to change, if institutions have not already changed their teaching system. Although policy for higher education is a national or even regional affair in most countries, this is such an international pressure that national governments cannot ignore the global development. In science, there has not been the same pressure, but as the borderline between engineering and science is blurred, it is likely that the same pressure will occur in science.

12.2 Problem-Based and Project-Based Learning

In engineering education, one of the answers to the societal demands for new skills has been to incorporate problem- and project-based learning (PBL). These changes are to be found at both course level and at system level as entire institutions decide to establish new teaching and learning systems. Ten years ago, it would have been possible to name the institutions that had implemented PBL to some degree at a departmental level, faculty level or as a total institutional approach. This is no longer the case, and nobody any longer has the full overview. All over the world, PBL has started to pop up on either a small or large scale in Australasia, Asia, Europe, Africa and South and North America.

The reasons for change to a full-scale PBL model at an institutional level have been multifaceted:

- address the new demands for learning outcomes
- create a more modern university profile that might attract students
- funding issues as PBL might decrease drop-out rates and improve the percentage of students who finish their study on time, and
- improve the quality of learning for students.

Along with a widespread use of PBL, various practices and models have occurred. The specific understanding of the PBL concept has also become fuzzier. Graaff and Kolmos (2003, 2007) argue that there will always be variations in the models used; especially when utilising PBL in various educational systems that represent a wide range of cultures, the very concrete models will and must be different. Therefore, it is not possible to define educational concepts by the concrete elements, but they have to be defined by the learning principles beyond the concrete practice. This will allow diversity on the concrete model level.

The two famous PBL models are the problem-based model (practiced at Maastricht and McMaster University – especially in health sciences) and the problem- and project-based model (practiced at Aalborg and Roskilde University – in subject areas such as engineering, science, mathematics, social science, and the humanities). The universities mentioned were established in the late 1960s or early 1970s and were organized around these educational approaches from the very beginning. The learning principles formulated for these two different models are more or less the same. Barrows (1996) stressed these elements as part of problem-based learning: the use of problems as a starting point for the acquisition and integration of new knowledge, that new information is acquired through self-directed learning, that it is student-centred, the use of small student groups, and teachers in the role of facilitators and guides. Almost the same elements were formulated beyond the problem-based and project-based models by Illeris (1976) as problem orientation, interdisciplinary learning, exemplarity towards overall educational objectives and teamwork.

At the model level there are quite a few differences. The problem-based model, as it is practiced at the university in Maastricht is based on thematic blocks of approximately 6 weeks. Each block focuses on a specific theme and the teachers have normally prepared cases for the students to work on. The students themselves choose to analyze one of the cases, which can be done either orally or in writing. It uses self-directed study groups, which usually meet once or twice a week to analyze and discuss the cases. Generally, the students use the seven-step procedure starting with clarifying, defining and analyzing the problem, finding the causes, searching for information and reporting the results. In the study group, each individual student presents his or her work to be discussed, and the group discusses who should continue with which tasks. The assessment system is based on an individual exam combined with individual and group-based formative assessment methods. The role of the teacher who attends the meetings is primarily to facilitate the learning process, in other words, to facilitate the group's work and internal communication (Schmidt and Moust, 2000).

The problem- and project-based model practiced at Aalborg University is quite different from the Maastricht model. The traditional Aalborg PBL model is based on problem- and project-based learning. The model was invented in the early 1970s as two new universities were established in Denmark: Roskilde (1972) and Aalborg University (1974). Right from the beginning it was decided that there should be a new educational approach even including the physical space. At Aalborg University, there are more than 1,200 small group rooms where students work on their projects, as a way of supporting the model (Kjærsdam and Enemark, 1994; Kolmos, 1996; Kolmos et al., 2004).

The model is used in all study programmes at Aalborg University within the Faculty of Humanities, the Faculty of Social Science, and the Faculty of Engineering, Science and Medicine. All project work is done in teams, and the same model is more or less followed from the first semester until the completion of a master's degree (tenth semester). The students submit one project per semester and during the study, the groups normally become smaller, starting with typically 6–7 students in the first year, and reducing to a maximum of 2–3 students in the final semester. In each semester the project and the majority of the courses must relate to the theme of the semester.

The understanding of the educational principles will vary from department to department, and these variations will affect aspects of the program. Typical elements that vary are:

- The semester themes that can be defined in different ways, e.g., describing various types of problems, or as themes covering certain subjects
- The choice of *projects*, which can be based on open or more controlled discipline formulations depending on the educational objectives. The open projects can be based on "real life problems" that start with students' formulation of problems
- The definition of a *problem*, depending upon the professional areas. In some programmes the problem must be a dilemma or a social discrepancy, in other programmes the problem will be what students are interested in investigating further
- The relation between *courses and projects*, depending on the knowledge, traditions and culture of the various departments. For example, there are large differences in the relation between courses and projects at the Faculty of Engineering, Science and Medicine, and the language study programmes at the Faculty of Humanities. At the Faculty of Engineering, Science and Medicine the students attend project courses, which support the objectives of the project and the project itself. At some of the languages studies the students are offered a number of courses and they are free to choose 2–3 courses and write their project within in the framework of these courses
- The extent of project *facilitation* and the method in which it is carried out
- The *group sizes*, which vary. There are more students in each group at the beginning of study. Group sizes also vary from department to department.
- Comparing the Maastricht and Aalborg models, there are substantial differences, which can be seen in the table below (Table 12.3).

The Aalborg model seems to be more student-centred and based on more open projects, whereas the Maastricht model seems to be teacher directed. The learning process is different – due to the submission of projects, students in the Aalborg model have to learn project management skills that can be transferred from one project to the other. The process for Maastricht students seems to be more structured from the very beginning. The team aspect is also different as the Aalborg students not only discuss the problem together, but have to come up with a common product. Finally, the assessment systems are different, both the formative system as well as the final exam.

Table 12.3 Comparison between Maastricht PBL model and Aalborg PBL model

	Maastricht	Aalborg
Problem	Cases defined by teachers: open and narrow	Problems defined by students or facilitators within a theme which can be open and narrow
Process	Seven jumps	Project management
Team aspect	Discussing together	Discussing and writing together
Assessment/exam	Individual progress testing Individual exam	Formative group assessment, Individual judgement in a team based exam

Although there are differences at the concrete model level, Graaff and Kolmos (2003, 2007) found that there are common learning principles that cross these two models. Based on an analysis of problem-based and project-based practices and the underlying learning principles, they have formulated PBL principles that can be captured in three approaches: learning, contents and social.

The *learning approach* in problem- and project-based learning means that *learning is organized around problems* and will be *carried through in projects*. It is a central principle for the development of motivation. A problem makes up the starting point for the learning processes, places learning *in context*, and bases learning on the learner's experience. It is a unique task involving more complex and situated problem analyses and problem solving because it is also *project-based*.

The *contents approach* especially concerns *interdisciplinary learning*, which may span across traditional subject-related boundaries and methods. It is *exemplary practice* in the sense that learning outcomes are exemplary to the overall objectives and support the relation between *theory* and *practice* by the fact that the learning process involves an analytical approach using theory in the analysis of problems and problem solving methods.

The *social approach* is team-based learning. The *team learning* aspect underpins the learning process as a social act where learning takes place through dialogue and communication. Furthermore, the students are not only learning from each other, but they also learn to share knowledge and organize the process of collaborative learning. The social approach also covers the concept of *participant-directed learning,* which indicates a collective ownership of the learning process and, especially, the formulation of the problem.

To formulate common learning principles for problem-based learning and for problem-based and project-based learning might cause critique as many researchers seek to maintain the differences. Savin-Baden (2003) and Prince and Felder (2006) argue that there is a substantial difference between, for example, problem-based and project-based learning. Whereas problem-based learning addresses an open learning process, project-based learning is more or less synonymous with a task-based approach. Behind this understanding is an interpretation of a project as a narrowly formulated task. One of the pioneers of the Aalborg PBL approach, however, defines a project as a complex, unique and situated task that will always involve an open approach (Algreen-Ussing and Fruensgaard, 1990). This definition of project-based learning includes a problem-based approach as the starting point for projects to find situations and problems to analyze and solve.

PBL as a teaching and learning model also has a lot in common with other educational approaches such as active learning, inquiry-based learning, experiential learning, co-operative learning, and case-based learning. These approaches are more or less based on the same learning principles, but do not fulfil all of them. Active learning is very often used in traditional classroom teaching, particularly in the USA, because it is easy to implement within the existing organizational framework for teaching and learning; however, it might not be based on team aspects or the same learning approach. PBL is a much more demanding method and often requires organizational changes in order to practice more complex projects.

12.2.1 Evidence of PBL

As there are quite a lot of differences among the different PBL models, it might be hard to draw the conclusion that PBL is the answer to the future challenges concerning engineering skills (Fig 12.1).

Fig. 12.1 Students at work in a PBL environment

Furthermore, a lot of research on PBL has been conducted, and this is not an attempt to give an overview, just to indicate some of the results. The research on PBL addresses different aspects of PBL:

- Management dimension
- Students' learning processes (undergraduates and graduates)
- Teaching processes (planning PBL, new teacher roles, assessment systems, etc.)
- Outcomes, skills and knowledge (graduates and employers)
- Change processes.

The management dimension concerns the financial implications of PBL. Several Danish evaluations show that Aalborg University, compared to other Danish institutions, have one of the lowest dropout rates among students and one of the highest percentages of students finishing their studies on time. In terms of the quality of teaching and learning, PBL institutions are often rated more highly than traditional teaching institutions.

Many authors have addressed students' learning processes and there are studies of students and graduates. Du and Kolmos (2006) explain why there is an improvement of the learning process through participating in team-based communities and by reflecting and experimenting with the practice. However, the learning experience remains tacit, if the students do not reflect and experiment with their own learning process (Kolmos, 1999). The formal as well as the informal learning processes motivate both women and men towards further learning. From the perspective of educational psychology, there have been several studies on motivation with the unambiguous conclusion that PBL increases students' motivation for learning (Schmidt and Moust, 2000; Thomas, 2000). This might be the most important finding: that PBL has an impact at the level of motivation. From a theoretical learning perspective, motivation is an important factor in the learning process – and if students are motivated, they learn more.

Much of the research on the teaching process of PBL has its roots in active research with the purpose of developing and improving PBL-systems and developing new conceptual understandings of the facilitator's role, e.g., the development of the tutor role (Savin-Baden, 2003; Savin-Baden and Wilkie, 2004; Hansen, 2004; Biggs, 2003). Most of the PBL research is to be found within this topic, but will not be reported here. Some of the research that has been conducted concerns improvement of PBL systems.

In terms of the development of skills, Dochy et al. (2003) have made a review of the literature from the 1990s on the evaluations of long-term effects of using PBL. Their main conclusion was that the use of PBL improves the development of transferable skills such as process competence. The impact on knowledge acquisition is missing or not significant. However, PBL students do not acquire less knowledge compared to students educated the traditional way.

Several studies come up with the same findings: that there is no significant improvement of knowledge acquirement, but a significant improvement of skills. These results concerning the acquisition of knowledge may have to be seen in the light that the concept of knowledge which is used for measurement is a concept from traditional education, so the PBL systems are compared to the values of the old system, not according to the more complex knowledge concept of PBL.

Nevertheless, there are results addressing the process, or transferable, skills and studies documenting that there is a significant improvement of that part. Faland and Frenay (2005) have conducted an empirical study of a transformation process at a particular institution. Their main conclusion was the same: that students do obtain process competence. Crosthwaite et al. (2006) document that the students' own perception of the achievement of skills had been significantly improved by PBL learning.

Schmidt and Moust (2000) have done a review of existing literature and concluded that PBL seems to have an effect on long-term retention of knowledge such as the remembering and understanding of various concepts.

Other studies are based on employers' response to education. Danish studies show that employers are very satisfied with candidates from the PBL institutions and note that these candidates are easy to integrate (Krogh and Rasmussen, 2004).

These studies are all made in different subject areas and evaluate the results of various PBL systems. The bottom line is that when you start to integrate collaborative and project-related elements into education, students achieve higher skill levels from the teaching and learning methods and do not learn a lesser amount of scientific knowledge compared to the traditional educational methods.

Furthermore, PBL might be a solution to the requirements of the global society. However, PBL is not implemented overnight. It is not an event; but rather a long, serious process of change from the traditional paradigm of learning to a new paradigm of collective, cognitive learning with the aim to achieve interdisciplinary knowledge for analyzing and solving problems.

12.3 The Institutional Implementation Process of PBL

Institutions all over the world implement PBL at a course level, program level and institutional level (Graaff and Kolmos, 2007). These transformation processes are often very challenging, and the question is how to do it. Fullan (2001, 2004, 2005) is one of the few authors who have developed models for the comprehension and organization of change in education. He emphasizes that the outcome is not only improved student learning, but also improved organizational capacity. The faculty is the ongoing factor, whereas the students leave once graduated. Change is a long process of continuous activities and improvements (Fullan 2005; Scott, 2003). It takes time, especially if the change not only concerns restructuring the scientific content, but also involves a cultural and conceptual change. This is in line with Henriksen et al. (2004) who argue that in order to understand organizational change, we need to understand the concept of reality. In order to comprehend reality, it is necessary to look into at least four elements: fact (documentation), logic (the core part of the constitution process), values (to describe the importance), and communication (as being a member of society and interpretation). Therefore, educational change should be analyzed and interpreted in a broader context, and values are an important part of the change process. Furthermore, each transformation process has its own story and thus change will always be contextualized.

To understand the process of change, Kotter's (1995) model for change is often used to illustrate phases at a more specific level. This model was developed in the management literature, but has been used as an analytical model for educational processes as well (Morgan and Roberts, 2002). Kotter works with eight phases:

1. Establishing a sense of urgency
2. Forming a powerful guiding coalition
3. Creating a vision
4. Communicating the vision
5. Empowering others to act on the vision
6. Planning for and creating short-term wins
7. Consolidating improvements and producing still more change
8. Institutionalizing new approaches

Kotter stresses the importance of urgency and the creation of visions. Normally, teachers do not experience any urgency. On the contrary, they feel confident and satisfied with existing teaching practices. Only a few staff members feel the need for change (Moesby, 2004). External reasons are most often the trigger for internal institutional change (Fig 12.2).

In Kotter's version, the vision consists of the leader formulating and communicating the vision to the faculty staff. As already pointed out in this article, governance, leadership and management are in a process of change: on the national level, towards more institutional autonomy, and on the institutional level, moving from a bottom-up approach to a top-down approach. This means that it might become easier to make decisions, but not necessarily to carry out changes. In educational settings, the role of the leader can be hard to define and fulfil, because leaders are often very good colleagues to their "employees." However, due to the new management systems in HE with appointed leaders, there might be more top-down decisions in this area and thus a phase in which it is necessary to involve faculty staff.

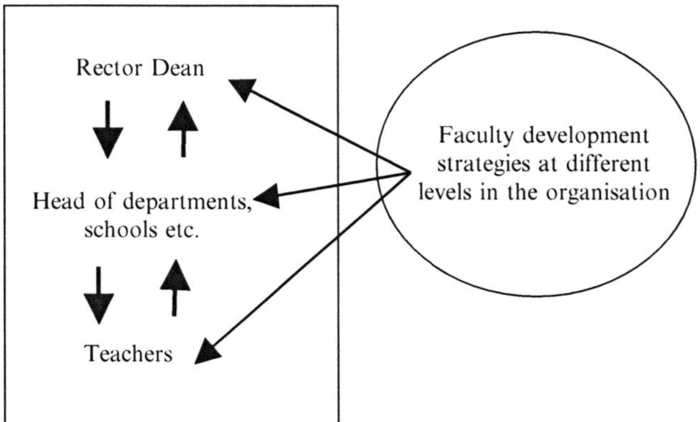

Fig. 12.2 Relation between faculty development units and the university organization

Researchers point out that all organizational levels have to become involved if the goal is successful change (Kolmos et al., 2004; Scott, 2003; Graaff and Kolmos, 2007). Bottom-up strategies are not efficient since change at a systemic level requires a decision at the top level. Top-down strategies are not efficient because they create resistance in the system. However, the two strategies taken together supplement each other and make change possible.

Furthermore, there is a need for change agents. If the change starts from the top level, change agents must be found among the involved faculty members. If the change starts from the bottom, change agents must be found in the top. The role of change agents is to motivate faculty staff and to lead the change process by constantly pushing for visions, exact plans, resources, strategies, etc. Each individual change agent should not cover all the responsibilities, but experience shows that drivers are necessary.

An important change agent is a faculty development unit, because they have the expertise and knowledge of other systems and new practices. Training is an important element, because it is necessary to establish a new educational practice. Faculty development units have to act at all levels, e.g., to join meetings for heads of departments, study boards, deans, etc., and promote awareness of the roles of both the faculty development units and the leaders at all levels in the university organization (Kolmos et al., 2001).

12.3.1 Two Stories of Change

In order to illustrate the described change processes, we will present two Danish stories of change within engineering education. These stories could have happened in many other countries – there are many stories to tell (Graaff and Kolmos, 2007; Kolmos, 2002). What is unique about Danish engineering education is that there is not just one individual institution implementing PBL, but rather all institutions over the last 10 years have implemented PBL on either a small or large scale. We have witnessed an emerging tendency to formulate models and systems, albeit of very different natures. In a research project on change, we interviewed eleven faculty staff at three different institutions. The eleven interviews at each institution were with former leaders, change agents and some faculty members that had to implement all of the ideas. All interviews have been transcribed and analyzed using a qualitative thematic method.

The two institutions are university engineering colleges and have similar characteristics: about 1,000 students, 100 faculty members and more or less the same programmes. The institutional change took place in two different years: institution A in 1998, institution B in 2002. Extensive institutional changes rarely occur without an external pressure, since many resources are involved, and the external cause (or urgency) was reduction in resources for both institutions. In general, the same educational–political processes have governed all Danish university engineering colleges, characterized by:

- Continuous reduction in resources, which is unfortunately an international trend
- Amalgamation of institutions – at the time when we collected data, both institutions went into new amalgamations
- Change in organizational conditions from democratically elected structures to appointed leaders and external boards. This is the real difference between the two cases, as they changed in two different years, under two different management systems
- Transition from technical college to university engineering college, where a certain percentage of staff have to be upgraded to master levels
- Joint educational development through the establishment of the Pedagogical Network for Danish Engineering Education (IPN). This network offered the same type of seminars and workshops to both institutions
- Previous experience with PBL, since interdisciplinary projects were integrated in the 1980s, which has earned the institutions their first project experiences. This was an important experience, because staff felt that PBL was not something completely new, but something that they already had in-house experience with.

12.3.1.1 Institution A

At Institution A, change took place under a democratically elected board of directors, which had no intention of causing division in the organization, but instead set realistic goals for the processes of change. In this way, they facilitated a bottom-up approach that led to a managerial decision on top-level (Table 12.4).

Table 12.4 Institution A's change process

Institution A
1996–1998: they joined several workshops held by IPN on project-based learning. Many teachers tried to experiment with project work
1998–1999: the decision was made stipulating that at least one third of the overall time should be used for project work. The decision was made by the senate, which at that time was a democratic forum, elected by faculty members
1998: rebuilding of the physical infrastructure to allow for group rooms
The internal staff development function was not well established at the institution. There were only sporadic, if any, follow-ups after the first workshops in 1996–1998
The interviewees could not remember any resistance among colleagues
There were no indications that the approach to teaching and learning changed. Projects are mainly regarded as an application of the knowledge that has been taught in lectures

Table 12.5 Institution A's curriculum structure

<table>
<tr><td>Semester 1-4</td><td>Course 5ECTS</td><td>Course 5ECTS</td><td>Course 5ECTS</td><td>Course 5ECTS</td><td>Project 10 ECTS</td></tr>
<tr><td>Semester 5</td><td colspan="5">Practicum</td></tr>
<tr><td>Semester 6</td><td>Course 5ECTS</td><td>Course 5ECTS</td><td>Course 5ECTS</td><td>Course 5ECTS</td><td>Course and pre project 10 ECTS</td></tr>
<tr><td>Semester 7</td><td>Course 5ECTS</td><td>Course 5ECTS</td><td colspan="3">Project 20 ECTS</td></tr>
</table>

The interviewees did not express any particular excitement about their change. This could be ascribed to several factors; for instance, the fact that a substantial period had passed since the change occurred, and therefore they were unable to remember it in detail. The institution had also started a new process of change initiated by a newly hired leader, who did not find the employees of the institutions properly prepared for changes.

Wide-scale project work was introduced and formulated as in the model in Table 12.5. A characteristic feature is the fact that there was no major merging of courses. A project worth 10 ECTS was introduced, though, in each of the semesters 1–4 and 6, and a project worth 20 ECTS was included in the last semester.

However, here we see a model that is not far reaching in its endeavours to implement a different educational style. The institution still holds a series of individual courses with individual exams and no attempt to integrate courses and projects.
It is important that this model was implemented at a time when there was still a democratically elected board of directors. This had a great influence on the process, since that type of structure limited how much the board could manage the process without clear support from the staff. Consequently, interviews with employees in this institution also reflect that they do not remember that there was any great resistance to the changes, because it was the employees' decisions to change. However, a few of the employees would have liked to carry through solutions that were more radical.

12.3.1.2 Institution B

In Institution B, the changes were implemented at a later stage when appointed leaders were in charge. Institution B consisted of three departments: one department had experience with project work to a wider extent; while the two other departments had more sporadic experiences and many faculty members were against any change. When a new board of directors was appointed, people from the department that had already adopted the project work dominated the board, and it was agreed that the entire institution was to develop a common model for teaching and learning (Table 12.6).

Table 12.6 Institution B's change process

Institution B
1996–2000: several training workshops on project work were held
Late 1990s: electronics and mechanics started to implement elements of PBL, but to a very different extent. Electronics went for nearly a full-scale model and had some very enthusiastic faculty members. The other department went for a more sporadic implementation
2001: a new leader was appointed from electronics
A top-down decision was made, and a group of change agents were asked to lead the actual transformation process
There was a strong internal staff development unit to support the idea
Finally, the committee in charge of the process of change formulated a number of visions, philosophies, and values to accompany the new models

The decision was a result of both a bottom-up approach and a top-down approach. It was a bottom-up approach because one department had already experienced a full scale PBL model; it was a top-down approach as the newly appointed leader made the decision to apply the model to the entire institution. If a democratic process had been allowed to form the basis for decision-making in the department that had not fully adopted PBL, the decision to implement the model might not have been as far-reaching.

The interviews from Institution B demonstrated that they were proud of their process of change. They all held their own understanding of the model and the underlying educational concept. At first, this phenomenon occasioned a great deal of reflection on the interviewers' parts, but along the way, this came to be interpreted as a positive expression of the personal internalization of the educational process of development, and that they had actually developed their own conceptual understanding of PBL.

Institution B also experienced resistance. Around 20% of the interviewees from this institution did not wish their interviews to be recorded, but they clearly expressed that they "just did as they had always done, only now it was named something new when in reality it was not new at all." The typical reply when asked why they kept doing what they were used to doing was "The students are unable to learn this profession in any other way." The leaders at institution B are well aware of the resistance and have chosen to ignore it this time around. As long as the resistance does not cause any student complaints, management will continue to disregard it, but at some point, it will have to be dealt with.

The curriculum model is described in Table 12.7. A characteristic feature of this process was the fact that smaller courses were grouped together in bigger professional units.

The process of change in institution B involved additional elements. In part, new goals for development of students' process skills were formulated in connection with the introduction of project work. In part, experiments were carried out involving the use of formative assessment methods in courses, with a particular focus on experiments concerning the timing of the relationship between courses and project.

Table 12.7 Institution B's curriculum structure

Semester 1-4	Courses	Courses	Project 10 ECTS pr. semester
Semester 5	Course	Course	Project 18 ECTS
Semester 6	practicum		
Semester 7	Project normally with a company 30 ECTS		

12.3.2 Comparing the Two Institutions

The two institutions have had quite similar conditions of change, but the processes of change have taken place at different times and under different managerial structures. This has given different basic conditions for decision-making.

Both cases demonstrate that top-down decisions at institutional level need to be combined with a pool of motivated staff to cause changes at a system level. A bottom-up approach with decentralized development in departments leads to variation within the institution, and it might be difficult to develop curriculum models at system level without top-down decisions. For universities in some countries, this conclusion might seem banal, but in a European context, this is a very important result.

Institution B also managed to develop a more holistically, coherent education. Through the process of change, the institution succeeded in giving most staff a vision for their own processes of change. This did not include all staff members, since approximately 20% of faculty was very negative towards the changes.

Looking at the results from the two cases in terms of Kotter's eight phases, more effort and attention could definitely have been devoted to two phases. The first phase lacking involves a sense of urgency, and no one in either institution really felt this. The faculties saw the changes as necessary because of financial reasons, not because the students would be able to gain new and different competences. Even though in Institution B there has been the formulation of new competences, the need for this was not the guiding factor for the individual employee. The common perception was that it was simply necessary to do.

The second phase that is missing is the vision. No vision was formulated in Institution A. In institution B, some visions were formulated, but both institutions tended to stop the process of change since no new energizers entered the process. Nonetheless, it is important to conclude that both institutions have developed models, and they have succeeded in institutionalizing the changes by, for instance, altering the physical infrastructure.

12.4 Perspectives

These two stories of institutional change could have happened anywhere in the world. Higher education is in a process of change, both at the level of governance and management and in the educational area. In Europe we have seen a lot of changes in these areas, partly in order to harmonize higher education, partly to address new global requirements. The same trend can be observed in the rest of the world: National governments are getting more and more oriented towards internationalization of higher education in order to secure mobility and sufficient number of academics for the knowledge society.

Along with these political changes come transformational processes at the institutional level. According to new market mechanisms, there is pressure on institutions to develop educational change. However, educational change is a challenging and energy-consuming process. The experiences with institutional change in Denmark do not differ very much from experiences in the southern part of Europe, in America, Australia and many other places. There are cultural and organizational differences, but the lesson for change is more or less the same: in order to manage institutional change, it is necessary to have both top-down and bottom-up processes, change agents, visions, realistic plans, qualified staff and so on. Therefore, the exchange of international experiences is an important part of the development of engineering and science education.

The requirement for new types of professional and process skills is global. Since engineers and scientists now have a global workplace, and education has to address these new types of requirements, PBL is one educational solution that has become widespread. In most parts of the world, PBL is integrated into courses. However, more and more institutions choose to develop their educational and pedagogical profiles, and PBL is implemented at a program, department, faculty or institutional level. By developing systems and organizations, it is possible to secure the implementation of PBL, which requires co-operation, project management, communication, problem identification and solving, and setting of objectives, followed by teaching, learning and assessment for these types of skills. This is one of the reasons that institutions want to change to PBL. Other reasons are the improvement of the learning process by the creation of motivation and improved retention rates, and establishment of institutional profiles for attracting more students.

In the future, more institutions will follow this trend. In order to support the global change to PBL, a new UNESCO Chair in Problem-based Learning in Engineering Education (2008) has been established at Aalborg University, Denmark. The chair has the overall objective of creating a global sustainable society of researchers, experts and institutions in PBL in order to facilitate the transformation process from traditional teaching to PBL. This Chair will promote an integrated system of research, training, information and documentation in the field of PBL in engineering and science – and provide evidence for the use of PBL in various subject areas. Experiences from engineering can directly be used in all areas of science such as computer science, nanotechnology, biotechnology, mathematics and physics. PBL is about learning and motivation for learning by the use of more student-centred learning principles rather than only scientific content principles. If science and technology want to recruit young people in the future, it is necessary to contextualize science and to use new more student-centred approaches.

References

ABET (2008). http://www.abet.org/. Retrieved 13 October, 2005.

Algreen-Ussing, H. & Fruensgaard, N.O. (1990). Metode i Projektarbejde. Aalborg: Aalborg University Press.

Askling, B. & Henkel, M. (2000). Higher Education Institutions. In M. Kogan, M. Bauer, I. Bleiklie & M. Henkel (Eds.), Transforming Higher Education – A Comparative Study (Chapter 5). London: Jessica Kingsley.

Assiter, A. (Ed.) (1995). Transferable Skills in Higher Education. London: Kogan Page.

Barnett, R. (1999). Reconfiguring the University. In P. Scott (Ed.), Higher Education Reformed. London: RoutlegdeFalmer.

Barrows, H.S. (1996). Problem-Based Learning in Medicine and Beyond: A Brief Overview. In L.A. Wilkerson & W.H. Gijselaers (Eds.), Bringing Problem-Based Learning to Higher Education: Theory and Practice (pp. 3–12). San Francisco: Jossey-Bass.

Biggs, J. (2003). Teaching for Quality Learning at University – What the Student Does. Buckingham: Open University Press, The Society for Research into Higher Education.

Bok, D. (2003). Universities in the Marketplace – The Commercialisation of Higher Education. Princeton, NJ: Princeton University Press.

Carter, J., Eriksen, K., Horst, S. & Troelsen, R. (2003). If Reform of Science is the Answer – What were the Questions? Copenhagen: University of Copenhagen, Center for Naturfagenes Didaktik.

Clark, B.R. (1998). Creating Entrepreneurial Universities: Organisational Pathways of Transformation. Amsterdam: Pergamon.

Crosthwaite, C., Cameron, I., Lant, P. & Litster, J. (2006). Balancing Curriculum Processes and Content in a Project Centred Curriculum – In pursuit of Graduate Attributes. In Education for Chemical Engineers, March 2006, Trans IChemE, Part D, no. 1, pp. 39–48 http://www.icheme.org/ECEsamplepaper.pdf.

Dochy, F., Segers, M., Van den Bossche, P. & Gijbels, D. (2003). Effects of Problem-Based Learning: A Meta-Analysis. Learning and Instruction, 13(5), 553–568.

Du, X. & Kolmos, A. (2006). Process Competence in a Problem and Project-Based Learning Environment. In Proceedings of the 34th SEFI Annual Conference: Engineering Education and Active Students, No. 34, July 2006, Uppsala University, Faculty of Science and Technology. Available at: http://www-conference.slu.se/sefi2006/.

EUR-ACE (2008). http://www.feani.org/EUR_ACE/EUR_ACE_Main_Page.htm http://www.pble.ac.uk/. Retrieved May 2008.

Faland, B. & Frenay, M. (Eds.) (2005). Problem and Project-Based Learning in Higher Education: Impact, Issues, and Challenges. Louvain-la-Neuve: Presses Universitaires de Louvain.

Friedman, T.L. (2005). The World is Flat – A Brief History of the Globalised World in the Twenty-First Century. London: Allen Lane.

Fullan, M. (2001). The New Meaning of Educational Change. New York: Teachers College Press.

Fullan, M. (2004). Leading in a Culture of Change: Personal Action Guide and Workbook. San Francisco: Jossey-Bass; Wiley.

Fullan, M. (2005). Leadership & Sustainability: System Thinkers in Action. Thousand Oaks: Corwin Press.

Gibbons, M., Limoges, C., Nowotny, H., Schwartzman, S., Scott, P. & Trow, M. (1994). The New Production of Knowledge. The Dynamics of Science and Research in Contemporary Societies. London: Sage.

Graaff, E. & de & Kolmos, A. (2007). Management of Change Implementation of Problem-Based and Project-Based Learning in Engineering. The Netherlands: Sense.

Hansen, S. (2004). The Supervisor in the Project-Organised Group Work Should Participate in Developing the Students' Project Competence. European Journal of Engineering Education, 29(3), 451–459.

Henriksen, L.B., Nørreklit, L., Jørgensen, K.M., Christensen, J.B. & O'Donnell, D. (Eds.) (2004). Dimension of Change – Conceptualising Reality in Organisational Research. Copenhagen: Copenhagen Business School Press.

Illeris, K. (1976). Problemorientering og deltagerstyring: oplæg til en alternative didaktik (Problem Orientation and Participation: Draft for an Alternative Didactic). Munksgaard, Copenhagen.

Kogan, M., Bauer, M., Bleiklie, I. & Henkel, M. (2000). Change and Continuity. In M. Kogan, M. Bauer, I. Bleiklie & M. Henkel (Eds.), Transforming Higher Education – A Comparative Study. London: Jessica Kingsley.

Kolmos, A. (1996). Reflections on Project Work and Problem-Based Learning. European Journal of Engineering Education, 21(2), 141–148.

Kolmos, A. (1999). Progression of Collaborative Skills. In Themes and Variations in PBL, Vol. 1, [Refereed Proceedings of the 1999 Bi-ennial PBL, Conference, 7–10 July 1999, Montreal: Canada] Callaghan, Australia: Problarc.

Kolmos, A. (2002). Facilitating Change to a Problem-Based Model. The International Journal for Academic Development, 7(1), 63–74.

Kolmos, A. (2006). Future Engineering Skills, Knowledge, and Identity. In J. Christensen, L.B. Henriksen & A. Kolmos (Eds.), Engineering Science, Skills, and Bildung (pp. 165–185). Aalborg: Aalborg University Press.

Kolmos, A. & Graaff, E de. (2003). Characteristics of Problem-Based Learning. International Journal of Engineering Education, 17(5), pp. 657–662.

Kolmos, A., Fink, F. & Krogh, L. (2004). The Aalborg PBL model. Aalborg: Aalborg University Press.

Kolmos, A., Gynnild, V. & Roxå, T. (2004). The Organisational Aspect of Faculty Development. In A. Kolmos, O. Vinther, P. Andersson, L. Malmi & M. Fuglem (Eds.), Faculty Development in Nordic Engineering Education. Aalborg: Aalborg University Press.

Kolmos, A., Rump, C., Ingemarsson, I., Laloux, A. & Winther, O. (2001). Organisation of Staff Development – Strategies and Experiences. European Journal of Engineering Education, 4(26), 329–342.

Kotter, J.B. (1995). Why Transformation Efforts Fail. Harvard Business Review, 73(2), 44–56.

Kjærsdam, F. & Enemark, S. (1994). The Aalborg Experiment – Project Innovation in University Education. Aalborg: Aalborg University Press.

Krogh, L. & Rasmussen, J.G. (2004). Employability and Problem-Based Learning in Project-Organised Settings at the University. In The Aalborg PBL Model: Progress, Diversity and Challenges (pp. 37–56). Aalborg: Aalborg University Press.

Moesby, E. (2004). Reflection on Making a Change Towards Project Oriented and Problem-Based Learning. World Transactions on Engineering and Technology Education, 2(3), 269–278.

Morgan, C.K. & Roberts, D. (2002). Herding Cats? Obtaining Staff Support for Curriculum Change and Implementation. International Consortium for Educational Development, Perth, July, 2002. Available at: http://www.csd.uwa.edu.au/iced2002/publication/toc.html.

Prince, M.J. & Felder, R.M. (2006). Inductive Teaching and Learning Methods: Definitions, Comparisons, and Research Bases. Journal of Engineering Education, 95(2), 123–138.

Savin-Baden, M. (2003). Facilitation Problem-Based Learning. Birkshire: Society for Research into Higher Education and Open Universtity Press.

Savin-Baden, M. & Wilkie, K. (Eds.) (2004). Challenging Research in Problem-Based Learning. Birkshire: Society for Research into Higher Education and Open Universtity Press.

Schmidt, H.G. & Moust, H.C.M. (2000). Factors Affecting Small-Group Tutorial Learning: A Review of Research. In D.H. Evensen & C.E. Hmelo (Eds.), Problem-Based Learning: A Research Perspective on Learning Interactions (pp. X–XX). Mahwah, NJ: Lawrence Erlbaum.

Scott, G. (2003). Effective Change Management in Higher Education. In EDUCAUSE Review, 38(6), X–XX.

Shuman, L.; Besterfield-Sacre, M. and McGourty, J. (2005). The ABET "Professional Skills" - Can They Be Taught? Can They Be Assessed? Journal of Engineering Education, Vol. 94, No. 1, pp 41–56

Sporn, B. (2003). Management in Higher Education: Current Trends and Future Perspectives in European Colleges and Universities. In R. Begg (Ed.), The Dialogue Between Higher Education Research and Practice: 25 Years of EAIR (pp. 97–107). Dordrecht: Kluwer.

Thomas, J.W. (2000). A Review of Research on Project-Based Learning. Autodesk Foundation. Available at: http://www.bie.org/pdf/researchreviewPBL.pdf.

UNESCO Chair in Problem-Based Learning in Engineering Education: http://www.ucpbl.net. Retrieved May 2008.

Part IV
Changes in Views of Science and Mathematics

Chapter 13
From Anomaly to Paralogy

The Post-modern Condition and its Consequences for University Science Education

Ole Ravn Christensen and Tom Børsen

In recent years natural science students' enrolment patterns have changed. Many new study-programmes have emerged with new hot names like nanotechnology, molecular biomedicine, medicinal chemistry, biotechnology, health mathematics, product and design psychology etc. These new study programmes transcend the traditional disciplinary borders of the classical scientific disciplines, so that we see how core knowledge from the old sciences like physics, chemistry, biology, mathematics etc. are mixed with each other and sometimes with humanistic disciplines and social sciences. With the new study programmes we say "[f]arewell to the old classifications, such as physics, chemistry, biology. Welcome to new ones, like GRAINN – short for genomics, robotics, artificial intelligence, neuroscience and nanotechnology [and] SHEE – the sciences of safety, health and environment plus ethics" as Jerome Ravetz has put it (2006, pp. 10–11). Hence, the contemporary tertiary natural science study-programmes face the challenge of coping with trans-disciplinarity.

One could, however, argue that the new study programmes are not new at all, and that the mixing of the knowledge of the old sciences has been a task undertaken by the disciplines of engineering (the applied sciences) for more than a century. Just because somebody makes up new names for what engineers are doing, does not mean that science and technology is changing fundamentally.

However, this argument rests on a division between pure (academic) and applied science. In this article we question this division and argue that the emerging new sciences represent a qualitatively new endeavour within the spheres of science and technology. We will demonstrate that a new scientific rationality is emerging, that fundamentally differs from that of the classical sciences. The new scientific logic is not primarily concerned with the expansion of the body of scientific knowledge, but is formed to fulfil new and strong external utility expectations. As the classical and the new trans-disciplinary scientific disciplines are fundamentally different one cannot construct new science study-programmes simply by mixing the educational

O.R. Christensen and T. Børsen
Aalborg University, Denmark and University of Copenhagen, Denmark

O. Skovsmose et al. (eds.), *University Science and Mathematics Education in Transition*,

modules of the classical sciences. The new trans-disciplinary study-programmes must differ qualitatively form those of the classical academic sciences.

In this chapter we set out to investigate the educational consequences of these transitions. Our claim is that classical academic science is undergoing changes that will eventually have an impact on university science education and we seek to explain what these changes amount to and how the impact on education may be conceived.

The starting point of our investigation will be two famous positions in the philosophy of science, namely Thomas Samuel Kuhn and Jean François Lyotard's different understandings of the processes underlying and causing scientific development. We will show how Kuhn's influential philosophy of science, first presented in the early 1960s, lacks the power to explain the dynamics of the trans-disciplinary sciences. Kuhn, especially in his earlier writings, is often accused of being relativistic with regard to scientific truth; and compared to earlier logical positivist standpoints his ideas may even be considered to have given birth to post-modern cultural science studies (Weinberg, 1998). However, we argue that this reading of Kuhn cannot stand alone and we want to point out that the Kuhnian conception of science is narrow and mono-disciplinary. Kuhn's understanding of academic science – like that of some logical positivists – defines scientific activity as the narrow endeavour of searching for scientific results within a specific scientific discipline (Pedersen, 1995). Kuhn gives an *internal* account of scientific development which points to the creation of a consensus on what is considered *true* among the members of a scientific community – and neglects external parameters.

In contrast to Kuhn's conception of scientific development we outline Lyotard's ideas. Lyotard's conception of post-modern scientific development takes into accounts the disciplinary mixture of many new fields of study, as well as the *external* relations and dependencies that scientific practices face today. According to Lyotard science does not mainly develop towards establishing a solid body of true knowledge but rather through pressures to become performative.

In this chapter we seek to identify how theories of science translate into ideas about the educational set-up at universities. We first examine Kuhn's and then Lyotard's conceptions of science with an emphasis on their views on the development and legitimation of science. Then, we discuss the educational consequences of a Kuhnian understanding of science in relation to university science education, and finally we discuss the consequences of Lyotard's proposed post-modern condition for university science education.

13.1 Kuhn's Conception of Science

In his famous book *The Structure of Scientific Revolutions*, Kuhn (1970) presents a socio-historical view on the development of the natural scientific disciplines. He perceives science as a social practice rather than an individual cognitive endeavour. Kuhn's account of the scientific development does not primarily aim at determining the validity of scientific claims – for example by showing how scientific knowledge can be perceived as "true" or that some scientific beliefs are more credible than other beliefs – but sets forth to map the socio-historical evolution of the natural sciences.

13.1.1 Conceptual Framework

According to Kuhn the scientific landscape consists of three types of sciences: normal science, revolutionary science and immature science. Immature science refers to a type of intellectual discussion climate from where all proper academic scientific disciplines originate. If one goes sufficiently far back in time all scientific domains were once situated in an intellectual context where many different accounts of natural phenomena, with more or less equal strength, competed in recruiting "believers". For a given discipline, at a specific and well-defined point in time, one of these accounts gained sufficient strength and intellectual capital to outdo competing views, hence forming a normal scientific discipline. Kuhn categorizes "the study of motion before Aristotle and of statics before Archimedes, the study of heat before Black, of chemistry before Boyle and Boerhaave, and of historical geology before Hutton" as immature (Kuhn, 1970, p. 15).

The "paradigm" is a central concept in Kuhn's philosophy of science. In one of its meanings a paradigm denotes a cognitive pattern, which a group of communicating scientists share. In the postscript to the second edition of *Structures* (1970) Kuhn clarifies the meaning of such a cognitive pattern – which he denotes as the disciplinary matrix – by splitting it into four elements: symbolic generalizations, metaphysical beliefs/model assumptions, values and exemplars (Kuhn, 1970, pp. 182–187):

Recognized formal equations make up the set of symbolic generalizations included in a discipline's matrix. Newton's laws of motion, or the three gas laws (Boyle's law, Charles' law, and Avogadro's law) constitute illustrative examples of symbolic generalizations (our examples)

Shared metaphysical beliefs such as "heat is the kinetic energy of the constituent parts of bodies" and "all perceptible phenomena are due to the interaction of qualitative neutral atoms in the void, or, alternatively, to matter and force, or to fields" (Kuhn's examples) are also part of a disciplinary matrix. This element of the disciplinary matrix one can think of as model assumptions

Values constitute a third element in the disciplinary matrix. Usually values are more widely shared among different scientific communities than either symbolic generalizations or metaphysical beliefs. Hence, Kuhn suggests that they generate a feeling of cohesiveness among natural scientists as a whole. Kuhn himself stresses scientific values like "scientific results should be accurate", "quantitative predictions are preferable to qualitative ones". He also states that "theories must permit puzzle-formulation and solution", and "be simple, self-consistent, plausible, compatible with other theories deployed"

By the notion of exemplars Kuhn refers to the exercises that students encounter during their scientific education. Exemplars are both practical (i.e., the tasks presented to students during laboratory work), and theoretical (the exercises found at the end of chapters in science textbooks). Technical solutions to scientific problems found in the journals that scientists study during their post-educational research careers are also considered as exemplars, as they show the scientists, by example, how their job is to be done.

A scientific (sub)-discipline is "normal" when its matrix' elements are left unchallenged by practitioners and surroundings. According to Kuhn, normal science is the most predominant form of academic science. Normal science progresses as normal scientists compete to solve the puzzles posed to them by the paradigm, by applying the symbolic generalizations to new particular situations, and thereby expanding the number of exemplars within their discipline's matrix. This dynamic is necessary to reach convergent scientific thought, which Kuhn considers the ultimate goal of normal science (Andersen, 2000).

13.1.2 Scientific Development

Normal scientists are sometimes unable to solve all scientific puzzles, and anomalies appear (i.e., situations where the symbolic generalizations fail to reduce the complexity of natural phenomena). When the number of anomalies become sufficiently high new practitioners become increasingly unsatisfied with the disciplinary matrix in power, and frustrated young scientists try to set up new and competing disciplinary matrices. Eventually they manage to formulate an alternative matrix that is strong enough to threaten, and maybe even over time replace, the old paradigm. Kuhn uses the phrase "revolutionary science" to denote an intellectual landscape, where an established paradigm or disciplinary matrix is being challenged by an emerging alternative one. In revolutionary periods two parallel paradigms exist, and compete regarding recruitment of new practitioners. Eventually one of the competing paradigms dies, as the followers retire without new young people taking their academic positions. A new scientific revolution has occurred when a new paradigm has replaced an old one.

Normal science deals with puzzle solving in contrast to revolutionary science that aims at the formulation of a new stable disciplinary matrix. Kuhn describes the history of science as a cyclic process: a long period of normal science is succeeded by a revolutionary period and eventually by a scientific revolution whereby a new normal scientific period is initiated. Both normal science and revolutionary science are necessary for scientific development, as "there is an essential tension implicit in scientific research. [If puzzle-solving] were the only mode of conducting science, science could lead to no fundamental innovation at all" (Andersen, 2000, p. 92).

According to Kuhn it is (practically) impossible for the individual scientists to be part of both the old paradigm – predominant before a scientific revolution has taken place – and simultaneously internalize the new paradigm that has taken over as the dominant cognitive pattern among practitioners. It is not possible for followers of the old paradigm to fully understand the proponents of the new paradigm, and vice versa. The two cognitive frameworks (the old and the new paradigm) are incommensurable. It *is* possible for scientists *over time* to change their conceptual framework, and hence internalize the new way of thinking that follows a scientific revolution; but it is a painful accommodative learning process of iconoclastic proportions. Such mental changes never happen over night.

13.1.3 Legitimation of Science

In his two books on Kuhn, Steve Fuller closely links Kuhn's philosophy of science to Vannevar Bush's social contract between science and society (Fuller, 2000, 2003). The contract says that society, through the state, finances academic science, and gives it autonomy to freely choose its research questions and methods, and to set up university curricula and recruitment mechanisms. In this way the state provides the fuel for the reproduction of academic science. In return for the fuel the state expects that the scientific communities make their knowledge production available as a free public resource (Bush, 1945).

The Bush line of reasoning is associated with what is known as the linear model of technological development. "This [model] locates science at the "upstream" end of a one-way process by which useful discoveries and inventions eventually flow down into the home, the shop, the hospital and the workplace" (Ziman, 2000, p. 15). In other words, the model says that pure scientific institutions – e.g., universities – produce and validate (via numerous examples of puzzle-solving) pure nomothetic knowledge (symbolic generalizations) that technological institutes, consultancy enterprises develop and deliver to smaller industries and entrepreneurs, who again are bought or enrolled in bigger companies. It is not possible to predict which particular piece of pure knowledge will be utilized later on. We end up with a clear-cut distinction between pure and applied science where the latter deals with the art of applying pure scientific knowledge to practical, but unique, problems.

When Bush's social contract is put into action, the linear model of technological development gains strength. Under such circumstances scientific communities need not (and ought not) to legitimize their particular activities as useful to funding agencies as research resources are provided by the state (of course researchers must show to their peers that they are scientifically competent). Furthermore scientific communities need to recruit new students, and hence new practitioners in order to survive (cf. Kuhn). The recruitment strategy of normal science cannot primarily refer to the societal utility of normal science, as pure science has no direct societal value, but must refer to some internal feature – like, science is fun or intellectually challenging.

The legitimation of science is therefore, in our conception of the Kuhnian framework of science, something that is connected closely to Bush's social contract between science and society. Science lives an isolated life supported by the state; it shares the products (symbolic generalizations) with the surroundings, and hereby fuels the industry's technology development.

13.2 Lyotard's Conception of Science

Jean-François Lyotard (1924–1998) is one of the leading figures in newer French philosophy. His ideas about post-modernity have greatly influenced many current intellectual debates. In this article we address his essay, *The Post-modern Condition: A Report on Knowledge*, which brought him fame outside France and made him a renowned philosopher around the globe[1]. This is also his most important work in which he reflects on science and brings about a conception of science that differs greatly from Kuhn's both with regard to his approach and the outcome of his analysis.

13.2.1 Conceptual Framework

In *The Post-modern Condition*, which was requested by and presented to the Conseil des Universitiés of the government of Quebec, Lyotard describes a move away from the modern era. The modern world's view is transforming into a post-modern framework of understanding and perceiving the world. The movement towards post-modernity is described as a transition in the attitude towards meta-narratives. In modernity meta-narratives were used to legitimize actions, whereas post-modernity is defined as a way of thinking where meta-narratives are rejected or "tranquilated" as Lyotard likes to depict their diffusion (Lyotard, 1992, p. 18).

Lyotard makes an important distinction between two types of discourse, narrative knowledge and scientific knowledge (Malpas, 2003, p. 21). By narrative knowledge he refers to the broad scale of stories we tell each other. The bedtime stories we tell our children before they fall asleep; our understanding of ourselves as revealed when we attend psychological therapy; and our political and ideological stand points about society are all narrative by nature, as they do not essentially need any legitimation. These stand in contrast to scientific knowledge, which is always in need of legitimation. In modernity legitimation takes place by referring to a so-called meta-narrative. Lyotard points out two such meta-narratives dominating modern thought:

- The idea that the history of scientific knowledge is a history of progress towards emancipation and social freedom.
- The understanding that scientific knowledge is progressing towards a unified encyclopaedic totality.

The first meta-narrative can be classified as a narrative about *justice* and *freedom*. Lyotard finds that this narrative begins with the French Revolution of 1789. Here, science is legitimized through its potential to emancipate humanity/the people from

[1] The French version of the essay was published in 1979 (Lyotard, 1979). The English translation appeared in 1984 (Lyotard, 1984).

material and social oppression. The second meta-narrative concerns *truth*. Here the scientist is seen as a romantic hero, who is making continued contributions to the construction of the unified building of scientific knowledge. Lyotard traces this meta-narrative back to the philosophy of Hegel institutionalized in the Humboldtian model of university. Here the function of the university is to retain the body of knowledge and explain and develop the principles and the foundations of knowledge in its totality. In the end it is Hegel's speculative spirit that legitimizes science (Lyotard, 1984, p. 33).

13.2.2 Legitimation of Science

Lyotard's central claim is that neither of these traditionally strong meta-narratives can any longer survive. In a later work, Lyotard states that the modern project "has not been forsaken or forgotten, but destroyed, "liquidated" (Lyotard, 1992, p. 18), due to internal inconsistencies (Brügger, 2001, p. 81). Lyotard states: "Simplifying to the extreme, I define post-modern as incredulity toward metanarratives" (Lyotard, 1984, p. xxiv). In that sense he describes the post-modern condition as one in which each narrative in society is left in its own particularity with regard to legitimation, scope of application, and interrelations with other narratives. No common goal of justice or common understanding of truth is perceived to exist.

What then legitimizes post-modern science when it is not a quest for justice, freedom or truth? In Lyotard's analysis of the post-modern condition a third type of legitimation of knowledge and scientific activities is identified. He refers to a scientific communal practice as a "language game" and distinguishes "between the denotative game (in which what is relevant is the true/false distinction) from the prescriptive game (in which the just/unjust distinction pertains) from the technical game (in which the criterion is the efficient/inefficient distinction)" (Lyotard, 1984, p. 46).

The first two games rest on the meta-narratives we have already touched upon. The third game – the *technical* game – values scientific knowledge in terms of its efficiency or performativity. It is concerned with other types of criteria for evaluating knowledge. Is this particular scientific knowledge efficient or not in fulfilling local aims? Does it entail the desired effects? Will it pay off?

The technical narrative of performativity reflects an instrumental rationality, as it only focuses on how one effectively reaches a given goal – it does not reflect on the aims themselves. Hereby the technical game differs from the denotative and prescriptive ones, which are both value rational. Their meta-narratives hold as an intrinsic element the criteria for evaluating specific scientific activities. Do they lead to new truths in the encyclopaedia of science respectively to emancipation?

According to Lyotard we have witnessed a gradual historical change from science being first and foremost legitimized by grand narratives (e.g., stating that scientific results are true or emancipative) to being legitimized locally through its performativity. The narrative of performativity can easily be connected to a number of small narratives in different ways. As there is no longer any single unifying goal of science one appears to be authoritative if one claims such a goal exists. The consequence of the rejection of "grand narratives" is not the total rejection of denotative and prescriptive justification strategies. But they can no longer be taken for granted, and they are reduced to little narratives ("petit récit") that function in subdomains.

In Lyotard's post-modern framework scientific activities that follow the rules of the denotative, respectively the prescriptive games are gradually being subordinated to the technical criteria of efficiency. Scientific development is therefore governed by research results' ability to perform and Lyotard uses the term "techno-science" to signify this state of affairs: "There is no denying the dominant existence today of techno-science, that is, the massive subordination of cognitive statements to the finality of the best possible performance, which is the technological criterion". (Lyotard, 1984, p. 77)

In other words techno-science should be understood as a specific scientific discourse on legitimation that cherishes efficiency and performance. This new quest for efficiency negates the "science for its own sake" dictum, as the technological criterion entangles any scientific work in a practical setting (in a company, in a grassroots organization, in a university context, in a political decision making process etc.). The quest for efficiency also negates the idea that science emancipates the whole of humanity, as humanity conceptually has been split up into many sub cultures. It is impossible to know what benefits the whole of humanity.

13.2.3 Scientific Development

According to Lyotard, this post-modern condition continuously prevents the crystallization of stable paradigms in science:

> Research that takes place under the aegis of a paradigm tends to stabilise [...] But what is striking is that someone always comes along to disturb the order of "reason". It is necessary to posit the existence of a power that destabilises the capacity for explanation, manifested in the promulgation of new norms for understanding or, if one prefers, in a proposal to establish new rules circumscribing a new field of research for the language of science. (Lyotard, 1984, p. 61)

Lyotard denotes this destabilization potential of science as *paralogy*. In order to have a fuller account of this fundamental notion we should consider how Lyotard links it to Wittgenstein's concept of a *language game*. Lyotard's method or procedure for understanding the post-modern condition is in general emphasizing a Wittgensteinian approach to language games. He makes three observations about language games (Lyotard, 1984, p. 10); here expressed in relation to a language game of a science:

- The rules of a science do not carry with them their own legitimation but are the object of a contract between the players of the game of this science. This means that the rules governing what is reasonable to do in, for example, physics or mathematics are not natural but determined by the members of the respective scientific communities

- The rules, by which a specific language game of science is played define the game. Some moves in the game are accordingly meaningful and others are not. Every utterance and action should be thought of as a move in the game
- Even an infinitesimal change of one rule in the scientific game alters the nature of the game. The language game is even open to changes as the result of legitimate moves enters in the game: how to follow a rule is never fully determined, and game players continuously "play" with the interpretation of a rule. It might change scientific precedent, when an existing rule is applied in a new setting.

Lyotard uses these three observations to clarify the definition of *paralogy* as new moves in a scientific language game (for example a formula is used in a new context) or as the invention of a new game of science (for example the recent construction of the language game of nanoscience). Hence, by paralogy Lyotard tries to pinpoint the fundamental ways of development in science under the post-modern condition. This development is partly fuelled by new moves on a small scale within an existing discipline and partly on macro level moves where new moves in the language game of science create or invent new fields of study.

In order to gain further insight into the meaning of Lyotard's notion of development through paralogy we may follow his description of *innovation*. It is important to observe that Lyotard does not equal paralogical development to innovation:

> Paralogy must be distinguished from innovation: the latter is under the command of the system, or at least used by it to improve its efficiency; the former is a move (the importance of which is often not recognised until later) played in the pragmatics of knowledge. (Lyotard, 1984, p. 61)

Lyotard links the efficiency/performativity criterion for scientific activities to the concept of innovation. Thinking in instrumental terms on making science efficient threatens the *authentic* development of science: the paralogical one that consists of making qualitative new moves in the language games of science.

Under the post-modern condition both performativity and paralogy are possible local legitimating criteria[2]. Therefore they can both act as the engine of scientific development. Lyotard is a spokesperson for development through paralogy as "[t]he logic of performance, of optimizing the system's overall performance, based on the criterion of efficiency, does violence to the heterogeneity of language-games" (Peters, 1995, p. 393). Whereas performativity tends to reduce science to instrumental reasoning and views science as a mere instrument for societal systems, paralogy brings forward new ways of thinking in a local scientific context. Performativity, in Lyotard's framework, potentially annihilates any scientific language game that does not benefit the efficiency of a system's performance. Paralogy increases the number of different games and hence the diversity and creativity of the overall scientific approach and development.

---

[2] Lyotard also mentions consensus achieved by open discussion as a third possible local legitimating criterion for scientific development. But he continuously criticises it by claiming that consensus cannot be attainted without oppression and exclusion of minoritarian and incommensurable games.

13.3 Normal Science Education

We have investigated Kuhn and Lyotard's different interpretations of the legitimation and development of science. In the following we shall try to draw some educational consequences of these two approaches. Both accounts focus on what scientists do, and how they are expected/need to legitimize their activities. Both Kuhn and Lyotard are specifically interested in the scientific process. Their theories can be seen as a move away from a focus on the products of science to a focus on the processes and the everyday practice of science. However they are in disagreement with regard to the dynamics of the scientific practice.

First we summarize two important characteristics of Kuhn's conception of science that have direct implications for the way science education should be executed – his focus on the *stability* of a discipline's established knowledge and the *isolation* of a given discipline from external interference. We contrast these focal points with Lyotard's conceptions and finally outline what has been called "Normal Science Education" – a perspective on university science education that is in line with Kuhn's theory of scientific development.

13.3.1 Stability Versus Instability

Kuhn's account of scientific development is essentially describing a modern condition for science where normal science develops according to a fixed logic of the paradigmatic game. Paradigms come and go, but do not essentially develop. Science is only involved in serious transitions and changes during rare cases of scientific revolutions where new paradigms emerge. The central point in Kuhn's philosophy of science is the construction of stable scientific knowledge that practically no scientific expert doubts: the goal is a state of universal consensus among the scientific experts. Kuhn cherishes the stability of scientific paradigms, and outlines how such stability is achieved and maintained. Scientists are expected to explore the consequences of the paradigm, not to reflect upon or change it.

In Lyotard's opinion, what characterizes contemporary science is something that can not be accounted for in a Kuhnian framework. The stable accumulative development of normal science is not sustainable under the post-modern condition because of the massive subordination under the technological criterion. Instead Lyotard suggests that science is best characterized as a paralogical endeavour. The rules of scientific language games (paradigms, if you like) change all the time. The assumptions and values that characterize what it means to be scientific are constantly negotiated in the pragmatics of science. Scientists are players in scientific games. But game rules are not fixed. Language games change all the time – they are instable.

13.3.2 *Isolation Versus Interaction*

In Kuhn's theory of scientific development the main idea is that a scientific paradigm is essentially isolated in its development and no external powers and language games influence this development. The development of science is thought of as isolated. In this conception of science society benefits from keeping scientific institutions (universities etc.) autonomous. What is thought to be the basic strength of science makes this isolation necessary. Society's needs enter only at the level of applied science.

In contrast Lyotard's account is based on the perspective that the legitimation of science is deeply interwoven with the demands of the social fields outside the realm of science. What counts as good science, is not the exclusive decision of the scientific communities. Instead external language games of politics, economics, religion etc. become influential in the formulation of research goals. Scientific institutions are required to engage in teleological dialogue with users of all sorts. In other words scientists need to have the ability to critically transcend their scientific language games – that is to see their own practice from other perspectives. The new sciences (e.g., biotechnology or nanoscience) can be understood as paralogical moves in that direction. They are scientific language games transcending the classical borders of scientific activities and by doing so new fields of research are formed with new methods of science that drive the scientific development.

13.3.3 *Normal Science Education*

Kuhn's focus on the stability and isolation of scientific development has educational implications. Here we outline what has been called "Normal Science Education" – a perspective on university science education that follows from Kuhn's theory of science. University science education plays an important part in Kuhn's philosophy of science. Students need to learn how to identify and solve puzzles, as this becomes their primary task when they themselves one day enter academia. The kind of university science education that prepares science students for puzzle solving has been given the name of "Normal Science Education" (NSE) by a Dutch research group (van Berkel et al., 2000).

NSE is probably the predominant form of teaching the classical academic disciplines of science in Denmark, and elsewhere at both the secondary and tertiary levels of science education. Kuhn's description of the classical sciences also includes an educational focus: university science education must ensure the production of good normal scientists. Academic science is very rarely presented to students as a body of knowledge with a turmoil history or competing current views on specific issues. In short, academic science is presented to university students as a set of rules for the scientific game that must be learned and followed without questioning.

NSE therefore leads to convergent scientific thinking. However, it also gives academic science a conservative flavour, since it promotes the reproduction of normal scientific communities via socialization of the up-coming generation of scientists. The socialization takes place by teaching students exemplary uses of the symbolic generalizations that take the form of puzzles. The puzzles are solved by, applying a given formula in the correct manor and the puzzles have one solution only. Another characteristic of NSE is the fictitious divide between science and its societal context, which decouples the sphere of science from external fields such as ethics and politics.

Hence, the Kuhnian account of university science education rests on the assumption that science aims at the establishment of stable, mono-disciplinary and isolated scientific thinking. However, this is not the only aim of science. Controversies can be epistemologically fruitful when the normal scientific methodology is incapable of capturing nature's complexity. Kuhn is aware of this – cf. his description of revolutionary science – though this sort of thinking ought not penetrate down to the educational level.

13.4 Science Education under the Post-modern Condition

In Lyotard's conception of science NSE serves no purpose other than to make science students familiar with a basic core of knowledge in the classical sciences. The idea that the stable knowledge of the classical disciplines should be more basic than other fields of knowledge hinges on the encyclopaedic narrative of science. What seems to be of the utmost importance in the contemporary state of science is, in Lyotard's view, the capacity to set up and use an efficient strategy in a particular context, that is, to solve a given problem.

> It should be noted, however, that didactics does not simply consist in the transmission of information; and competence, even when defined as a performance skill, does not simply reduce to having a good memory for data or having easy access to a computer. It is a commonplace that what is of the utmost importance is the capacity to actualise the relevant data for solving a problem "here and now," and to organise that data into an efficient strategy. (Lyotard, 1984, p. 51)

Data and information is manifold in the post-modern era. Lyotard speaks of this situation as "perfect information" as opposed to a situation where you (for example the teacher) have the upper hand in the game by having access to more information than the other players (for example the students).

> But in games of perfection, the best performativity cannot consist in obtaining additional information in this way. It comes rather from arranging the data in a new way, which is what constitutes a 'move' properly speaking. [...] It is possible to conceive the world of post-modern knowledge as governed by a game of perfect information, in the sense that data is in principle accessible to any expert: there is no scientific secret. [...] what extra performativity depends on in the final analysis is 'imagination,' which allows one either to make a new move or change the rules of the game. (Lyotard, 1984, p. 52)

Lyotard points to the need for fostering *imagination* in science! Without imagination neither the production of performative nor paralogical knowledge is possible. These two types of knowledge production are described as processes of arranging "data" in new ways that change the rules of the game or constitute new moves in science. Fostering imagination is therefore also related to the ability to connect spheres of data previously disconnected by the traditional disciplinary landscape of the Humboldtian university model. If the task of science education is not only to pass on mono-disciplinary information but also to promote paralogical thinking in science it should also secure the ability to transcend the borders of a given discipline. Education should,

> [...] include training in all of the procedures that can increase one's ability to connect the fields jealously guarded from one another by the traditional organisation of knowledge. [...] In Humboldt's model of the University, each science has its own place in a system crowned by speculation. (Lyotard, 1984, p. 52)

Under the post-modern condition the traditional disciplinary matrix is undermined because the pressure towards new moves in science erodes this matrix. Aiming for performative or paralogical knowledge production is inconsistent with teaching inside the narrow frame of Humboldtian scientific disciplines. Urgent and complex problems can only rarely be solved inside the frame of only one discipline. Most real world problems are in fact transdisciplinary and Lyotard therefore expects the classical ordering of the sciences as well as of science teaching to gradually give in to this pressure.

The transitions in the disciplinary organization have implications for the role of the teacher. Because of the perfect information situation (Lyotard of course had no knowledge of the spread of the internet in 1979) he argues that the authority of the mono-disciplinary scholar will be challenged by the superior imagination of interdisciplinary teams.

> But one thing that seems certain is that [...] the process of delegitimation and the predominance of the performance criterion are sounding the kneel of the age of the Professor: a professor is no more competent than memory bank networks in transmitting established knowledge, no more competent than interdisciplinary teams in imagining new moves or new games. (Lyotard, 1984, p. 53)

If the prime focus for science education is to nurture imagination (performative or paralogical thinking) in science education, it cannot consist of only receiving information about a tradition of what is considered bullet-proof knowledge within a given discipline. This was the original task of "the professor".

We find Lyotard's perspectives on university education inspiring inputs to the educational strategies of universities in coping with the trans-disciplinary sciences but also the classical science programmes of physics, mathematics etc. Let us summarize what this inspiration might look like by formulating some principles for science education's curricula development under the post-modern condition. We propose four important principles that deal with both the matter of content and the best possible learning milieu for students, as well as the role of the university teacher.

13.4.1 Open-Ended Problem Solving

Students should learn to handle open-ended problems, not only to solve one-ended puzzles. In NSE the idea is to control the outcomes of strategies for solving a scientific problem, whereas the opposite is what could promote paralogical thinking. Open-ended and situated problems unavoidably require paralogical solutions, as new moves will be needed to find a scientific strategy for solving exactly this particular problem in a reasonable way. Paralogical thinking could thrive in dealing with this type of problem because there is no "true" or "singular" solution to an open-ended problem. In an educational environment that presents to students unsolved open-ended questions, discussions on scientific method automatically surface as an extremely important part of learning a scientific discipline because there is no directly applicable method for answering the question.

Problem-Based Learning – or just PBL – is a well known, and in several universities well-established, way of handling open-ended problem solving educationally (Kolmos, this volume). A case-based approach is another way of handling open-endedness. In contrast to the NSE it is not enough to present students with basic puzzles where only one methodological approach is allowed and only one solution is the right one. This is obviously counterproductive in making students imaginative and creative in their work with scientific problems.

13.4.2 Contextualization

We find that PBL and case-based learning environments are in thread with Lyotard's general account on the status of knowledge in highly developed societies. However, the problems and questions that students are asked to work on within these settings could be extremely theoretical. This, however, in our opinion could be countered by supplementing PBL and case studies with a desire to contextualize the problems and questions. Students should be acquainted with problems situated in a particular real world context rather than only reproducing canonical and abstract knowledge.

In contrast to the NSE, the performativity criterion that (according to Lyotard) presses the development of all sciences forwards means that it will be essential for any future scientist to be working in close relation with ongoing real life issues, problems or visions. Theory is important but could be reinterpreted as "important when needed" in relation to a contextualized problem instead of only being considered important in itself.

13.4.3 Trans-Disciplinarity

If problems and questions that students work on are contextualized and open-ended the possible ways of finding a methodological approach will very often include trans-disciplinarity. The solution strategy will combine knowledge from more than one discipline in order to give the best possible solution to a given research question. The strategy is poly-paradigmatic rather than mono-paradigmatic. It is valued by its effort to combine the insights of several fields of research in relation to a particular problem or question that doesn't fit the disciplinary establishment and turn them into a strategy for thinking about this particular problem or question in this particular context.

In contrast to the NSE, it is very important for students to be familiarized with radically different types of being scientific than those of their own field of study. Only by knowing what other scientific approaches look like will it be possible for students to learn to navigate in the field of science. The NSE has a very strong tendency to isolate each science and especially to seclude them from the sciences of other faculties, which makes a truly trans-disciplinary scientific approach of the highest quality impossible from the outset, as well as from the wider social, political and financial contexts.

13.4.4 Teams of Teachers and Supervisors

In order to bring imagination into students' learning processes the role of the teacher must change. First of all, teams of teachers consisting of researchers with different educational backgrounds are needed. Teams of teachers can build the trans-disciplinary environment that is needed to both qualify and help the construction of a combined trans-disciplinary approach to a given problem or question. They must also show students in practice what science is all about, how research is done properly when trying to solve complex problems or answering complex questions involving different scientific paradigms and many methodological considerations.

In contrast to the NSE concept teachers cannot be thought of as only passing on information to students. Students need to be engaged in creating new moves within a given scientific discipline. Only by showing students how research can be done under highly particular circumstances can they improve their thinking about what it means to do scientific research.

13.5 Final Discussions

Different conceptions of what legitimizes scientific activities and what constitutes scientific development point towards different educational set-ups in science education. Here we have presented two conceptions of science that in many ways are in opposition to one another. Kuhn's conception of science is mono-disciplinary, encourages stability and considers scientific activities as a fairly secluded enterprise from the remaining spheres of society. Lyotard, on the other hand, is eager to explain the transition of the status of knowledge in highly developed societies.

Knowledge production is trans-disciplinary – what counts as relevant knowledge is not stabile – and demands an immediate cooperation with external spheres of society to gain legitimate status.

Inspired by these differences we have sought to contrast the educational settings that are inherent in these two conceptions of science. Kuhn's conception of science provides us with a framework for understanding university science education as Normal Science Education where puzzle solving and the socialization into a mono-disciplinary paradigm are essential ingredients. Lyotard, on the other hand, sees university education as developing towards dealing with real world problems and questions.

We suggest a number of principles that can guide the design of new trans-disciplinary study-programmes. The principles are derived from Lyotard's analysis of contemporary science and their relevance hinges upon the correctness of this analysis in describing the conditions for doing science today.

The article, however, highlights another very important concern about university science education. The way we as scientists as well as the way in which politicians, industry etc. conceive science is intrinsically connected to the way university science education is structured: what kind of content is considered essential, what is expected from teachers and students, etc. From our point of view university science education should be considered as being in a transitional phase as a result of the changed status of knowledge in contemporary society. We could think for example of the ways in which we generally interpret society as a "knowledge society", our economy as a "knowledge economy", and how "modus 2 research" takes place not (only) at universities but in private companies. This change – whether it is best characterized as a post-modern condition or something different – can be expected to have vast implications in the future for the university as an institution, and as part of the university, science education will not be left unaffected.

References

Andersen, H. (2000). Learning by Ostension: Thomas Kuhn on Science Education. *Science & Education*, 9, 91–106.

Brügger, N. (2001). What About the Postmodern? The Concept of the Postmodern in the Work of Lyotard. In R. Harvey & L.R. Schehr (Eds.), *Jean-François Lyotard: Time and Judgment (Yale French Studies, 99)* (pp. 77–92). New Heaven: Yale University Press.

Bush, V. (1945). Science the Endless Frontier – A Report to the President by Vannevar Bush, Director of the Office of Scientific Research and Development. Washington: United States Government Printing Office. Retrieved May 2008, http://www1.umn.edu/scitech/assign/vb/VBush1945.html.

Fuller, S. (2000). *Thomas Kuhn: A Philosophical History of Our Times*. Chicago: University of Chigaco Press.

Fuller, S. (2003). *Kuhn vs. Popper: The Struggle for the Soul of Science*. New York: Icon and Columbia University Press.

Kuhn, T.S. (1970). *The Structure of Scientific Revolutions 2nd edition*. Chicago: University of Chicago Press.

Lyotard, J.F. (1979). *La condition postmoderne: Rapport sur le savoir*. Paris: Minuit.

Lyotard, J.F. (1984). *The Postmodern Condition: A Report on Knowledge*. Manchester: Manchester University Press.

Lyotard, J.F. (1992). *The Postmodern Explained: Correspondence 1982–1985*. Minneapolis: University of Minnesota Press.

Malpas, S. (2003). *Jean-François Lyotard*. New York: Routledge.

Pedersen, S.A. (1995). Kuhns videnskabsfilosofi, dens udvikling og betydning. In T.S. Kuhn (Ed.), *Videnskabens revolutioner*. Copenhagen: Fremad.

Peters, M. (1995). Education and the Postmodern Condition: Revisiting Jean-François Lyotard. *Journal of Philosophy of Education*, 29(3), 387–400.

Ravetz, J. (2006). *No-Nonsense Guide to Science*. Rotherham: New Internationalist Publications.

van Berkel, B., de Vos, W., Verdonk, A.H. & Pilot, A. (2000). Normal Science Education and its Dangers: The Case of School Chemistry. *Science & Education*, 9, 123–159.

Weinberg, S. (1998). *The Revolution that didn't Happen*. New York: The New York Review of Books.

Ziman, J. (2000). *Real Science, What it is, and What it Means*. Cambridge: Cambridge University Press.

Chapter 14
Modernities, Sciences, and Democracy

Sandra Harding

The "modern" in "modern science" is a relatively uninterrogated and untheorized concept within the sciences and in the philosophy, sociology, and history of science. This is so today at a time when other aspects of Western sciences have been fruitfully explored in critical and illuminating ways (see Christensen and Hansen, Chap. 13; and Skovsmose, Chap. 15). In particular, the exceptionalism and triumphalism characteristic of Western attitudes toward our sciences have been explicitly criticized and purportedly abandoned by many of the scholars working in science studies fields. By exceptionalism is meant the belief that Western sciences alone among all human knowledge systems are capable of grasping reality in its own terms – that these alone have the resources to escape the human tendency to project onto nature cultural assumptions, fears, and desires. By triumphalism is meant the assumption that the history of science consists of a history of achievements – that this history has no significant downsides. According to this view, Hiroshima, environmental destruction, the alienation of labour, escalating global militarism, the increasing gap between the "haves" and the "have nots," gender, race, and class inequalities – these and other undesirable social situations are all entirely consequences of social and political projects, to which the history of Western sciences makes no contribution. Such conventional Euro centric assumptions can no longer gather the support either in the West or elsewhere that they could once claim.

In recent decades a huge amount of literature on modernity has emerged from the social sciences and humanities. Stimulated by the massive shifts in local and global social formations during the last half of the Twentieth Century, and by the post-modern response to such changes, social theorists, literary and other cultural critics, and, especially, historians have debated the uneven and complex origins, nature, and desirable futures of modernity, modernization, and modernism. Such controversies about modernity are first and foremost about a culture's relation to its past and its possible futures. They arise as ways of asking what went wrong, and

Reprinted from Social *Philosophy Today*, Vol. 22 (2007) Ed. John Rowan, with permission of the author and The Philosophy Documentation Center (Charlottesville, VA, USA).

S. Harding
University of California, Los Angeles, USA

O. Skovsmose et al. (eds.), *University Science and Mathematics Education in Transition*,

what needs to be corrected. The last half a century has witnessed the global decline and fall of belief in the unquestionably legitimate authority of the white, bourgeois, male. Contributing to the epistemological, economic, political, and cultural rubble left by his demise – or, at least, deflation – have been compelling and influential counter-histories of social relations between the races, classes, genders, and within colonial and imperial eras.

In light of such extensive recent discussions of modernity and its woes, one might wonder whether yet another engagement with the topic could be worthwhile. I think there are two reasons to do so. Most of these accounts do not have modern science and technology clearly in focus, nor do they seem to think that gender relations are relevant to either modernity's crisis or its possible successor. First, the science and technology issue. When these authors are from the humanities or classical social theory and its successors in the West, and whether they are politically on the Right, Left, or Centre, science and technology are usually to be found off at the periphery of such accounts. Modernity for them is about exclusion of the influence of religion and kinship in forms of government and citizenship, economy, and education, and about a shift from past to future in social orientation. Such exclusion makes the creation of autonomous, rational institutions, including those of value-free modern science and, consequently social progress possible. Since these scholars are largely unfamiliar with the critiques of exceptionalist and triumphalist science indicated above, they often treat modern sciences as if they played no role in whatever economic, social and political ills lead them to question modernity.

Yet interrogating what is meant by the modernity of Western sciences, and what have been the consequences and will be the likely futures of commitments to modernity in scientific institutions, their cultures and practices, is a more important intellectual and political task than such accounts reveal or comprehend. Such a project poses frustrating questions, which challenge familiar ethical and political assumptions, and even seem critical of the psychic framework that well-intentioned academics bring to such a project. Do we have the right to try to answer such questions? Can we, especially those of us who are U.S. citizens, contribute to creating the polycentric democratic political spaces, in Egyptian economist Amin's (1997) phrase, called for by so many critics of the West's modernity who want to encourage the design and emergence of desirable successors to the West's global hegemony? On the other hand, what are the consequences of our neglecting to engage such issues?

A small handful of critics and defenders of modernity, its political realities and promises, have directly focused on the natural sciences and their technologies. In today's world, they ask, do Western sciences promote or retard the growth of the democratic social relations and social progress, which have been taken to be distinctive marks of modernity? Some have argued that these sciences and their philosophies in some respects actually block important directions in the growth of scientific knowledge and toward social progress – though they have different visions of social progress. Here I look at the consideration of these issues by three such critics of modernity who focus on the sciences: the French ethnographer and philosopher of science Latour (1993), the German sociologist Beck (1992, 1999), and the Indian science studies intellectual Nandy (1988).

The analyses of these three are especially interesting in several respects. First, I selected each to represent a distinctive focus in science studies. Latour is an ethnographer and philosopher, one of the founders and continuing contributors to mainstream European-American science studies. Beck brings a background in the German environmental movement and in sociological theory to thinking about science's role in "risk society." Nandy has been a continuing contributor to postcolonial science studies and especially to its critique of the disastrous recent effects of Western sciences in the Third World. Second, while all are critics of modernity, all find postmodernism an unattractive alternative. Third, all three argue that science has become a kind of governance which illegitimately bypasses democratic processes. However, fourth, in contrast to postmodern critics of modernity and of philosophies of modern science, all three are optimistic about the possibilities for transforming the sciences into ones that are politically accountable for their practices and consequences and can contribute to social progress, and that are epistemologically less "underdeveloped." All three call for more science, though they want different kinds of sciences than those favoured in the contemporary West. They each strategise about how to democratise science in the service of a radically democratized social order, and how to do so by strengthening and expanding the reach of the scientific impulse. This set of commitments and projects makes it difficult to categorize any of them as having fully modern or fully postmodern commitments.

There is one more feature shared by all three. They are all gender blind – the second reason to critically examine the assumptions and promises of modernity and of its possible successors. Yet accounts of modernity from the social sciences and cultural studies reveal that gender relations are something like the proverbial 800-pound gorilla standing in the living room of modernity but invisible to all but feminist scholars. In ignoring gendered aspects of both their objects of study and of their own accounts, all three deeply undermine the chances of success of their own projects. Those of us concerned with social justice need for their projects to succeed, so this lacuna requires attention.

The first section below outlines a few dimensions of current debates over modernity. The following ones sketch out main claims of these three theorists. Section 5 suggests that these theorists' relations to modernity are even more complex than they have acknowledged in that their criticisms are themselves enlivened, for both better and worse, by central projects of the modern ethos, which presumably they would reject. The final section briefly identifies some of the gender issues that these and other participants in debates about modernity, science, and political theory need to address.

14.1 What Counts as Modern?

To begin, "modern" can refer to a temporal era such as the one following the European medieval era, or it can refer to substantive constituents of a society, its structures, practices, and discursive commitments or worldview (Wittrock, 2000).

The temporal notion currently is used in the West with three distinct referents. For philosophers and many historians of science, modern science begins with the scientific revolution of Copernicus, Galileo, Boyle, Harvey and Newton, and modern philosophy begins with Hobbes and Descartes. The early modern philosophers often engaged with implications of the new features of the world which new sciences such as astronomy revealed, and they thought about the shifts in European social formations which they were experiencing. They thought about the new experimentalism in the sciences, and about the new science movements of their day (Van den Daele, 1977; Shapin and Schaffer, 1985).

Yet some historians of science would date the emergence of fully modern sciences later, in the bourgeois revolutions of the Eighteenth Century and the industrial revolution of the late Eighteenth and early Nineteenth Century. They are concerned especially with the increasing power of scientific technologies. Modernization theorists, who produce the second kind of temporal notion of modernity, draw especially on this kind of history of science. Those concerned with modernizing traditional societies, for example in the Third World development policies of national and international agencies and institutions after the Second World War, always focus on transferring to underdeveloped societies, as they were characterized, Western scientific and technological rationality in manufacturing, health care, agriculture and other economic sectors. They take Western forms of modernization to be the only ones, as did their Nineteenth Century forerunners such as Marx, Durkheim, and Weber. For such theorists, as well as for some of their critics, modernization is identical to Westernization. Modernization means Western modernization, and science refers only to Western science.[1] Like Nineteenth Century theorists of modernity such as Marx, Durkheim, and Weber, modernization theorists have expected a gradual homogenization of global societies as Western forms of modernity disseminate around the globe. By now, those Third World development policies are widely criticized for further immiserating the majority of the world's poorest citizens whom such policies were supposed to benefit (Amin, 1997; Sachs, 1992; Escobar, 1995). Feminist work has been an important part of this critique, (Mies, 1986; Shiva, 1989; Sparr, 1994). Moreover, while modernity is now a global condition shaping how all societies engage with the world around them (Wittrock, 2000), the expected homogenization of societies around the world has not occurred (Amin, 1997; Eisenstadt, 2000).

Finally, for literary and cultural theorists, modernism is the late Nineteenth and early Twentieth Century movement which follows romanticism. T.S. Eliot, James Joyce, Picasso and Seurat are modernists. This literary and cultural movement has been the focus of what is perhaps the most developed analyses of the gender of modernity (e.g., Felski, 1995; Jardine, 1985), though particular aspects of modernity have long been the topic of feminist sociologists and political theorists. "Postmodernism," also, can refer to any one of these three eras. No wonder discussions of "the modern" among people from different disciplines can get confusing.

Deciding just when such temporal eras begin and end, however, requires the specification of substantive criteria of the modern, which some particular era does or does not meet. Thus the temporal notion collapses into or depends upon the

[1] Yet see Skovsmose's interesting discussion (Chap. 15).

disputed substantive criteria (Wittrock, 2000). Substantive conceptions focus on the emergence of a differentiated social structure with independent political, economic, religious/moral, educational (including scientific) and family structures, and such democratic institutions as representative government and a free press. They usually centre the presence of a secular worldview, the idealization of universal instrumental rationality, and a social orientation toward the future rather than toward the past. They also include several kinds of contradictory tendencies, such as the insistence on universal reason, yet also recognition and even toleration of the pluralism of rationalities, and a critical and self-critical attitude along with severe restrictions on the appropriate targets of such criticisms. And these substantive conceptions contain unarticulated tensions such as the promotion of a continuing global dissemination of the modern alongside tacit approval of a continual reconstitution of modernity's other, the traditional – a point to which we return. (Eisenstadt, 2000; Wittrock, 2000).

This is the kind of complex and conflicted background against which Latour, Beck, and Nandy undertake their critiques of modernity and their examinations of the tensions between scientific practices and philosophies, on the one hand, and particular modern projects, on the other hand. This background enables them to revise modern sciences in illuminating ways. It also enables us to understand their work somewhat differently from how they understand it.

14.2 Latour: Where Is a Science for Our World of Hybrids and Networks?

Latour argues that modernity and its sciences have an ontology problem. They conceptualize our knowledge of nature as separate from matters of our interests, of justice, and of power, though it is in fact inseparable.

> On page six [of my daily newspaper], I learn that the Paris AIDS virus contaminated the culture medium in Professor Gallo's laboratory; that Mr. Chirac and Mr. Reagan had, however, solemnly sworn not to go back over the history of that discovery; that the chemical industry is not moving fast enough to market medications which militant patient organisations are vocally demanding; that the epidemic is spreading in sub-Saharan Africa [...] [H]eads of state, chemists, biologists, desperate patients and industrialists find themselves caught up in a single uncertain story mixing biology and society.
>
> (Latour, 1993, pp. 1–2)

> We live in an incommensurable mix of nature, politics, and discourse. Yet no one seems to find this [story] troubling. Headings like Economy, Politics, Science, Books, Culture, Religion and Local Events remain in place as if there were nothing odd going on. The smallest AIDS virus takes you from sex to the unconscious, then to Africa, tissue cultures, DNA and San Francisco, but the analysts, thinkers, journalists and decision-makers will slice the delicate network traced by the virus for you into tidy compartments where you will find only science, only economy, only social phenomena, only local news, only sentiment, only sex... By all means, they seem to say, let us not mix up knowledge, interest, justice and power. Let us not mix up heaven and earth, the global stage and the local scene,

> the human and the nonhuman. "But these imbroglios do the mixing," you'll say, "they weave our world together!" "Act as if they didn't exist," the analysts reply.
>
> (Latour, 1993, pp. 2–3)

Latour argues that the world we experience consists of networks linking aspects of nature, cultural legacies, states and nations, agencies, institutes, corporations, official and unofficial policies, de facto practices, mechanisms and other artefacts, and even deities. Reality consists of such hybrid networks. Yet modernity requires a representation of reality consisting only of images of purified objects. It delinks nature from culture, appropriate policies and practices from deities, and it delinks agencies, institutes and corporations from mechanisms and other material artefacts. Modernity, its epistemologies, philosophies of science, and its sciences represent a world of broken networks and dismembered hybrids that is not the one in which we live or about which we want explanations. Its sciences are intentionally isolated from the reality that needs explanation, and that practice itself is a culturally distinct one. We do not live in the modern world that the epistemologies and philosophies of science of modernity imagine in their claims to transcend cultural values. We, and our sciences, have never been modern, he proclaims. He implies that we, and our sciences, are as historically specific, as much in the thrall of our own culturally shaped reality, as any other culture and its knowledge-system.

Where, Latour asks, are sciences of these networks and hybrids? Not in the laboratories or field sites. Rather such sciences have been developed in the disciplines of science studies, he argues. Its histories, sociologies, ethnographies, and textual studies of moments in the history of Western sciences focus on the relations between knowledge-seeking and the social, cultural, economic, political and even psychic projects of an era (e.g., Biagioli, 1999). Latour is not a fan of feminist contributions to this field, but they have been significant (e.g., Braidotti et al., 1994; Haraway, 1989, 1991; Harding, 1986, 1991, 2003). Moreover, we can note, though Latour does not, that, such studies have been expanded by postcolonial scholars to include scientific, instead of triumphalist and Eurocentric, investigations of other cultures' knowledge-systems and of relations between the knowledge projects of the West and those of other cultures. While Latour appears to be unaware of the increasingly large literature with just such a focus that has been developing in these extensions of science studies over the last three decades (Harding, 1998a; Hess, 1995; Petitjean et al., 1992; Sachs, 1992; Sardar, 1988; Selin, 1997), he is nevertheless critical of the West's traditional negative attitude toward other cultures. And his account opens the door to understanding the necessity of re-evaluating other cultures' knowledge systems, though he does not himself step through that door or, more likely, even see the value of such a move.

Latour does not wish to abandon the West's Enlightenment project, in contrast to some other critics of Western philosophies of science. Instead, he proposes that we redefine the Enlightenment to exclude its vision and practices of modernity. Thus he is opposed to both the illusion and what he understands as the ideal of modernity. However, he is not anti-modern, he says, since the modern never has existed; he is only against taking as real the imagined world of modernity's sciences. And he is critical of postmodernists because, he says, they simply give up and enjoy the confusion of the present moment without trying to improve the sciences or their faulty ontology:

> Postmodernism is a symptom, not a fresh solution [...]. It senses that something has gone awry in the modern critique, but it is not able to do anything but prolong that critique, though without believing in its foundations. Instead of moving on to empirical studies of the networks that give meaning to the work of purification it denounces, postmodernism rejects all empirical work as illusory and deceptively scientistic. Disappointed rationalists, its adepts indeed sense that modernism is done for [....]
>
> (Latour, 1993, p. 46)

Latour proposes that we think instead in terms of the "a-modern." More competent sciences as well as the political work of constructing democratic social relations both require reuniting aspects of the world which modernity keeps sundered. "Half of our politics is constructed in science and technology. The other half of Nature is constructed in societies. Let us patch the two back together, and the political task can begin again" (p. 144). The project will be difficult. But "[t]he task of our predecessors was no less daunting when they invented rights to give to citizens or the integration of workers into the fabric of our societies" (p. 145).

Thus Latour links failures of the modern political project to failures of its knowledge project, though he is vague about just what the former project and its failures have been. Advancing democratic social relations as well as restoring the environment, another important political project, both require a scientific study of kinds of objects around us that modernity has banished from view. It is the recently emerging field of science studies that has developed the resources to engage in such work, he argues. Such studies bring systematic scientific assumptions and methods to the description and explanation of the hybrids and networks that constitute reality.

In a later book Latour illuminatingly expands his criticism of the standard conceptual framework of modern sciences, which he delineated in the 1993 study. He also takes on the task of trying to redefine "the political," recognizing that if sciences and what we know of nature are inevitably infused with social and political assumptions, interests, and desires, then revising science requires revising politics also. I recommend the part of this book on the faulty standard philosophy of science, but I think that the way he develops "the political" is excessively contained by distinctively French commitments to "republican democracy." At least he understands the need to take on such a project.

14.3 Beck: The Incomplete First Modernity of Industrial Society

Ulrich Beck argues that a "risk society" has emerged alongside modernity's industrial society. Modern sciences and technologies have enabled industrial society to create terrifying global risks. Now we live in a context of generalized employment insecurity. We fear the kind of economic crises that have sunk national economies virtually overnight. We fear pandemics such as AIDS, SARS, or Asian Bird Flue which can quickly spread around the world with little warning, little possibility of immunity, and often few known remedies. New kinds of environmental destruction seem daily to threaten our lives and the natural resources upon which human life depends. Damaging forms of radiation, whether from armaments, power plants, or work sites, seem impossible to predict let alone to eliminate or control. "Everywhere pollutants and toxins laugh and play their tricks like devils in the Middle Ages. People are almost inescapably bound over to them. Breathing, eating, dwelling, wearing clothes – everything has been penetrated by them" (Beck, 1992, p. 73). Little appears to be known about the effects of genetically modified foods, yet it seems impossible to stop agribusiness from producing and selling them. And there are the dangers from second-hand smoke, mad cow disease, urban crime and violence, the unregulated global firearms trade, and a host of other contemporary phenomena – including, now, biological, chemical, and military terrorism – all of which are enabled through modernity's sciences and technologies. However isolated from cultural, social, and political influences scientists may imagine their work to be, the consequences of their work today are always already embedded in cultural, social, economic, and political possibilities for those with the power to turn such possibilities into actualities.

Who is to blame? Everyone and thus no one seem responsible for these risks. No one is held accountable for them. It is impossible to prove with a reasonable degree of certainty that the toxic industry upstream, the tobacco industry, or agribusiness's use of pesticides and chemical fertilizers was responsible for a particular pattern of increased cancer rates, or that blame for urban violence or terrorism should be placed on the hugely expanded, profitable, and minimally regulated arms industry. Scientific technologies seem out of control politically. Thus we all seem doomed to have to continue to countenance these increased threats to life and health. We live in an environment of manufactured uncertainties and institutional irresponsibility.

Beck argues that this risk society is not the consequence of failures of modern industrial societies but of their successes. Beck's question is "How can the risks and hazards systematically produced as part of modernization be prevented, minimized, dramatized or channelled?" (p. 19). Modern sciences, their epistemologies and philosophies of science have permitted the development of only semi-modern societies, he proposes. Moreover, these sciences and their philosophies are not likely to look kindly on the kinds of transformations necessary to make possible fully-modern societies and their more self-critical, reflexive, sciences. They will resist those developments of sciences that stand a chance of eliminating or even diminishing the extent and power of the risk society. The problem is that modern sciences profit too much from risk society. "[S]cience is one of the causes, the medium of definition, and the source of solutions to risks, and by virtue of that very fact it opens up new markets of scientization for itself" (p. 155). The actual and perceived threats of the risk society are created by modern sciences and their technologies. However, it is also modern scientific and technological experts who are asked to analyze and measure for us these risks. Finally, it is just such experts who are then called upon to "solve" the threats of the risk society. The risk society is knowledge-dependent. Thus modern sciences and technologies are profiteers dependent upon their monopoly on the production of truths, and thus on the continuation of the very risks their epistemologies and philosophies of science made possible in the first place. Here is another way in which we can see how science has become a form of governance, a way of "ruling" our lives. Decisions that will affect how we live (and die!) are made on scientific and technological grounds that bypass the democratic processes to which political decisions are supposed to be subjected.

How are we to escape this dreadful scenario? Beck's solution calls for completing the modernity, which he argues industrial society only began. He conceptualizes the present moment as a break or shift in modernity. The risk society is the consequence of incomplete modernization.[2]

We have experienced only the first, industrial modernity, he argues. Only more strongly reflexive practices can open up scientific and technological decisions to appropriate democratic rule. We need the second, reflexive, modernity that can provide more extensive scientific knowledge seeking, broader rationality and greater objectivity. By "reflexive" he means here self-critical.

> [...T]he framework of risk society again connects what have been strictly discrete areas: the question of nature, the democratisation of democracy and the future role of the state[....] Risk society demands an opening up of the decision-making process, not only of the state but of private corporations and the sciences as well"
>
> (Beck, 1999, p. 5)

A "democratization of democracy" is needed, he argues. Beck pointed out why we should expect resistance to such an opening up of decision making. Yet he has another insight, which leads him to be optimistic about the possibility of such a transformation: it can occur through the expansion of science. There are three ways in which such a process is already under way, he points out. One is the dissemination of the scientific attitude and practices into the systematic study of sciences and technologies themselves: the development of "sciences of science" in the field of critical science and technology studies. This is one way in which the production of scientific knowledge has escaped the monopoly of laboratory and field scientists. Scientists do not and should not be given the "last word" on the nature of their own enterprise, the culture of science in which they are the natives. After all, one could point out, they always recommend against such a practice with respect to other groups of natives. Beck cites Latour's work as an example of this kind of critical science and technology studies – while disagreeing with Latour's claim that "we

[2] Here one could argue against Beck that the risk society is the result of a central failure of modernity, namely its restrictions on the modern project, its imposed limits on the scientific impulse, and its devaluation of the ethics and politics of science. Beck could respond that he means to point to how the modern project restricted its own goals to precisely the kinds of successes which have produced these problems.

have never been modern" (Beck, 1999, pp. 150–152). We need a reflexive modernity, he argues, one that can scientifically describe, understand, and explain its own principles and practices, which industrial modernity cannot.

A second way in which the risk society itself has inadvertently expanded scientific practice is that it makes us all participate in the production of reliable knowledge. We must do so because the first modernity proliferates experts who continually disagree with each other. "...[E]xperts dump their contradictions and conflicts at the feet of the individual and leave him or her with the well intentioned invitation to judge all of this critically on the basis of his or her own notions" (Beck, 1992, p. 137). Consequently, "science becomes more and more necessary, but at the same time, less and less sufficient for the socially binding definition of truth" (p. 156). Are smoking and second-hand smoke harmful or not? Do vitamins and herbal remedies improve our health and longevity? Which ones and how many should we take? Should we increase our intake of fish, since fish oils seem to provide protection against various forms of cancer, or decrease the amount of fish in our diet since fish seem to pass on environmental toxins? Should we take the radiation and chemotherapies recommended by Western biomedicine to stop the growth of cancers, or the herbal remedies recommended by German pharmacologists? Will the benefits of genetically modified foods outweigh their risks? Is global warming increasing, and if so what should we do about it? The proliferation of conflicting expert opinions on such topics requires that each of us must figure out how to make informed decisions about our own nutrition, health, and safety and that of our children and other dependents. We are forced to become part of the production of scientific knowledge. The production of scientific knowledge escapes the monopoly of scientist experts in this second way.

A third expansion of science can be found in the new emphasis on the importance of everyday experience to the formation of new scientific questions. Beck refers to this as a "science of questions." We can note, though he does not, that daily experience comes in different social forms for the rich and the poor, for men and women, for colonizers and their objects of colonization. Thus the new social movements representing the oppressed groups in these dichotomies have produced important new questions for scientific research; they have produced sciences for oppressed groups, rather than primarily to serve the interests of militarists, profiteers, colonizers, and male supremacists – an issue to which we shall return.

All three cases of the expansion of scientific processes contribute to a democratization of science in which previously inaccessible political processes are made observable. Like Latour, Beck inveighs against the division of the world into nature, culture, and discourses. There is no pure nature or pure culture, let alone disembodied and de-cultured representations of either. Like Latour, Beck also rejects postmodernisms since they simply abandon the crises of the risk society:

> A new kind of capitalism, a new kind of economy, a new kind of global order, a new kind of society and a new kind of personal life are coming into being [...] This is not "post modernity" but a second modernity [...]"
>
> (Beck, 1999, p. 2)

> Where most postmodern theorists are critical of grand narratives, general theory and humanity, I remain committed to all of these, but in a new sense. To me the Enlightenment is not a historical notion and set of ides but a process and dynamics where criticism, self-criticism, irony and humanity play a central role [... M]y notion of "second reflexive modernity" implies that we do not have enough reason [...] in a new postmodern meaning to live and act in a Global Age [...] of manufactured uncertainties.
>
> (Beck, 1999, p. 152)

Like Latour, Beck is also optimistic. Needed transformations of the first modernity's sciences and social relations are already underway. Beck's social theory is more robust and comprehensive than Latour's; it provides more clues to how and where progressive transformations can and, perhaps, are already occurring. This project is pursued in later books, and especially in *The Reinvention of Politics: Rethinking Modernity in the Global Social Order* (Beck, 1997).

Beck's controversial writings have had great influence on German and European social theory. He has also been criticized on a number of grounds, most notably for underestimating the continuing power of class (Draper, 1993; New 1994), especially in global contexts, for essentializing science and failing to sufficiently analyze its political/economic content (Boyd, 1993), for conflating the reality and the perception of risk, and for regarding as general a situation which is more distinctive to Germany than to other parts of the world. He has also been charged with overemphasizing the rift between a first and second modernity, and for taking the ideology of modern science for its reality. He responds to some of these criticisms directly in the last chapter of *Word Risk Society* (Beck, 1999), and indirectly in such other writings (see, e.g., Beck, 1997). Whatever the limitations of his account, he has conceptualized in provocative ways important questions about how contemporary science functions as a kind of governance of our daily lives behind the back, so to speak, of democratic political processes. He also offers alternatives to both the current politics of modernity and its sciences and to postmodernism's fatalism. He points to valuable ways in which scientific expertise is already escaping the monopoly of natural scientists.

Both Beck and Latour mention science and Western modernity in their global contexts, and criticize Western attitudes toward other cultures. Indeed, Beck's 1999 book focuses on this topic. Yet neither exhibits the kind of deep grasp of this issue that can be found in postcolonial science and technology theorists. There are deep differences between various tendencies in this now rich literature, (see, e.g., Harding, 1998a, 1998b and Hess, 1995 for overviews of central themes in these writings, and Harding, 2006, 2008 for further discussion of such issues). Yet Ashis Nandy articulates some themes common to most of them.

14.4 Ashis Nandy and Postcolonial Science and Technology Studies

Postcolonial science and technology studies have produced a distinctive vision of modern Western sciences and their philosophies. The origins of this field can be found in attempts to re-evaluate objectively – that is, outside the familiar Eurocentric exceptionalist and triumphalist framework – indigenous scientific and technological traditions in ex-colonized societies. It can also be found in criticisms of the imperial and neo-colonial character of the new Third World development policies in the 1950s, and the role of Western sciences and technologies in such policies and subsequent practices. Development was from the beginning conceptualized as the transfer of Western sciences and technologies, and their distinctive rationality, to the so-called underdeveloped societies. Thus the practices and philosophies of Western sciences and technologies would be implicated in the successes and failures of Third World development. This point has been virtually invisible to Western philosophies of science, which still focus for the most part on what happens in laboratories in Europe and the United States. These postcolonial science and technology studies emerged as a distinctive field in English-language writings by the early 1980s, and so have now accumulated more than two decades of books, articles, journals, conferences, manifestos, websites, and a significant presence in ongoing projects of United Nations organizations as well as other national and international institutions and agencies. There has always been at least a sprinkling of Western activists and scholars involved in these projects.[3]

Meanwhile, studies of "multiple modernities" have appeared (Daedalus, 2000). Interestingly, modern sciences and technologies, Western or not, are barely mentioned in this work. Yet many of its assumptions and arguments also appear in postcolonial science and technology studies. One shared point is that modernity is not the same as Westernisation. At least some non-Western societies are clearly modern, with differentiated social institutions, multiple political parties, representative government, a free press, or some combination of these, as well as highly developed technology and science sectors (consider, for example, India and South Korea). Yet they look very different from modern Western societies. Often their sciences look different, for they overtly combine elements of modern Western sciences and of traditional knowledge systems, including their local values and interests. Many of the elements the theorists of modernity presumed were requisite for modernization are missing in these cases. For example, they often do not give up central elements of their traditional identities. Their sciences are often openly infused with traditional religious and cultural assumptions and projects.[4] Thus, in spite of increasing global economic, political, and cultural linkages, global

---

[3] A few of the important works here are Brockway (1979), Groonatilake (1984), Haraway (1989), Headrick (1981), Hess (1995), Joseph (1991), Kochhar (1992–93), Lach (1977), McClellan (1992), Nandy (1988), Needham (1954), Petitjean et al. (1992), Philip (2003), Prakash (1999), Reingold and Rothenberg (1987), Sabra (1976), Sachs (1992), Sardar (1988), Selin (1997), Shiva (1989), Third World Network (1988), Watson-Verran, and Turnbull (1995). See also Harding (1998a).

[4] I say "openly" in contrast to the widespread but unacknowledged permeation of Christian assumptions in European/U.S. sciences and their philosophies, as historians of science point out. (Needham, 1969; Noble, 1992; 1995) And, of course, Western sciences and their philosophies also contain distinctively Western economic, political and social assumptions beyond the religious ones.

homogenization has not been the consequence of the dissemination of modernization, its sciences and their rationality. Instead we have a world of "indigenous modern knowledge systems." Indeed, Latour and Beck, like the postcolonial theorists, both identified distinctively Western features of Western modernity's sciences. Are these "incomplete modernities" (Ortiz, 2000) or simply different modernities? Does regarding them as incomplete invariably retain Western standards for modernity?

Here the focus will be on just one leading figure in the postcolonial science studies debates, the Indian psychologist and science theorist and activist, Ashis Nandy. His discussion of Western modernity's sciences expresses kinds of criticism widely articulated in the various streams of postcolonial science and technology studies. Nandy's work has made him a well-known and controversial intellectual in India as well as in the West. *The Intimate Enemy: Loss and Recovery of Self Under Colonialism* (Nandy, 1983) is the book most familiar to Westerners. He is the author or editor of some 13 other books, including one on racism in the West, reflections on Adorno and Marcuse, on environmentalism in India, and many other topics. In most of these writings he has focused on knowledge systems. As one observer puts his main point:

> How is it, as Nandy has asked, that the notions of modernity, science, development, and instrumental rationality have come to predominate in our understanding and ordering of the modern world, and what have been the consequences of imposing, largely through the mechanisms of colonialism, nationalism (the purported opposite of colonialism), and now the nation-state, these supposedly natural categories, upon the entire world?
>
> (Lal, 2000, p. 7)

With this background in mind, I focus here on the particular issues he raises in his essay *Science as Reason of State* (1988). These bring into focus some themes shared with Latour and Beck's criticisms of science and modernity and some that are distinctive to postcolonial science studies.

In this essay Nandy focuses on the fact that the Indian Government developed modern sciences and technologies, after the end of British colonisation, as a way of claiming the power and potential global influence of the Indian state. India saw the establishment of state-of-the-art scientific and technological institutes and the training of huge numbers of Indian scientists and engineers as the way for India to excel at activities highly valued by and politically advantageous to the West, and to insert itself into the highest level of global politics.

Today India has the third largest scientific and technological workforce in the world. Large segments of it are located in "off-shore" research elsewhere in Asia for U.S. corporations, in "silicon Valley" in California, in U.S. and European university science departments, and other science and technology sites in the West.

Nandy points out that like the space programs in the U.S., the Indian science and technology initiative clearly had military and commercial benefits. Also like the space programs, it also served nationalism, increasing the global status of the Indian state. Such a triumph justifies the means through which it was achieved which could not stand the scrutiny of democratic processes. That is, this state project was not the result of the democratic political processes to which the Indian government is supposedly committed. The status of science and technology was used as an excuse for the Indian state to install political programs – science and technology programs – that bypassed democratic decision processes. Science became a "reason of state." But Nandy is not just making another "external" criticism of modern science. Instead he asks what it is about modern science itself that permits it to invite the state to develop and use scientific knowledge outside democratic processes. He is concerned not only about modern science as practiced in India, but also in the West, of course.

This question about the nature of science is a crucial one, Nandy argues. Through the application of scientific and technological expertise justified by appeals to social progress, violence is permitted around the globe against nature, cultures, and the Third World. Purportedly progressive scientific and technology projects destroy the environments upon which the world's poorest citizens directly depend for their daily subsistence, and upon which all of us, and the generations following, depend. They deny cultures' democratic participation in decisions that have powerful effects upon their living conditions.

> [M]odern science has the capacity within it to sustain a culture of science, which is incompatible with democratic governance as well as with the democratic rights of those who are turned into the subjects of modern science and technology.
>
> (Nandy, 1988, p. 10)

Scientific projects produce a massive silence about the future of the Third World, which is being further immiserated through the practices and consequences of scientific and technological projects. He argues that in the imagination of Western states and their scientific and technological thinking, apparently the Third World is to be discarded. It is to exist for the extraction of raw materials, to serve tourism, as the recipient of toxic industries and toxic dumps, and as a source of labour in such projects there and to meet labour needs in the West, (see also Amin, 1997). We can add that it also serves as a consumer of the Western arms industry. Perhaps "exploited" is more accurate than "discarded" to characterize a segment of the world that is part of global networks so vital to Western interests; nevertheless, Nandy's point is well taken.

Particularly disturbing to Nandy is "[...] the manner in which the link between science and violence in India has been strengthened by forces within the culture of Indian science, forces which in other cultures of science in some other parts of the world have been either less visible or less powerful" (p. 6). The Indian science establishment,

> [...]on its own initiative, has taken advantage of the anxieties about national security and the developmental aspirations of a new nation to gain access to power and resources [...T]he privileged among Indian scientists have often been the most vigorous critics of civil rights group struggling for protection against the hazards of a callous nuclear establishment. (p. 6)

Thus it was not some totalitarian external power that forced the scientific establishment to state agendas. Rather, the scientific community used state and popular anxieties to advance its own status.

Nandy calls for the "repoliticization" of science. By this he means the establishment of an ongoing political audit of the effects of science projects by those external to the sciences and to the states at issue.

> [T]he intellectual challenge is to build the basis of resistance to militarisation and organised violence, firstly by providing a better understanding of how modern science or technology is gradually becoming a substitute for politics in many societies, and secondly by defying the middle-class consensus against bringing the estate of science within the scope of public life or politics. (p. 10)

In this respect, protest movements and their critical science studies writings can help to make "every man his own scientist" – to democratise decisions about what kinds of scientific and technological knowledge to seek and who will receive the benefits and bear the costs of such projects. India needs to re-examine and reedit its own scientific traditions, incorporating aspects of Western sciences where desirable. Similarly, Western societies need to engage in such a project with respect to their own traditions.

India and other Third World societies could provide models for the West in this respect, Nandy argues. India has had some six centuries of interaction with the West, two of which were as colonies of England. It has a deep familiarity with Western knowledge systems as well as, of course, with its indigenous systems.

> [...] the Indic civilisation today, because it straddles two cultures, has the capacity to reverse the usual one-way procedure of enriching modern science by integrating within it significant elements from all other sciences – pre-modern, non modern and postmodern – as a further proof of the universality and syncretism of modern science. Instead of using an edited version of modern science for Indian purpose, India can use an edited version of its traditional sciences for contemporary purposes. (p. 11)

Thus India, "by virtue of its bicultural experience, manages to epitomize the global problem of knowledge and power in our times" (p. 11). Its strategies for resisting the substitution of science for politics by democratizing science and technology, and for strengthening a modernity which is not identical to Western modernity, can serve as an important model or social experiment for Western societies.

14.5 Still Modern, Still Traditional

Each of these theorists of modernity and its sciences has drawn to our attention distinctive aspects of the current crisis. Each criticises the limitations of the modernity and its sciences that we have had. Each criticises the inability of postmodernism to provide solutions to these limitations. Each focuses on the tensions between the commitments of Western societies to democratic processes of governance, on the one hand, and, on the other hand, the authoritarian ways that scientific and technological decisions are made, decisions that "govern" our lives in powerful ways. Each calls for democratizing science in the sense of extending critical scientific practices to the examination of sciences themselves, and providing an "audit" of which groups bear the costs and which receive the benefits of the practices and consequences of particular scientific projects.[5] Each proposes ways to transform, update for the present world, and democratize modernity and its sciences, while Nandy also suggests that such projects in the Third World can serve as models for the Western ones. All three are optimistic about the possibilities for successfully engaging in such immensely challenging projects.

One could argue that the ways in which these theorists relate their concerns to issues about modernity, pro and con, are not the most important or valuable parts of their rich analyses for those interested in relations between systems of knowledge and of democratic governance. This may be so. Yet I suggest that because issues about how a culture thinks about modernity are issues about its relation to its past and to possibilities for the future, we should resist the temptation to dismiss the modernity issues as irrelevant to the scientific and political projects of these theorists. Moreover, it is interesting to note that the relationship of their accounts to modernity remains more complex than their pronouncements about it acknowledge.

14.5.1 Still Modern

For one thing, in several significant senses of the term all three of these analyses are fully inside modernity's program. Of course Beck already frames his analysis as demonstrating the need for an even more completely modern program than the first modernity, industrial modernity, embraced. And Nandy calls for multiple modernities. As several commentators on Nandy's work have pointed out, Nandy's relation to modernity is far more complex than a simple rejection or affirmation can convey (Lal, 2000). Yet all three theorists' projects can be understood as centrally modern ones in more conventional ways. This becomes clear once we keep firmly in mind the insistence of Nandy as well as of others that modernization does not have to – indeed, should not – mean only Westernisation (see, for example, Philip, 2003; Prakash, 1999; Eisenstadt, 2000).

Consider how all three accounts exhibit a good part of the classical set of substantive characteristics of modernity indicated earlier. All three emphasize the importance of active, human agency in transforming the sciences we have. All appeal

[5] It is useful to refer to this extension of critical scientific practices as 'politicizing science' by returning issues of governance to the realm of public discussion and political decision. Yet it would be as reasonable to refer to the processes by which scientific institutions come to monopolise scientific and technical decisions that have political consequences as the politicisation of science. Such processes surreptitiously redefine as merely scientific or technical issues fundamental political issues of how we shall live our lives, who shall receive the benefits and who bear the costs of scientific and technological work, which, incidentally (for financing is not the main issue here), is itself financially supported by the citizenry. The scientific monopolisation of what are fundamental political issues removes from the public realm discussion of how we shall be governed. I borrow this point from Proctor's (1988) discussion of how the Nazis politicized science.

to the reasonableness, the rationality, of their proposals as a motivation for voluntary action. All three stress the importance of bringing the sciences and their philosophies under the control of commitments to a rational and democratic social order. Thus all three are committed to social progress. All three call for an expansion and intensification of science's self-critical attitude and practices, specifying distinct aspects of science that need criticism. All articulate a historical consciousness, insisting on conceptions of sciences as fully inside not only the conventional intellectual histories, but also social, cultural, economic, and political histories.

Moreover, from its origins modernity has promoted pluralism and protest movements. All three specifically assert the value of multiple, culturally distinctive sciences. And all are engaged in precisely the kind of protest, and protest movements, that have been prominent in the history of modernity (whether or not one fully agrees with the agendas of these movements). Furthermore, all three contribute to contemporary struggles over the appropriate definition of the realm of the political. Contrary to the standard view that politics occurs only in and about formal governmental structures, these theorists argue that expert scientific decisions are inherently political, for they disenfranchise those groups which don't count as experts from participating in decisions about the basic conditions of their lives – conditions which determine whether they will live or die.

Finally, one could argue that all three represent an unfortunate, but not accidental, tendency in modernity in that they ignore women's issues and, in particular, women's relation to modernity and its sciences. They treat women as if they existed outside the realm of the social and political. In their accounts we could be forgiven for thinking that perhaps women are just a natural phenomenon. And these theorists are writing not in the Nineteenth Century when they would have had to struggle to produce the kinds of feminist analyses which emerged three decades ago, but in the 1990s, when gender analyses are to be found virtually every day in the newspapers and other media. This is not to say that these authors have no appreciation at all of feminist work. Indeed Latour gestures to Donna Haraway's work, Nandy has voiced support for some feminist issues, and Beck evidences a most fulsome appreciation of the political importance of women's movements (Beck, 1997). Yet none engage with feminist science or modernity critiques; none take women to be agents of the transformations of science of interest to them.

Thus it appears that these three severe criticisms of conventional forms of modernity are still fully within the historical ethic and practice of modernity in a number of ways – a claim to which perhaps none of the three would object. Yet the plot thickens, for it turns out that modernity's projects are always dependent on the persistence of pre-modern skills and practices.

14.5.2 Still Pre-Modern

The power of appeals to the modern always depend upon the continued discursive construction of its opposite, the traditional. "The modern" is always defined in relation to this partner. Or, to be more precise, modernity continually discursively recreates the power of the threat of the traditional as its own rationale. This is one way in which modernity does and must continually reproduce the pre-modern as its necessary Other which is both outside it and yet inside its discursive framework.

But it is not the only way. Modernity also hides its reliance on actual pre-modern skills and practices. Modernity cannot make its way in the world without the skilled, "expert" work, one could say, of people who can and will continuously translate, network, or otherwise link modernity's conceptions and activities to traditional ways of living in and interacting with the world around us. In this way also modernity perpetuates the pre-modern inside itself, so to speak. Or, to put the point another way, modernity can only exist within an unacknowledged cocoon of pre-modern skills and activities. A couple of examples can clarify this issue.

In an earlier essay I argued that if we take the standpoint of the economically and politically most vulnerable women in the Third World – those who have been a main target of First World so-called development aid – we can come to understand an aspect of modern scientific rationality which is hard to see from the perspective of those of us who most benefit from it here in the First World (Harding, 1998b). This is that modern scientific philosophies and modernization practices can only succeed if they continually reproduce social practices which they associate with the pre-modern. Pre-modern cultures and their knowledge-seeking practices are inside modernity and its sciences in this sense.

Consider how this phenomenon became visible to feminists engaged in the debates over women's work, environmental destruction, and Third World development. Early feminist criticisms of development policies argued that women around the world were being left out of development. The kind of scientific and technological training which made possible greater access to agricultural and manufacturing jobs for export production had been restricted to men. This was because Europeans and Americans in the development agencies and their funders perceived such jobs as men's work regardless of which gender had performed traditional forms of agricultural and manufacturing labour in Third World societies. Modern production was consistently gendered as masculine. Thus men were drawn into modernized work in distant cities or plantations, leaving women, children, the old and the sick behind to survive on their own largely without benefit of adult male economic or social assistance. The early feminist argument went that women, too, should be educated so that they could access such means of livelihood. Modern work did not have to be gendered. Androcentric development policies were unfairly leaving women behind with access only to pre-modern forms of work and community, and under conditions of deprivation at that.

However, it soon became apparent that there was another story to be told about such development processes. Mies (1986) argued that development policies succeeded, when they did, only because they legitimated the appropriation of both women's and peasants' land rights. They did so through processes much like the enclosing of the "commons" in industrializing Europe and the United States in which public grazing and agricultural lands were transferred to private owners. Moreover, the labour of women and peasants was appropriated from subsistence work for their own kin and communities into the capitalist agricultural and manufacturing export economies favoured by Northern corporations and international agency sponsors. Mies pointed out that the Marxists were wrong; "primitive capital accumulation" had not ended by mid-Twentieth Century, but rather continued through Third World development policies and practices. The families and communities dependent on women's domestic labour were left even more the responsibility of women alone, who now had to replace through low-paid labour for local and international industries, as well as through sex work and domestic labour far from home, the economic contributions that had been provided to them, their dependents and communities through their work and their men's work on their own lands.

Thus the claimed social progress of modernization could only be achieved through social regress for women and peasants. Women and peasants lost their pre-modern rights and ways of living, though their pre-modern labour was and remains a necessary part of modern scientific agriculture and manufacturing. Indeed, someone must reproduce the next generation of agricultural and manufacturing labourers. The new forms of pre-modern labour of women and former peasants were created through modernization processes; they are not an anachronistic residue of pre-modern conditions. Beck made a similar point about women's continuing contemporary "feudal" household labour in the North, modernity transformed a form of oppression and exploitation that existed outside capitalism into a different one that was central to the successes of capitalism (Beck, 1992).

Third World development policies and practices thus exemplified the pattern earlier detected by historians whereby women's political and economic regress is often an enabling condition for "progress for humanity." Kelly-Gadol (1976) had drawn attention to the persistence of this phenomenon, focusing in particular on how women consistently lost legal rights, economic power and social status at moments of so-called democratic progress from Fifth Century B.C. Greece through to the European industrial revolution and the Jacksonian era in the U.S. In spite of the benefits "social progress" has brought to some groups of women, on balance women in every class tend to lose social status and material benefits in such processes. Women's losses thus constitute a "contribution" to so-called human progress.

This reproduction of the pre-modern within modernization processes has now also been described with respect to the introduction of modern sciences and technologies themselves into traditional societies. For example, Japanese philosopher and historian of technology Murata (2003) describes how the creativity of an advanced sector of modern technology has depended upon the restriction to culturally acceptable patterns of the flexible ways it may be interpreted or linked to local practices. This adaptation is provided by the traditional technology sector in Japan.

> One of the most conspicuous characteristics of the modernization process in Japan is the dual structure of its socio-technical network with an advanced sector of modern technology and a parallel domestic sector of traditional technology. The advanced sector functions as if transferred technology guides and determines the way of modernization. In reality, however, the advanced sector interacts with the domestic sector, where traditional technology plays a role of instrumental rationality, decreasing the gap between the two sectors sufficiently that advanced technology becomes adapted to local practices. Through this interaction, the scope of flexivity is restricted, the process is channelled in a certain direction, and rapid and continuous adaptation and development of technology becomes possible, (Murata, 2003, p. 263).

The very abstractness and generality of modernity's sciences and technologies permit them to be interpreted in many different ways and thus to be practiced, applied, or interpreted in many different cultural contexts. But in each case, such abstract and general principles must be connected to local cultural resources, values, and interests. This task can only be done through pre-modern "craft labour." In such accounts we can see another way in which the intellectual and social progress achieved by modern scientific and technological rationality in fact depends upon the continued nourishment of pre-modern, traditional local modes of thought and practice within modern practices.

My point here is, first, that it is a mistake to think that modern ways of thinking and acting could ever completely replace pre-modern ways, as the classical modernization theorists supposed. Thus the spread of modernization globally does not have a homogenizing effect on cultures, as the theorists predicted. Rather, the advance of modernization requires the continual culturally local adaptation and realignment of pre-modern thinking and acting. They do not disappear, but rather adjust themselves to "the modern" as they adjust their particular, local part of "the modern" to culturally distinctive values and interests.

Modernity's commitment to trans-cultural abstractness and generality insures that it can only be practiced in ways pre-modern thinking and practice can interpret. This realization provides yet another reason to question the kind of temporal specification of modern societies, which locate them after traditional ones. Rather, it appears that the former emerge within or alongside traditional societies. Moreover, to the extent that the three science theorists discuss adopting the temporal specification of modernity, they fail to engage with modernity's persistent and necessary dependence upon tradition skills and practices.

14.5.3 Two Missing Foci

This brings us to two issues under-engaged in these accounts. One is the role already being played by existing practices of social justice movements, such as some environmental movements, anti-racist movements, and feminist movements in developing and putting into practice precisely the programs recommended by these theorists. Of course Nandy speaks from the history and practices of the lively postcolonial science movement and points to its importance in agitating for political audits of science and technology programs. And Beck (1997) emphasizes the importance of environmental movements. Yet critical examinations of the gender of modernity (Felski, 1995; Gole, 2000; Jardine, 1985; Philip, 2003) and of androcentrism in science and technology projects and their philosophies, North and South, are virtually unacknowledged in the thought of all three theorists. (See earlier citations). Gender analyses have charted the distortions and political damages created by the hyper-masculinization of modernity and its scientific and technological projects. Third World feminist criticisms of development policies have specifically focused on their androcentrism as well as their racism, colonialism and imperialism. These identify how such androcentrism further disempowers not only the half of the Third World constituted by women and girls, but also the other half, the male half, which depends upon the knowledge and labour of women to maintain families, the elderly and sick, and the community relations and resources upon which men and boys depend, as well as to produce cash income. The range and power of gender analyses is too big a topic to pursue further here, but its necessity should be noted.[6] The "modernities" on which Latour, Beck, and Nandy focus are distinctively gendered, and the remedies to their scientific and political limitations must directly engage with these gendered aspects of the democracy to which these theorists aspire.

Another absence from these accounts is recognition of the relatively recent change in the economic status of the production and management of information. This kind of work has moved to the "base" of the global political economy thanks to the development of electronic media, especially the Internet (Castells, 1996). The production and delivery of goods (food supplies, manufacturing components, retail store stocks), as well as scientific research, and international environmental policy are now organized through the Web, e-mail, and cell phones. So are terrorism, criminal activity, and progressive political protests. New configurations of scientific communities have emerged following the 1989 end of the Cold War and the huge migration into expertise-hungry private industry of scientists whom university laboratories no longer had funds to employ (Gibbons et al., 1994; Nowotny et al., 2001). New forms of capitalist production are complicating the traditional understandings of modernity's rationality in yet further ways (Kleinman and Vallas, 2001). A large-scale map of possible future directions of capitalist expansion and of possible effective forms of resistance to it are required for effective planning for democratic sciences in democratic societies (e.g., Amin, 1997). The West's prevailing philosophies of science, modernity, and democracy are poorly equipped to enable democratic policies for directing these new social relations. Latour, Beck, and Nandy have provided helpful clues for how we might begin to rethink central aspects of this new production and management of information, but none provide the conceptual resources to grasp such phenomena in maximally inclusive and effective ways.

14.6 Conclusion

In conclusion, there is a tendency among both defenders and critics of modernity in the humanities and in social theory to neglect or undervalue the role that modern sciences have played in advancing modernity and modernization projects, and to misunderstand the relation between the modern and the traditional. Reflecting on

---

[6]The relations to feminism of Latour and Nandy deserve more critical assessment than I have here given them, as some readers will be aware. On the other hand, Beck is more eloquent in his overt appreciation for the power of the women's movements than I have indicated. Nevertheless, he does not engage with their work on science or modernity. He doesn't see them as making any distinctive contribution to the kind of project he is advancing.

the arguments of these three critics of modernity, Latour, Beck and Nandy, enables us to appreciate how historically specific forms of knowledge-production always are co-constituted with specific forms of governance and of public politics. Today, when the production and management of information have moved to the base of the global political economy, and when sciences' power to govern behind the back of democratic political processes is at its all-time high, we certainly cannot afford to ignore identifying fully the actual links between scientific practices and democratic principles, nor can we ignore the need to debate in maximally democratic spaces goals and strategies for change. We cannot afford to turn our backs on our own actual and desirable embeddedness in a modernity which has consistently defined itself against the feminine and the traditional, both of which are firmly lodged, and necessarily so, in its very core.

Well-educated students in the sciences and mathematics now need a far richer understanding of how scientific and technological research and practices function in the changing world around us. Conventional philosophies of science, and the science and mathematics education curricula and pedagogy which they still guide, lead us only to an obsession in research with a mythical past that never could, even in principle, deliver the scientific and social benefits it promised.

References

Amin, S. (1997). *Capitalism in the Age of Globalisation: The Management of Contemporary Society*. New York: Zed Books.

Beck, U. (1992). *Risk Society: Towards a New Modernity*. London: Sage.

Beck, U. (1997). *The Reinvention of Politics: Rethinking Modernity in the Global Social Order*. Cambridge, UK: Polity.

Beck, U. (1999). *World Risk Society*. Oxford: Blackwell.

Biagioli, M. (Ed.) (1999). *The Science Studies Reader*. New York: Routledge.

Boyd, W. (1993). Risk Society: Towards a New Modernity. *Economic Geography*, 69(4), 432–436.

Braidotti, R. et al. (1994). *Women, the Environment, and Sustainable Development*. Atlantic Highlands, NJ: Zed.

Brockway, L.H. (1979). *Science and Colonial Expansion: The Role of the British Royal Botanical Gardens*. New York: Academic Press.

Castells, M. (1996–1998). *The Information Age: Economy, Society & Culture* (vol. 1–3). Oxford: Blackwell.

Draper, E. (1993). Risk, Society, and Social Theory. *Contemporary Sociology*, 22(5), 641–644.

Eisenstadt, S.N. (2000). Multiple Modernities. *Daedalus*, 129(1), 1–30.

Escobar, A. (1995). *Encountering Development: The Making and Unmaking of the Third World*. Princeton: Princeton University Press.

Felski, R. (1995). *The Gender of Modernity*. Cambridge, MA: Harvard University Press.

Gibbons, M., Limoges, C., Nowotny, H., Schwartzman, S., Scott, P. & Trow, M. (1994). *The New Production of Knowledge: The Dynamics of Science and Research in Contemporary Societies*. Thousand Oaks: Sage.

Gole, N. (2000). Snapshots of Islamic Modernities. *Daedalus*. 129(1), 91–117.

Goonatilake, S. (1984). *Aborted Discovery: Science and Creativity in the Third World*. London: Zed.

Haraway, D. (1989). *Primate Visions: Gender, Race, and Nature in the World of Modern Science*. New York: Routledge.

Haraway, D. (1991). *Simians, Cyborgs, and Women*. New York: Routledge.

Harding, S. (1986). *The Science Question in Feminism*. Ithaca: Cornell University Press.
Harding, S. (1991). *Whose Science? Whose Knowledge? Thinking from Women's Lives*. Bloomington: Indiana University Press.
Harding, S. (1998a). *Is Science Multicultural? Postcolonialisms, Feminisms, and Epistemologies*. Bloomington: Indiana University Press.
Harding, S. (1998b). Gender, Modernity, Knowledge: Postcolonial Standpoints. *Is Science Multicultural? Postcolonialisms, Feminisms, and Epistemologies* (pp. X–XX). Bloomington: Indiana University Press.
Harding, S. (Ed.) (2003). *The Feminist Standpoint Theory Reader: Intellectual and Political Controversies*. New York: Routledge.
Harding, S. (2006). *Science and Social Inequality?: Feminist and Postcolonial Perspectives*. Edwardsville: University of Illinois Press.
Harding, S. (2008). *Sciences from Below: Feminisms, Postcolonialisms, and Modernities*. Durham: Duke University Press.
Headrick, D.R. (Ed.) (1981). *The Tool of Empire: Technology and European Imperialism in the Nineteenth Century*. New York: Oxford University Press.
Hess, D.J. (1995). *Science and Technology in a Multicultural World: The Cultural Politics of Facts and Artifacts*. New York: Columbia University Press.
Jardine, A. (1985). *Gynesis: Configurations of Woman and Modernity*. Ithaca: Cornell University Press.
Joseph, G.G. (1991). *The Crest of the Peacock: Non-European Roots of Mathematics*. New York: I.B.Tauris.
Kelly-Gadol, J. (1976). The Social Relation of the Sexes: Methodological Implications of Women's History Signs. *Journal of Women in Culture and Society*, 1(4), X–XX.
Kleinman, D.L. & Vallas, S.P. (2001). Science, Capitalism, and the Rise of the 'Knowledge Worker': The Changing Structure of Knowledge Production in the United States. *Theory and Society*, 30, 451–492.
Kochhar, R.K. (1992–1993). Science in British India (parts I and II). *Current Science*, 63(11), 689–694; 64, 55–62.
Lach, D.F. (1977). *Asia in the Making of Europe* (vol. 2). Chicago: University of Chicago Press.
Lal, V. (Ed.) (2000). *Dissenting Knowledges, Open Futures: The Multiple Selves and Strange Destinations of Ashis Nandy*. New Delhi: Oxford University Press.
Latour, B. (1993). *We Have Never Been Modern*. Cambridge, MA: Harvard University Press.
McClellan, J.E. (1992). *Colonialism and Science: Saint Domingue in the Old Regime*. Baltimore: Johns Hopkins University Press.
Mies, M. (1986). *Patriarchy and Accumulation on a World Scale: Women in the International Division of Labor*. Atlantic Highlands, NJ: Zed.
Murata, J. (2003). Creativity of Technology and the Modernisation Process of Japan. In R. Figueroa & S. Harding (Eds.), *Science and Other Cultures: Issues in Philosophies of Science and Technology* (pp. X–XX). New York: Routledge.
Nandy, A. (1983). *The Intimate Enemy: Loss and Recovery of Self Under Colonialism*. New Delhi: Oxford University Press.
Nandy, A. (1988) Science as a Reason of State. In A. Nandy (Ed.), *Science, Hegemony, and Violence: A Requiem for Modernity* (pp. X–XX). New Delhi: Oxford University Press.
Needham, J. (1954) *Science and Civilisation in China* (7 vols). Cambridge, UK: Cambridge University Press.
Needham, J. (1969). *The Grand Titration: Science and Society in East and West*. Toronto: University of Toronto Press.
New, C. (1994). Class Society or Risk Society? *Radical Philosophy*, 66(56).
Noble, D. (1992). *A World Without Women: The Christian Clerical Culture of Western Science*. New York: Knopf.
Noble, D. (1995). *The Religion of Technology*. New York: Knopf.
Nowotny, H., Scott, P. & Gibbons, M. (2001). *Re-Thinking Science: Knowledge and the Public in an Age of Uncertainty*. Cambridge, UK: Polity.

Ortiz, R. (2000). From Incomplete Modernity to World Modernity. *Daedalus*, 29(1), 249–260.

Petitjean, P., Jami, C. & Moulin, A.M. (Eds.) (1992). *Science and Empires: Historical Studies About Scientific Development and European Expansion*. Dordrecht: Kluwer.

Philip, K. (2003). *Civilising Natures: Race, Resources and Modernity in Colonial South India*. New Delhi: Orient Longman.

Prakash, G. (1999). *Another Reason: Science and the Imagination of Modern India*. Princeton, NJ: Princeton University Press.

Proctor, R. (1988). *Racial Hygiene: Medicine Under the Nazis*. Cambridge: Harvard University Press.

Reingold, Nathan, and Marc Rothenberg, eds. (1987) *Scientific Colonialism*. Washington: Smithsonian Institution Press.

Sabra, I.A. (1976). The Scientific Enterprise. In B. Lewis (Ed.), *The World of Islam*. London: Thames and Hudson.

Sachs, W. (Ed.) (1992). *The Development Dictionary: A Guide to Knowledge as Power*. Atlantic Highlands, NJ: Zed.

Sardar, Z. (Ed.) (1988). *The Revenge of Athena: Science, Exploitation, and the Third World*. London: Mansell.

Selin, H. (Ed.) (1997). *Encyclopedia of the History of Science, Technology, and Medicine in Non-Western Cultures*. Dordrecht: Kluwer.

Shapin, S. & Schaffer, S. (1985). *Leviathan and the Air Pump*. Princeton: Princeton University Press.

Shiva, V. (1989). *Staying Alive: Women, Ecology, and Development*. London: Zed.

Sparr, P. (Ed.) (1994). *Mortgaging Women's Lives: Feminist Critiques of Structural Adjustment*. London: Zed.

Third World Network (1988). *Modern Science in Crisis: A Third World Response*. Penang, Malaysia: Third World Network.

Van den Daele, W. (1977). The Social Construction of Science. In E. Mendelsohn, P. Weingart & R. Whitley (Eds.), *The Social Production of Scientific Knowledge* (pp. X–XX). Dordrecht: Reidel.

Watson-Verran, H. & Turnbull, D. (1995). Science and Other Indigenous Knowledge Systems. In S. Jasanoff, G. Markle, T. Pinch, & J. Petersen (Eds.), *Handbook of Science and Technology Studies* (pp. 115–139). Thousand Oaks: Sage.

Wittrock, B. (2000). Modernity: One, None, or Many? European Origins and Modernity as a Global Condition. *Daedalus*, 129(1), 31–60.

Chapter 15
Towards a Critical Professionalism in University Science and Mathematics Education

Ole Skovsmose

Descartes is often referred to as the first modern philosopher, as he introduced a new way of looking at knowledge and nature. This new way is initiated by a universal doubt that allows Descartes to question what has been presented as knowledge, regardless of what authority has claimed it to establish fundamental truths.

Modernity can also be related to scientific and industrial developments by paying special attention to the fact that the so-called Scientific Revolution was followed by an Industrial Revolution. Naturally, there are no direct causal links between the two revolutions as many elements, including non-scientific ones, contributed to the Industrial Revolution. But it is still important to observe that Modernity relates to scientific and technological development, which introduces new forms of production.

Modernity is characterized by certain political developments including the advancement of democratic ideas. As part of the French Revolution, ideas of freedom, equality and fraternity entered the scene and signified a new political outlook. This revolution composes part of the modern era, regardless of the fact that a political barbarism was soon to follow this revolution providing the first raw material for Madame Trousseau's production of masks of famous heads.

The great discoveries made an initial step into the modern area, and Modernity marched along with colonization of the so-called New World. Slavery and racism accompanied the exploration of modern principles of governing and the emerging democratic outlook. This brings highly critical elements into the glossy self-portrait of Modernity. An ultimate destruction of the rosy picture is found in the *Holocaust and Modernity*, where Bauman (1989) portrays the Holocaust as a possible aspect of modern rationality, and not as a simple pathological event in the flow of history.

Finally, one could claim that modernity has come to an end, and there is no lack of suggestions for what to call the epoch we are entering, or about to enter: post-modernity, late modernity, liquid modernity, risk society, reflexive modernity, information society, informational society, network society, hyper-complex society, knowledge society, learning society, Modus-2 society.

Ole Skovsmose
Aalborg University, Denmark

O. Skovsmose et al. (eds.), *University Science and Mathematics Education in Transition*,

When we take a look at science and mathematics education, in particular as they are organized in a majority of universities and higher educational institutions all over the world, it appears to many (including me) that this education is to a great extent still framed according to the outlook of Modernity. As a consequence, an important challenge facing science and mathematics education is to consider what it could mean to move beyond the assumptions of Modernity.

15.1 Assumptions of Modernity

What then should be counted as the assumptions of Modernity? I will impose some strong limitations and address only some assumptions concerning knowledge and science. I will concentrate on aspects of "modern rationality" as this is expressed in science and mathematics. By doing so, I will try to prepare for a critique of rationality as part of a critique of Modernity.[1] I will pay attention to three assumptions, namely the assumptions of *progress*, *neutrality* and *epistemic transparency*. The assumption of progress claims that the development of human knowledge and of science in particular is the true motor of progress in all aspects of life. The assumption of neutrality claims that science and mathematics are socio-political neutral entities. The assumption of epistemic transparency states that it is possible to provide an overall clarification of what counts as knowledge.

One could wonder to what extent these assumptions are mutually consistent. In fact, it appears that the assumptions of progress and neutrality contradict each other. A motor of progress can hardly be called neutral. However, it is important to observe that we need not expect that such overall paradigmatic claims be consistent. We can deal with different discourses which, depending on the situation, could be articulated. The assumption of progress is most convenient when one has to portray the usefulness of science. At present, this assumption operates forcefully in "external" discussions – not least of all when requesting additional funding. The assumption of neutrality, however, serves perfectly well as part of an intrinsic discussion about what topics to include in a specific curriculum. With reference to neutrality, it can be claimed that ethical issues are of no relevance for the development of scientific professionalism, and that they therefore need not be included in any university curriculum.

15.1.1 The Assumption of Progress

One can assume that the very idea of progress is a defining characteristic of the modern age, and such a postulate is presented by Bury in *The Idea of Progress* (Bury, 1955). He claims that it is only during Modernity that the notion of progress becomes part of philosophic considerations as well as of cultural and political

[1] See also Skovsmose (2005, 2006).

outlooks. This notion was foreign in the Middle Ages, when the church interpreted life on earth as a period with no particular value of its own, as the whole purpose of life before death was to prepare for salvation and life after death. If we follow Bury's interpretation, the idea of progress is a defining aspect of Modernity.[2]

The Scientific Revolution, as established through the works of Copernicus, Kepler, Galilei, Descartes, Newton and many others, becomes a paradigmatic illustration of the idea that science by its very nature is progressing. From then on, this idea became a defining element of scientific self-understanding. The idea that society makes progress is more complex, for instance expressed through considerations of new forms of governing, as formulated by Locke in *Two Treatises of Government* (Locke, 1967). What I refer to as the assumption of progress states that the two forms of progress are related: science-in-progress ensures progress-in-general.[3] This assumption forms an integral part of many philosophic worldviews. One example is found in Comte's positive philosophy, which establishes progress as the central notion for understanding the history of ideas and society.

The rationality of science has been presented as the experimental method, the inductive method, or the hypothetical-deductive method. Whatever the interpretation of scientific rationality has been, the assumption of Modernity has been the same: By means of scientific rationality, every kind of superstition and dogmatism can be confronted. Rationality has to be developed and sharpened in its scientific forms. Obstacles to this purification have to be identified and eliminated. The application of this rationality needs no particular justification, as it is the motor of progress.[4]

15.1.2 The Assumption of Neutrality

The assumption of neutrality is clearly presented by Hume, who in the concluding lines of *An Enquiry Concerning Human Understanding* from 1748 states:

> If we take in our hand any volume [...] let us ask: Does it contain any abstract reasoning concerning quantity or number? No. Does it contain any experimental reasoning concerning matter of fact and existence? No. Commit it then to the flames: for it can contain nothing but sophistry and illusion.[5]

Formulated differently: Hume finds it impossible to deduce a statement including "ought" or "ought not" from statements including "is" or "is not." There is no logical path from the descriptive to the normative.

---

[2] Nisbet (1980) modifies this interpretation by pointing out that the notion of progress has been presented long before the modern period itself. Furthermore, we should never forget that progress was permanently questioned during the modern period, for instance by Rousseau.

[3] For a careful analysis of the assumption of progress, see Christensen (2003).

[4] Thus, Dewey presented the scientific method as a form of thinking, which was applicable not only in sciences, but also in education where it could ensure a 'progressive education' and democratic forms of life. See, for instance, Dewey (1966).

[5] Hume (1975, p. 165).

A particular elaboration of this idea is presented by logical positivism.[6] The claim is that science, developed in the proper way, is a value-neutral enterprise. (That mathematics is neutral is a given, although for different reasons, namely its pure formal nature.) If science has to be established in the proper way, it has to be cleansed from all elements, which could include political priorities, religious outlooks, metaphysical perspectives, or personal opinions: such elements are foreign to science.

The principle of verification became a basic tool for purifying science. By means of this principle, according to logical positivism, one can distinguish sentences with meaning from sentences without meaning.[7] This is important as the grammar of natural language allows the formulation of sentences which are correct from a grammatical point of view, but which have no meaning from a scientific point of view. According to the principle, a sentence has meaning if and only if it can be verified. Meaningfulness does not presuppose that the statement is true. The statement "there are 369 days in a year" is meaningful although it is false. It is meaningful because it is possible to outline a procedure according to which one can clarify its truth-value. Furthermore, verification must be understood as "in principle, possible to verify." The formulation of the principle of verification was soon elaborated in many different variations.[8] A main difficulty in this analytical development of the principle was caused by the status of the principle itself. If we apply the principle to itself, the principle of verification is meaningful if and only if it can be verified. And as it appears to be impossible to identify any observation that could either support or disqualify the principle, it appears to be meaningless according to its own formulation. This paradox was acknowledged by logical positivism as a tricky one, and the discussion of status of the principle of verification became rather heated.

This, however, did not affect the possibility that the principle in fact became applied. The principal target was metaphysics in all its variations. Any formulations referring to "idea" or "form" in a Platonic sense went to the waste bin as "nonsense," together with formulations referring to "imps," "angels," "God's mercy," Freudian notions in general, Kantian categories elaborated through transcendental consideration, Heidegger's discussion of "existence," and much more. The principle of verification revealed them all as nonsense – that type of nonsense, which according to logical positivism, had haunted philosophy for too long. Natural language was too charitable as it included formulations that were grammatically correct but nevertheless meaningless.

The principle of verification was applied to ethics, and Ayer's book *Language, Truth and Logic*, published first time in 1936 after attending meetings of the Vienna Circle during the year 1933, provides a provocative delineation of the nature of ethics. Although ethical statements can be formulated in natural language, one

[6] See, for instance, Ayer (1959) and Stadler (2001).

[7] See Carnap (1959).

[8] Thus a more elaborated formulation could be: a proposition p has meaning if and only if it is possible to make an observation, which would change the probability of p being true (see Hempel, 1959).

should not take this as a guarantee of any epistemic significance. As one cannot point out procedures for verifying a statement containing words like "ought to" or any other ethical notions, such statements are devoid of any epistemic content. They can only be seen as expressions of emotions, similar to "buh" and "hurra." No such expressions of sympathy and antipathy belong in science, and in this way, ethics was eliminated from the whole scientific discourse. Ethics could represent "nothing but sophistry and illusions," as already pointed out by Hume.

The idea of logical positivism moved rapidly world wide, and after the Second World War, logical positivism became a broadly accepted interpretation of science, which could be referred to as neo-positivism. Although the principle of verification was not maintained, the elimination of ethics (including any socio-political issue) from the scientific domain became a defining element of scientific self-understanding. And during the 1950s and 1960s, neo-positivism dominated the Anglo-American interpretation of science. The assumption of neutrality turned into scientific common sense.

15.1.3 The Assumption of Epistemic Transparency

That it is possible to provide an overall clarification of what counts as knowledge does not imply that it is simple to obtain knowledge. On the contrary, it can be the most demanding task to identify particular bits and pieces of knowledge. The assumption of epistemic transparency only claims that the overall definition of knowledge can be stated in a fairly simple and transparent way.

The search for epistemic transparency is one defining elements of Descartes' modern philosophy. According to him, methodological doubt draws us into a situation where we are able to do away with beliefs and assumed knowledge. Whatever it is possible to doubt, we doubt. And whatever we doubt, we remove from our (supposed) stock of knowledge. Not much remains. According to Descartes, only one statement: *Cogito, ergo sum.* On this basis, however, it is possible to build up knowledge by following some well-defined logical steps. This is a sublime presentation of epistemic transparency.[9]

However, Descartes' presentation of epistemic transparency was programmatic only. The Scientific Revolution had initiated a proliferation of scientific insight and formulations, which might, however, include very many different, also non-scientific, elements. A notion like "force" could easily include metaphysical elements. What about the conception of "effect at a distance?" A tremendous attempt to re-establish epistemic transparency in science was carried out by logical positivism. As an essential supplement to the principle of verification, they tried to outline the structures of a language that could contain exactly the empirically meaningful statements. Carnap made a huge effort to clarify the nature of a language adequate for expressing statements of scientific significance.[10] At the same time, this language

[9] See Descartes (1993).

[10] See, for instance, Carnap (1937)

should ensure that no other statements could be formulated as syntactically correct statements. As a consequence, a syntactic demarcation between sense and non-sense would coincide with the semantic demarcation provided by the principle of verification. Carnap's candidate for such a language was to be found in mathematics and logic. In this way we get to the idea that mathematics can be interpreted as the language of science. What can be formulated in this language represents what is relevant in science.

In order to ensure further epistemic transparency, it becomes important to identify the logical architecture of scientific theories. According to logical positivism, mathematics and natural science are organized according to two different structures. While mathematics could be clarified in terms of deductive systems, science must be clarified as an inductive system. At least this was the first proposal elaborated by logical positivism, and here it was possible to draw directly on the whole empirical tradition represented by Bacon, Locke, Berkeley, Hume and Mil. An inductive system must start with "facts," formulated in the so-called protocol statements, and from these facts, through induction, one could reach the general laws of science. Protocol statements were the empirical foundation for the theoretical superstructures.

It soon became apparent, however, that it was no simple task to clarify what a protocol statement might be like.[11] An identification of such statements depended on many factors, theoretical ones. Furthermore, outlining logic of induction brought about many difficulties. The notions, protocol statements and induction turned out to be extremely complex and obscure. It became necessary to make a different attempt to establish epistemic transparency. Popper made a proposal in terms of the hypothetical-deductive structure of science.[12] This would dispense completely with inductive logic. Furthermore, the need for an elaborated set of protocol statements providing the foundation of the scientific system was also dispensed with, at least to a large extent. Bold guesses were the start of deduction in a hypothetical-deductive system; and from these guesses one could deduce implications, which could be falsified through observations. So some kind of empirical statements were needed, but now they emerged at the end of a deduction, and not as an extensive input to inductive processes. If the deduced implication was not falsified, it was not possible to draw any radical consequence with respect to the bold guesses. They could be maintained and we could continue to consider them our best guesses. If the deduced implication was falsified, however, the guesses had to be abolished and new guesses had to be established. This led Popper to the insight that scientific theories are constituted by bold guesses, which are not (yet) falsified. It is always possible that they might become falsified. And it is the task of scientists to try to falsity theories. The reason is that bold guesses, which have survived attempts of falsification, *might* be true. Through this presentation of scientific development in term of conjectures and refutations, Popper's philosophy of science turns into a master-piece of epistemic

[11] See Neurath (1959).

[12] See Popper (1965, 1972).

transparency. Both mathematics and science are organized in deductive systems, but while mathematics starts with axioms, natural sciences starts with bold guesses.[13]

15.2 Modern Conceptions of Learning Science and Mathematics

I see modern science and mathematics education as an education that, in one way or another, reflects the assumptions of progress, neutrality and epistemic transparency. The idea of progress serves as an extrinsic organizing (and protecting) principle, which confirms that science and mathematics, and therefore science and mathematics education, are socially important. The principle of progress serves an extrinsic purpose, which could somehow be summed up as: the more the better. The assumption of neutrality represents an intrinsic organizing principle claming that the curriculum has to be organized according to structures and insights of the particular scientific subjects. There is no need for ethical elements in a scientific curriculum. The assumption of epistemic transparency implies that it is possible to provide a clarification of the essential content of a curriculum in terms of simple organizing principles.[14] The three assumptions are naturally mixing, and as a paradigmatic example, I will comment on "modern mathematics."

"Modern mathematics" refers to the New Math Movement that emerged in the late 1950s. It grew from developments in mathematics research in the late nineteenth century and matured as an elaborated perspective on mathematics in the first half of the twentieth century, which aimed at stabilizing epistemic transparency within mathematics. Thus, according to formalism, the nature of mathematical knowledge can be described in the following way: a mathematical theory presupposes a formal language for which the basic units are symbols. Symbols can be organized in sequences, and some of these sequences can be counted as formulas. Some formulas are established as axioms. The rules of deduction state when a formula is a consequence of other formulas. A proof can be described as a sequence of formulas with the following property: any formula in the sequence must either be an axiom or a consequence of some of the previous formulas in the sequence (according to stated rules of deduction). A theorem, then, is any formula that occurs as the last formula in a proof. Truth of a mathematical statement (a formula) means provability of this statement within a certain formalized theory.[15]

Bourbaki developed a more advanced conception of formalism, a pseudonym for a group of predominantly French mathematicians.[16] According to formalism, it seems to make sense to investigate implications of any axiomatic system, but Bourbaki eliminated this relativism by identifying some formal structures as mathematical mother structures, and it was shown how different mathematical theories

---

[13] See also Hempel (1965).

[14] The presentation in Bruner (1960) is paradigmatic for this assumption.

[15] See Curry (1951) for an outline of a formalist philosophy of mathematics.

[16] See Bourbaki (1950) and Dieudonné (1970).

can be described in terms of these structures. Bourbakism brings structures of pure mathematics into focus and, in doing so, presents a conception of mathematical transparency characteristic of Modern Mathematics.

In Europe the main initiative for the New Math Movement can be referred to the Royaumont seminar, which took place in 1959. The seminar was introduced with a lecture from the mathematician Marshall Stone. His essential point was that the teaching of mathematics was going to be recognized as the true foundation of the emerging technological society. Here we find a clear expression of the assumption of progress with a specific reference to mathematics, as mathematics education will have the task of carrying "the ever heavier burden of the scientific and technological super-structure which rests upon it" (OEEC, 1961, p. 18). This is one line of argument linking "modern mathematics" with the assumption of progress.

The main presentation at the Royaumont seminar was given by Dieudonné who, more that any other, represented Bourbakism. He outlined a mathematics curriculum that left behind the classic organization in topics like algebra, analyses, number theory, etc. The new organizing principles were defined through the basic structures of mathematics.[17] And Dieudonné did not make any reference to the overall considerations made by Stone. In fact, Stone's introduction came to serve only as a decoration for the task of defining a new curriculum in mathematics. In this way, the Royaumont seminar illustrates not only the assumption of progress, but also the assumption of neutrality. The discourse of progress can serve an external function, while neutrality becomes expressed in a curriculum discourse that only refers to the mathematical content as defined through its logical structures. After the Royaumont seminar, modern mathematics education spread world wide, and dominated a variety of curriculum reforms.

Piaget has pointed out that the structures of mathematics can be reached through a natural development of the child, as the epistemic structures of the child somehow anticipated the mother structures of mathematics as pointed out by Bourbaki.[18] In this way the epistemic transparency of structuralism became transposed into a learning theory. I see Piaget's genetic epistemology as an example of a modern theory of learning, i.e., an interpretation that reflects modern assumptions about knowledge. This interpretation has inspired the development of many different learning theories. Radical constructivism provides a particular interpretation of Piaget.[19] This idea was brought strongly into the community of mathematics education in the middle of the 1980s. However, this radical form of constructivism did not provide movement beyond the assumptions of modernity.[20]

Dewey made a strong case for progressive education. Here we find a direct celebration of the importance of the scientific method, which in Dewey's interpretation

[17] See OEEC (1961).

[18] See Beth & Piaget (1966).

[19] See Glasersfeld (1995).

[20] With the emergence of social constructivism, this private outlook started to become modified, and initiatives to leave the modern paradigm in education did emerge. See, for instance, Ernest (1998).

was experimental and inductive.[21] And we find a nice example of epistemic transparency: even though scientific knowledge is complex in details, it is straightforward to formulate the scientific method. This method ensures progress in science, and also in education. And better education guides us towards a better society, where the democratic spirit prevails. The scientific method becomes identified as the nucleus of progress in all human affairs. In the work of Dewey, the assumption of progress stands out in a powerful way. Many educators have established combinations of Piaget's and Dewey's perspectives, and in this way a variety of modern conceptions of learning have developed.[22]

15.3 Beyond the Assumptions of Modernity

What could it mean to move beyond the assumptions of Modernity? I will outline some considerations that add question marks to the assumptions of progress, neutrality and epistemic transparency. First, the idea of progress has been disputed by the many catastrophes and almost-catastrophes associated with scientific and technological development. Second, the idea of neutrality appears doubtful as the production of scientific knowledge is now taking place in many organizations, companies and institutions, which are different from the traditional sites of scientific knowledge production, the universities. Knowledge production is going on the market, and this does not appear to be a neutral matter. Third, the assumption of epistemic transparency has been challenged through the analyses of knowledge-power complexities, demonstrating that a particular outlook, including a seemingly scientific one, can be an expression of particular but powerful interests.

In short: we will consider whether or not we are dealing with *myths* of progress, neutrality and epistemic transparency. If so, it becomes an important task to develop science and mathematics education without subscribing to these myths.

15.3.1 Risk Production: The Myth of Progress

The notion of risk society and world risk society has been developed by Beck (1992, 1999), and these notions have brought new dimensions to the way we interpret the interaction between science and society. While the notion of risk society was originally developed with reference to Western societies, in particular with reference to the introduction of atomic energy, it has been further developed in scope to reflect processes of globalization. Here I will present the ideas in general form (see also Harding, Chap. 14).

---

[21] See Dewey (1938, 1963).

[22] The notion of "experiential learning" as developed by Kolb (1984) being one example.

What is a *risk*? Mathematically speaking the risk, R, associated with an event A, can be symbolised as $R(A)$. It can be calculated as $R(A) = P(A)C(A)$, where $P(A)$ refers to the probability of A occurring, and $C(A)$ refers to the consequences of A occurring. In particular, with respect to the planning and construction of atomic power plants, estimates of $P(A)$ have been attempted. Beck's initial analysis of the risk society was related directly to the problem of carrying out this estimation. It is also complicated to estimate $C(A)$. What should in fact be considered? The cost of building a new reactor? The consequences for the environment? And for how long a period should we consider these consequences? What about the human consequences? How does one consider consequences in terms of deformity of the unborn? What unit should be applied when we describe $C(A)$? Could human consequences be expressed in terms of money – in terms of insurance costs, for instance?[23]

How do we *experience risks*? One can think of risks as propensities, as something that could happen (sooner or later); so risks are not straightforwardly experienced phenomena. Nevertheless, they form part of our life conditions. One can consider risks in all aspects of life. One can consider risks as associated with health: What is the risk of working in a polluted environment? What is the risk of undergoing a particular operation, or of not being operated on? What is the risk associated with a particular investment? What is the risk for a company of moving production to a different country? What are the ecological risks that are then associated with production? Is it possible to export risk? The "we" that is submitted to risks can refer to very many different groups. The globalized process of distributing industries and types of production the world around also includes a distribution of risk. Different groups of people could come to face radically different forms of risk.

This brings us to the notion of *risk production*. In the modern era, science, industrialization and production have been considered humankind's best protection against hostile nature. According to the modern outlook, one can see the development of the whole technological and scientific expertise as a way in which humanity tries to cope with such nature-caused catastrophes. We can try to protect ourselves against floods by constructing dams; we can try to protect farming against dry seasons by digging channels. A very long modern legend outlines how the complexity of industry and techno-science support us in the struggle against nature. However, the discourse of the risk society addresses the doubleness of production. Catastrophes and dangers cannot simply be seen as the work of nature. They can also be produced by science and science-based industry. In other words, catastrophes form part of human production. We are producing the catastrophes we ourselves are going to face. Risk production finds resources in science, including mathematics. Here we find a new role for science. The possibility of using atomic energy was identified through science. And only through science could one imagine such a

[23] The notion of risk is closely related to the notion of chance, where the consequences are seen as being attractive. We could also talk about risk-change society. In the following, however, I continue to talk about risk, and in some cases, the connotation of chance.

possibility. To formulate it more generally, a wide range of technological possibilities can only be established through science, and such new possibilities include new forms of risk. We have to reconsider the assumption of progress carefully. It might be that the progress ensured through science brings us into new risk environments. Instead of seeing science as the motor of progress, it also – through the new possibilities it establishes – brings us into new risk structures.

The world risk society might be characterized by a new form of unequal distribution of risks and opportunities in an entangled mix. It might be that scientific development can be related to progress, in some sense of progress, in some areas, and within some cases. But it might as well be that scientific development represents a step backwards in other aspects. During the time of Modernity, we find examples in literature of counter-stories to the assumption of progress. In *Doctor Jekyll and Mister Hyde*, published first time in 1886, Stevenson provided a simple metaphor for the doubleness of science. However, it might be necessary to take a step further beyond the Jekyll-Hyde duality with respect to science. It might be that a risk situation has to be explored, not through a good-bad duality, but through a much more complex conceptual network: in terms of possibilities, dangers, uncertainties, doubts, implications for some groups and implications for other groups.

15.3.2 Science on the Market: The myth of Neutrality

In *An Inquiry into the the Nature and Causes of the Wealth of Nations*, published in 1776, Smith presented the "invisible hand" as a metaphor signifying the logic of the free market. This hand ensures that supply and demand find an equilibrium that guarantees balanced prices and, in the end, maximum wealth and well being for all. In this way, the idea of progress was beautifully symbolized by the actions of the invisible hand. Obviously, the best one can do is to leave this hand in peace, undisturbed, to lead us towards the paradise ahead.

To the extent that science has joined the market place, one can raise the following questions: Can we be sure that supply will be of the proper quality? And what kind of knowledge would in fact be in demand? What is the relationship between, on the one hand, what has been referred to as scientific qualities and, on the other hand, the market qualities of scientific knowledge? For instance, is it a "valuable" thing to discover for the second time something that has already been discovered? From a non-commercial point of view, the duplication of scientific discoveries does not make sense. From a market perspective, however, this could make very good sense. For instance, discoveries in the field of medicine are eagerly duplicated, which is essential in order for the different companies to operate in a market with many patents in place. So one should not be surprised to find that the invisible hand of the market is promoting qualities, which are not in accordance with classic qualities in science (see also Christensen & Hansen, Chap. 13). One must also notice that the invisible hand might not be that invisible. It might be a guided hand, expressing the free will of certain companies.

I have tried to condense the observations about the new mode of knowledge production into the following three comments.[24] First, one finds *new sites* for knowledge production. The universities represent the classic site, and their task as formulated by Wilhelm von Humboldt, is to produce universal knowledge, the principal point being that this production should take place within a research community that is somehow protected. However, now the sites for knowledge production have expanded in many directions, meaning that we find a variety of companies, institutions and organizations involved in knowledge production. This applies with respect to pharmaceuticals, information technology, biotechnology, nanotechnology, etc. This variety establishes a completely different and much more flexible network for research. It also implies that research is taking place at sites that can be overloaded with different business interests

Second, the *structuring interests* for knowledge production become much more complex. According to the modern outlook (the modern myth) knowledge production was grounded in some sublime epistemic interests: the search for new insight that could fit into constantly growing stock of human knowledge. Research should be independent, as it, by its very nature, is able to promote its own proper qualities. In *Knowledge and Human Interest*, Habermas (1971) outlined a more complex picture by pointing out different knowledge guiding (if not constituting) interests, namely a technical, a hermeneutic and an emancipatory interest. Although such an analysis brings the notion of interest into the scope of knowledge production, one could go much further. Instead of considering three knowledge-constituting interests, one could consider knowledge supporting interest, knowledge guiding interest, knowledge limiting interest, knowledge obstructing interest, etc. The notion of interest, when elaborated in particular sites, brings us directly into an analysis of knowledge and power.[25] The complex interaction of knowledge and power makes up part of the modern development, as particularly pointed out by Foucault. Nevertheless, the myths of progress, neutrality and epistemic transparency were elaborated through this same period.

Third, the *criteria of quality* for knowledge production are changing. Following for instance, Popper's conception of knowledge development in terms of conjectures and refutations, it is possible to describe criteria for quality in a simple way. These criteria can be expressed with reference to how a conjecture-refutation dialectic is supported. However, when knowledge has gone on the market, many new criteria

[24] Considerations about knowledge on the market have in particular been addressed through the notion of the Mode-2 society as presented in Gibbons et al. (1994).

[25] An explicit expression of the idea that knowledge production can be guided by a particular interests is found at the home page of the Danish Ministry of Research, Technology and Development, where one of the Ministry's guiding aims is stated as to ensure that "Danish research is relevant and the motivating power for the ongoing development of welfare. As a consequence the Ministry will help to strengthen the research within areas which are of strategic importance for the Danish society". (Ministry of Research, Technology and Development, 2006, my translation, from the homepage: http://videnskabsministeriet.dk-/site/forside/forskning. (Last visited 23 November 2006) Such a formulation represents a direct rejection of the Humboldtian tradition. It indicates that science, as far as Denmark is concerned, is placed on the market.

mix with classic ones. Quality criteria are developed with reference to commercial interests: What is it possible to introduce into the market? What is it possible to patent? Thus, quality criteria get expressed in terms of competitive power, which could refer to companies, regions and to nations. Eventually, classic sets of quality criteria are annihilated by a market complexity. Again, one must consider that scientific knowledge was on the market long before it was realised that this was the case.

One conclusion to be drawn from these observations concerning sites, interest and quality of research is that any idea about the neutrality of sciences is a myth. It is a myth produced as part of the modern outlook.

15.3.3 Framing and Action: The Myth of Epistemic Transparency

Popper's presentation of the dialectics between conjectures and refutations can be seen as the last (almost desperate) attempt to maintain the myth of epistemic transparency. Initially, Feyerabend was in favour of Popper's project, but he later became a strong critic, ridiculing the devastating simplicity in the Popperian research programme.[26]

One important step further beyond the assumption of epistemic transparency was taken by Kuhn who emphasized that scientific discoveries appear to be governed by many non-rational elements, while, according to Popper, the dynamics of the conjecture-refutation follows a proper logical pattern.[27] In particular, Popper assumed that when a theory was falsified, logically speaking, it was also falsified in scientific practice. Kuhn, however, pointed out that a falsification brings about many attempts to save the falsified theory, as the theory might be deeply embedded in paradigmatic assumptions. Although falsified, such theories stay on. During the period of normal science, scientific rationality protects the paradigm from attacks caused by falsifying observations. Only in revolutionary periods are paradigmatic assumptions addressed. This means that research is not a simple rational process; it involves many metaphysical priorities.

This opens up the discussion for further considerations inspired by Foucault's work. He analyses how a range of extra-scientific factors constitute the conceptual framing of research. He opens up research formed through Modernity as representing particular perspectives and interests. A seemingly general and neutral research agenda can include many, much more specific agendas, brought into operation through the way the "object" of research and the "objective" of research are discursively constructed. In fact the object comes to reflect the objective in a much more profound way than normally assumed.[28]

[26] See Feyerabend (1975, 1987).

[27] See Kuhn (1970).

[28] Instead of paradigm, Foucault (1994) talks about *epistemé* as signifying the framing of research.

Knowledge is *framed* through complex processes, which might be revealed through a careful genealogy. According to Kant, it is part of the human condition that we cannot obtain any knowledge without categories. Furthermore, Kant finds that these categories have an eternal nature. They represent a universal and a-historical epistemological framing. I agree that knowledge is framed, but that the framing categories are reliable (not to say given a priori) is to me an illusion. The framing of what we present as knowledge is structured through a multitude of processes, which is demonstrated by both Kuhn and Foucault. To address the framing, one could try to clarify how perspectives and priorities have been incorporated as a scientific outlook.[29]

Knowledge also means *action*. This means that one is not concentrating on what "facts" knowledge might represent, but instead on how knowledge becomes a constituting part of actions. In another context, I have specified different elements of mathematics in action.[30] Mathematics can serve as a resource for all kinds of technological innovations. I have tried to be more specific about this by addressing different dimensions of mathematics in action: mathematics as an important element of technological imagination; mathematics as a resource for hypothetical reasoning; mathematics as part of processes of design and decision-making etc.

As a consequence of these observations about framing of knowledge and knowledge in action, I find that the assumption of epistemic transparency is a myth. The framing of scientific concepts is complex, and the scope of scientific knowledge brought into action cannot be delineated in a simple way.

15.4 A Critical Professionalism?

When the assumptions of progress, neutrality and epistemic transparency become recognized as myths, science and mathematics education face new challenges. Here I will concentrate on university education, although many of the comments may have broader applications. I will address three issues with respect to the professionalism that university mathematics and science education might prepare students for. I use the word "professionalism" to cover all kinds of job-functions as specialists, teachers, advisors, technicians, and researchers, based on university studies.

First, I want to consider the conceptual *framing* of the scientific content that is basic to professionalism. What does it mean to address how scientific theories have developed through a historical process as part of a complex theoretical, cultural and metaphysical conceptual framing? Second, I want to consider the *action*-dimension of knowledge. What kind of technological actions may future professionals come to take part in? Third, I want to address *ethics* as part of the professional outlook:

[29] Lakatos (1976) provides an analysis, which illustrates the internal complicity of generating mathematical concepts. Jammer (1957) documents the complex genealogy of the notion of force.

[30] See Skovsmose (2005).

What kind of reflections are relevant as part of the developed professionalism? Such questions may prepare university sciences and mathematics education for establishing what I will call a *critical professionalism*.

15.4.1 Framing

It is important that the myth of epistemic transparency is not replicated. In fact there are many ways of propagating this myth. The sanitized presentation in lectures and in lecture notes of definitions of concepts in science and mathematics is one of them.

Let us, as an example, consider the sequence in calculus of the specifications of: limit, differentiation, integration, the fundamental theorem of calculus, etc. This sequence, when elaborated in terms of definition–theorem–proof–definition–theorem–proof–definition–, etc. appears to many students to be established by some kind of magic. But one can try to get behind this magic by asking: what is the reason for the fundamental theorem of calculus being identified as fundamental? In what sense is it fundamental? When did it become fundamental? What is the point of introducing limits? When were limits identified as important? Why are there no infinitesimals in present day infinitesimal calculus? What was the point of eliminating infinitesimals? One could consider: what is the connection between these mathematical notions and other more philosophic world pictures? As outlined, for instance, by Leibniz in his *Monadologie*, where the infinitesimals assume a life of their own, and seem to compose the whole world. The development of calculus is connected to the formulation of the deterministic worldview. In fact one can find many connections between mathematics, science, astronomy, cosmology and theological positions.[31] This illustrates the conceptual complexity from which the fundamental notions of calculus emerge. An insight in such complexities makes part of a critical professionalism.

The notion of force can be introduced in physics as a simple concept through Newton's second law: $F = mg$. However, the notion of force forms part of a long conceptual story that can be traced back to antiquity.[32] Here I limit myself to pointing out the importance of the notion with respect to the formulation of the heliocentric worldview. It was not a straightforward empirically based task for Copernicus to introduce this view. He had, however, studied Neo-Platonism, according to which the sun is God-like, and in Greek philosophy he could find formulation of heliocentric worldviews. What could in fact be more natural than God taking up a central position in the universe? When the earth is removed from its fixed position, some movements call for explanations. What sense to make of the idea that the earth is "swung" around the sun in a tremendous circle (which Kepler later pointed out to be an ellipse)? Descartes had introduced whirls in his world picture, and maybe they could bring about some circulation. Descartes, however, had also formulated

[31] This has, for instance, been carefully pointed out by Tarnas (1991).

[32] See Jammer (1957).

the idea of inertia and that the earth was somehow "falling" towards the sun. How then should one think of the connection between the sun and the earth? Conceptually speaking, one had to think in terms of a very strong but invisible cable; something that could keep the earth in its orbit by permanently making it "fall" towards the sun. The notion of force seems to be necessary, and Newton brought all observations together in a nice conceptual pattern. However, accepting this pattern also means giving up the ability to specify the "nature" of force. Instead, one has to live with the conception of "working at a distance" as being part of a worldview.

Scientific notions emerge from a complex set of ideas and presumptions. The clear-cut definitions presented in much mathematics and science education are only the final outputs of a complex process which has no similarity with the conceptual architecture presented in most textbooks. Scientific concepts are developed in complex historical processes of tentative formulations, reformulations, reconsiderations, etc. An understanding of this complexity is an important feature of any critical professionalism.[33]

In university science and mathematics education, it becomes important to address relationships between conceptions within the field and conceptions and ideas from other fields. It is important that scientific concepts become recognized as parts of a bigger conceptual landscape, and how they have emerged through a complex process including many extra-scientific ideas, assumptions and priorities.[34] Science and mathematics have a particular reference to Modernity, as the Scientific Revolution provides a portal to the modern worldview. But as already mentioned, Modernity also means colonisation, suppression and exploitation. And this includes colonization of ways of thinking and doing. Such an insight also forms part of any critical professionalism within the domain.

15.4.2 Action

Knowledge can be brought in to action, which also applies to science and mathematics. But this is ignored in much university education.

By means of knowledge, also mathematical knowledge, one becomes enabled to construct new technological possibilities. Thus, modern cryptography emerged from number theoretical considerations. The relevant theoretical insight had been at hand for a long while, but only during the 1960s was it realized that one could organize coding and decoding in a completely different way than ever before. As a consequence, the number theoretical insight expressed in Euclid's Theorem, the

[33] The initiative of introducing philosophy of science (in Danish: *Fagenes Videnskabsteori*) as an obligatory ingredient in all university education in Denmark can be seen from this perspective. For a careful discussion of the introduction of *Fagenes Videnskabsteori* see Christensen (2003).

[34] The whole discussion of gender issues in mathematics and science education takes departure in this observation. See also Harding (Chap. 14)

Prime Number Theorem, Euler's Theorem, and Fermat's Little Theorem acquired significance well beyond the limits of pure mathematics. Furthermore, a principal knowledge for developing the new schemes of coding concerns the difficulties of formulating an efficient algorithm for factorising.[35] Such number theoretical notions and results have one type of significance within number theory, but they obtain a completely different significance within cryptography. The mathematical notions become included in different universes of meaning. Philosophically speaking, one can claim that the meaning of a notion is linked to its use. This means that the meanings of mathematical notions reflect the space of actions of which they make part. And different spaces could provide them with different meanings. I find it important that students experience the possible meanings of scientific concepts when they are recognized as resources for action. For me this is an important element of a critical professionalism.

One can see Problem-Based Learning (PBL) as one possible approach of bringing the action dimension of knowledge into focus. In Denmark, one initial step in the history of PBL was the establishment of Roskilde University in 1972, and of Aalborg University in 1974. These two universities were explicitly organized around the ideas of problem-based learning, interdisciplinarity and project-organized studies. These ideas were also reflected in the physical construction of the universities. The buildings contained many more offices than classic universities, as each group of students had a room of their own as their working place. The students' working day could include lectures, supervision and project work. All of this established a new working atmosphere at the universities: it was thought of as a shared working place for both staff and students.[36] The PBL-framework also helped to define project work in science and mathematics. The PBL-framework emerged from a radical interpretation of education of the late 1960s and the 1970s. It was emphasized that knowledge, not least science and mathematics, could be acted out in a variety of socio-political endeavours.

The realization of the PBL at Aalborg University was soon accompanied by certain pragmatism in the identification of what problems to address. Initially, in particular in the humanities and the social sciences, the problems were often identified with clear political references, while in mathematics and science, they were interpreted in more technical terms. The whole PBL-approach became elaborated in study-plans, and it came to form part of routines for examination. In fact a new professionalism was developed on a grand scale, and it is now well documented that candidates in mathematics, science and engineering from Aalborg University are often better received in private companies than candidates from other universities. The PBL-professionalism is really useful in many work-situations. The transfer from PBL to PBW (problem-based working) is experienced

---

[35] See Skovsmose and Yasukawa (2004).

[36] For initial presentations of problem orientation and project work see Kolmos et al. (2004); and Kolmos (Chap. 12). For a presentation of project work in mathematics at the university level, see Vithal et al. (1995).

as being efficient. Thus, PBL-students are accustomed to working in teams, to addressing problems that are not well defined and reformulating them in more specific ways, to solve conflicts in a group, etc. In many situations, the political sensitivity which was originally associated with the PBL-approach has been eliminated. Instead one can see a strong tendency to develop PBL-within the functionality of the market.

The PBL-approach can provide a blind professionalism that can be embodied and applied in any work situation, regardless of the socio-political and economic context. Therefore, it becomes an important challenge for the development of a critical professionalism to reveal the action dimension also of mathematics and science and at the same time to stress that no professionalism can assume that progress is intimately linked to these forms of knowledge.

15.4.3 Ethics

Knowledge in action means action, and actions call for reflections. Actions do not take place in a value-free space. No actions are "good in themselves." This also applies to science- and mathematics-based actions. They can be as risky, surprising, questionable, suppressing, unfair, marvellous and generous, etc. as any other form of action. I see that notions of actions and ethics are closely connected. Actions demand ethical considerations.

I do not see ethics as morality, so to me ethics does not mean assuming certain "proper" moral positions. I see ethics as having to do with structures for analyzing and reflecting on actions. In particular, I find it important to address some of the conflicts and controversies included in such reflections. Let me illustrate with an example. Science has now made it possible to intervene in life processes, in the form of genetic manipulation. According to a *teleological* position in ethics, one has to judge the ethical aspects of an action by the implication of the action. So how does one consider implications of actions that are, in this way, deeply rooted in technological inventions? How can one identify the implications of a certain form of genetic manipulation, when most of the implications only take the form of possible implications? They cannot (yet) be experienced. And how can such implications be measured? What unit of measurement should be used? In fact it appears that a teleological approach in ethics is impossible to manage. Instead one could consider a *deontological* approach, where at least some elements of the ethical potion are based on certain a priori formulated rules, i.e., rules that are also applicable in case we have no knowledge of possible consequences. However, such an approach brings the ethical discussion into a different difficulty: where to look for a priori rules applicable to science-based actions?

I do not see any simple solution to such difficulties concerning ethical argumentation, but I find that an awareness of such difficulties should be part of mathematics and science education. Students should come to see that mathematics-and science-based technological initiatives raise ethical issues with no straightforward solutions.[37] Eriksen (2003) has presented the notion of ethics-anchored dialogue in university chemistry education, which can illustrate some important elements of what it could mean for an educational practice to open the ethical dimension. It is important that students be invited into a discussion of ethical issues related to a particular issue (see Eriksen, Chap. 2). Eriksen presents the issue of the development of DDT as a possible issue for such a dialogue. This highly poisonous material was produced in the beginning of the twentieth century. However, its poisonous effects were thought of as being the most useful quality of the material. It could serve to fight noxious insects and vermin. Originally one did not conceptualize any of the consequences later to be experienced, as the DDT in accumulated form turned out to have disastrous environmental implications. A critical professionalism must include ethical reflections on actions based on the particle scientific domain in question.

The DDT-example makes it possible to address some fundamental ethical challenges. One can raise the idea of the risk society, as the example illustrates the difficulty of identifying risks, and therefore of carrying out an ethical evaluation based on a teleological principle. One can also see how a science-based risk production might have taken place long before the notion of risk society was coined. A straightforward notion of science-based progress becomes questioned, and so does the idea of the neutrality of science. The exploration of ethical issues can be anchored in an example, but a main educational point is that it is the dialogue between teacher and students and among students that opens the way for the ethical insight. As a consequence, I find that ethics-anchored dialogues compose an important part of university science and mathematics education.

15.5 Uncertainty as Conclusion

According to the assumptions of modernity, mathematics and science can be seen as sublime representatives of scientific rationality, and the best one can do is to ensure that this rationality gets the best possible space for developing, according to its own intrinsic dynamic.

Denying the assumption of progress, however, does not imply that we must assume that scientific rationality leads to catastrophes. Although we might find that trust in scientific rationality has brought about threatening risk configurations, we cannot assume that negating scientific rationality may establish a safe distance to such configurations. Leaving the modern conception of science means to question (and not to abolish) the rationality of science, and therefore this rationality needs to be addressed by a critique. Such a critique could include: an analysis of the framing of the rationality; an identification of what actions such a rationality could bring about; and ethical reflections on such actions.

---

[37] There is huge development of applied ethics. See also Hansen (2002).

Trying to formulate a critique of rationality immediately brings us to the question: from what perspective and from what epistemic resources could one formulate a critique of rationality? In philosophy one finds many forms of critique. Critical Theory criticized the positivist position, and in this way it expressed a critique of rationality. The problem, however, in establishing a critique, as suggested by Critical Theory, is that it may assume the existence of a position from which one can formulate a critique. *Making* a critique means *assuming* a critical perspective. One can assume a perspective from where one can criticize the functioning of the free market and how knowledge is operating within a capitalist economy. One can assume a perspective from where one can identify knowledge-guiding interests included in so-called technical disciplines like science and mathematics. One can try to establish a firm foundation for such a critique. The tricky point is that the more one tries to elaborate a solid foundation for a critique, the more one digs into the modernist assumption that knowledge (and therefore also critique) need to be established on a firm foundation.

What does it mean, therefore, to give up searching for a foundation for a critique of scientific rationality? A critique of rationality is needed, but adequate sources for such a critique are not identified. Scientific rationality appears to be a powerful and constructive force; at the same time, we seem to be in a weak position to criticize this force. We come to face uncertainty as a general epistemic condition. Is the implication, then, a harsh form of relativism, which in fact could make the modern position just as reliable as any other position? Moving into relativism, does not, however, mean that we have to assume absolute relativism. To me it means to move beyond the assumptions of modernity and acknowledge that uncertainty becomes a defining element of a critical professionalisms.

References

Ayer, A. (Ed.) (1959). *Logical Positivism*. New York: The Free Press.

Ayer, A. (1970). *Language, Truth and Logic*. London: Victor Gollancz.

Bauman, Z. (1989). *Modernity and the Holocaust*. Cambridge: Polity.

Beck, U. (1992). *Risk Society: Towards a New Modernity*. London: Sage.

Beck, U. (1999). *World Risk Society*. Cambridge: Polity.

Beth, E.W. & Piaget, J. (1966). Mathematical Epistemology and Psychology. Dordrecht: Reidel.

Bourbaki, N. (1950). The Architecture of Mathematics. *The American Mathematical Monthly*, 57, 221–232.

Bruner, J.S. (1960). *The Process of Education*. Cambridge: Belkapp Press.

Bury, J.B. (1955). *The Idea of Progress: An Inquiry into its Origin and Growth*. New York: Dover.

Carnap, R. (1937). *The Logical Syntax of Language*. London: Routledge and Kegan Paul.

Carnap, R. (1959). The Elimination of Metaphysics Through Logical Analysis of Language. In A. Ayer (Ed.), *Logical Positivism* (pp. 60–81). New York: The Free Press.

Christensen, O.R. (2003). *Exploring the Borderland: A Study on Reflections in University Science Education*. Ph.D. Thesis. Aalborg: Department of Education, Learning and Philosophy, Aalborg University. Available at: http://www.learning.aau.dk/index.php?id = 28.

Curry, H.B. (1951). *Outlines of a Formalist Philosophy of Mathematics*. Amsterdam: North-Holland Publishing.

Descartes, R. (1993). *Meditations on First Philosophy in Focus. Edited and with an Ontroduction by Stanley Tweyman*. London: Routledge.

Dewey, J. (1938). The Relation of Science and Philosophy as a Basis for Education. *School and Society*, 47, 470–473. [Reprinted in R.D. Archambault (Ed.). (1964), *John Dewey on Education: Selected Writing* (pp. 15–19). Chicago: The University of Chicago Press].

Dewey, J. (1963). *Experience and Education*. New York: Macmillan.

Dewey, J. (1966). *Democracy and Education: An Introduction to the Philosophy of Education*. New York, London: The Free Press [first published 1916].

Dieudonné, J.A. (1970). The Work of Nicholas Bourbaki. *The American Mathematical Monthly*, 61, 134–145.

Eriksen, K.K. (2003). *The Role of Reflectivity in Tertiary Chemical Education*. Ph.D. Thesis. Copenhagen: The University of Copenhagen.

Ernest, P. (1998). *Social Constructivism as a Philosophy of Mathematics*. Albany: State University of New York Press.

Feyerabend, P. (1975). *Against Method*. London: Verso.

Feyerabend, P. (1987). *Farewell to Reason*. London: Verso.

Foucault, M. (1994). *The Order of Things: An Archaeology of the Human Sciences*. New York: Vintage Books.

Gibbons, M., Limoges, C., Nowotny, H., Schwartzman, S., Scott, P. & Trow, M. (1994): *The New Production of Knowledge: The Dynamics of Science and Research in Contemporary Societies*. London: Sage.

Glasersfeld, E.V. (1995). *Radical Constructivism: A Way of Knowing and Learning*. London: Falmer Press.

Habermas, J. (1971). *Knowledge and Human Interests*. Boston: Beacon Press.

Hansen, T.B. (Ed.) (2002). *The Role of Philosophy of Science and Ethics in University Science Education*. Göteborg: NSU Press.

Hempel, C.G. (1959). The Empiricist Criterion of Meaning. In A. Ayer (Ed.). *Logical Positivism* (pp. 108–129). New York: The Free Press.

Hempel, C.G. (1965). *Aspects of Scientific Explanation*. New York: The Free Press.

Hume, D. (1975). Enquiries Concerning Human Understanding and Concernning the Principles of Morrals. Oxford: Oxford University Press.

Jammer, M. (1957). *Concepts of Force: A Study in the Foundations of Dynamics*. Cambridge: Harvard University Press.

Kolb, D.A. (1984). *Experiental Learning: Experiences as the Source of Learning and Development*. Englewoord Cliffs: Prentice Hall.

Kolmos, A., Fink, F.K., Krogh, L. (Eds.) (2004). *The Aalborg PBL-Model: Progress, Diversity and Challenges*. Aalborg: Aalborg University Press.

Kuhn, T.S. (1970). *The Structure of Scientific Revolutions*. Chicago: University of Chicago Press.

Lakatos, I. (1976). *Proofs and Refutations: The Logic of Mathematical Discovery*. Cambridge: Cambridge University Press.

Locke, J. (1967). *Two Treatises of Government*. Cambridge: Cambridge University Press [first published 1689).

Neurath, O. (1959). Protocol Sentences. In A. Ayer (Ed.), *Logical Positivism* (pp. 199–208). New York: The Free Press.

Nisbet, R.A. (1980). *History of the Idea of Progress*. New York: Basic Books.

OEEC (1961). *New Thinking in School Mathematics*. Paris. OECD.

Popper, K.R. (1965). *The Logic of Scientific Discovery*. New York: Harper and Row.

Popper, K.R. (1972). *Conjectures and Refutations: The Growth of Scientific Knowledge*. London: Routledge and Kegan Paul.

Skovsmose, O. (2005). *Travelling Through Education: Uncertainty, Mathematics, Responsibility*. Rotterdam: Sense.

Skovsmose, O. (2006). Challenges for Mathematics Education Research. In J. Maasz & W. Schloegelmann (Eds.), *New Mathematics Education Research and Practice* (pp. 33–50). Rotterdam: Sense.

Skovsmose, O. & Yasukawa, K. (2004). Formatting Power of 'Mathematics in a Package': A Challenge for Social Theorising? *Philosophy of Mathematics Education Journal*. Retrieved May 2008. Available at: http://www.ex.ac.uk/~PErnest/pome18/contents.htm.

Smith, A. (1776). An Inquiry into the Nature and Causes of the Wealth of Nations. Retrieved May 2008. Available at: http://www.adamsmith.org/smith/won-index.htm.

Stadler, F. (2001). *The Vienna Circle: Studies in the Origins, Development and Influence of Logical Empiricim*. Vienna: Springer.

Stevenson, R.L. (1987). *Dr. Jekyll and Mr. Hyde*. New York: Signet classic.

Tarnas, R. (1991). *The Passion of the Western Mind: Understanding the Ideas that have Shaped Our World View*. New York: Random House.

Vithal, R., Christiansen, I.M. & Skovsmose, O. (1995). Project Work in University Mathematics Education: A Danish Experience: Aalborg University. *Educational Studies in Mathematics*, 29(2), 199–223.

Name Index

Subject Index

Printed in the United States
129708LV00003B/263/P

9 780387 098289